Second Workshop on Numerical Modeling in MHD and Plasma Physics (MHD-PP-2019)

AA007642

Methods, Tools, and Outcomes

Journal of Physics: Conference Series Volume 1336

Moscow, Russia
10 - 11 October 2019

ISBN: 978-1-5108-9992-6
ISSN: 1742-6588

Printed from e-media with permission by:

Curran Associates, Inc.
57 Morehouse Lane
Red Hook, NY 12571

CURRAN ASSOCIATES INC.
proceedings
.com

Some format issues inherent in the e-media version may also appear in this print version.

This work is licensed under a Creative Commons Attribution 3.0 International Licence.
Licence details: http://creativecommons.org/licenses/by/3.0/.

No changes have been made to the content of these proceedings. There may be changes to pagination and minor adjustments for aesthetics.

Printed with permission by Curran Associates, Inc. (2021)

For permission requests, please contact the Institute of Physics
at the address below.

Institute of Physics
Dirac House, Temple Back
Bristol BS1 6BE UK

Phone: 44 1 17 929 7481
Fax: 44 1 17 920 0979

techtracking@iop.org

Additional copies of this publication are available from:

Curran Associates, Inc.
57 Morehouse Lane
Red Hook, NY 12571 USA
Phone: 845-758-0400
Fax: 845-758-2633
Email: curran@proceedings.com
Web: www.proceedings.com

TABLE OF CONTENTS

IOP Conf. Series: Journal of Physics: Conf. Series **1336** (2019) 011001 doi:10.1088/1742-6596/1336/1/011001

Preface

Brad Gibson[1], Igor Kulikov[2], Igor Chernykh[2]

[1]University of Hull, Cottingham Road, Hull HU6 7RX, United Kingdom
[2]Institute of Computational Mathematics and Mathematical Geophysics SB RAS, 6-th Lavrentyev avenue, Novosibirsk, Russian Federation

mhd2019@sscc.ru

The fields of magnetohydrodynamics and plasma physics provide the nexus for a powerful array of multidisciplinary fields. Diverse fundamental and applied topics meet here, ranging from astrophysics, thermonuclear fusion, and earth sciences, through to bleeding-edge mathematical algorithm developments, including particle-in-cell, smoothed particle hydrodynamics, and adaptive mesh refinement approaches, each underpinned by rapid advances in high performance computing technology. The Second Workshop on Numerical Modeling in MHD and Plasma Physics (MHD2019), held 10-11 October 2019 at the Keldysh Institute of Applied Mathematics, in Moscow, brought together 40 leaders in their respective fields, under the guidance of a Program Committee spanning five countries. The overarching goal of the Workshop was the breaking down of artificial barriers which exist between disciplines, in order to share best practice in these diverse fields. By doing so, the associated up-skilling of all the participants could only benefit all those involved, as well as their respective staff and students; the timeliness of such a philosophy should not be underestimated, given the exponential growth in impact of Big Data across all of the disciplines represented at MHD2019. The success of MHD2019 is reflected in the breathtaking array of work highlighted in these proceedings, and bodes well for the continued success of the workshop series in the future.

We gratefully acknowledge the financial support of the Budget Project of Institute of Computational Mathematics and Mathematical Geophysics SB RAS No. 0315-2019-0009

Content from this work may be used under the terms of the Creative Commons Attribution 3.0 licence. Any further distribution of this work must maintain attribution to the author(s) and the title of the work, journal citation and DOI.
Published under licence by IOP Publishing Ltd

IOP Conf. Series: Journal of Physics: Conf. Series **1336** (2019) 011002 doi:10.1088/1742-6596/1336/1/011002

Peer review statement

All papers published in this volume of *Journal of Physics: Conference Series* have been peer reviewed through processes administered by the proceedings Editors. Reviews were conducted by expert referees to the professional and scientific standards expected of a proceedings journal published by IOP Publishing.

Content from this work may be used under the terms of the Creative Commons Attribution 3.0 licence. Any further distribution of this work must maintain attribution to the author(s) and the title of the work, journal citation and DOI.
Published under licence by IOP Publishing Ltd

IOP Conf. Series: Journal of Physics: Conf. Series **1336** (2019) 012001 doi:10.1088/1742-6596/1336/1/012001

A Mathematical Model of Tornado

M B Gavrikov[1,2] and A A Taiurskii[1]

[1] Keldysh Institute of Applied Mathematics, Miusskaya sq., 4, Moscow, 125047, Russia
[2] Bauman Moscow State Technical University, ul. Baumanskaya 2-ya, 5, Moscow, 105005, Russia

E-mail: mbgavrikov@yandex.ru, tayurskiy2001@mail.ru

Abstract. We propose a mathematical model of tornado in the framework of continuum mechanics. According to this model, the swirling upward air motion in a tornado can be explained even without taking into account the Coriolis force or the vertical convection of warm air. Our results show that for a tornado to appear two factors are necessary: a powerful rotational motion of air in the upper atmosphere and the growth of air pressure from the center to the periphery of an expected region of tornado. The characteristics of air flow in tornado can be found by solving a boundary value problem for a nonlinear parabolic integro-differential equation for an unknown complex-valued temperature. We formulate this problem, propose an algorithm for its numerical solution, and examine its stationary solutions by the stabilization method.

1. Introduction

The phenomenon of tornado plays an important role in the life of many regions of the Earth because of its disastrous consequences. The prediction and control of this phenomenon requires the study of the flow of air masses in a tornado. There is still no common understanding of its causes. There is a widespread opinion [1–4] that it is generated by the vertical movement of warm air, due to the local heating by the sun of parts of the earth's or water surface and the air masses adjacent to them: ascending air flows first lead to radial, and then vortex air movements at the Earth's surface owing to the Coriolis force. Our study suggests another possible mechanism of air motion in a tornado. It does not rely on the Coriolis force or vertical convection of warm air, but is based on small-scale phenomena in atmospheric air, in particular, its viscosity. It should be observed that the effective hydrodynamic viscosity of air in a real tornado significantly increases due to the presence in it of a large amount of water or sand dust (depending on whether it is formed above the surface of water or land). The reasons for the occurrence of a tornado, according to our research, are, firstly, the excess pressure of air in its periphery over the pressure in its central part, and secondly, the presence of a powerful vortex motion in the upper atmosphere. In this case, we show that the angular velocity of the vortex motion is directly related to the periphery-center pressure drop. Thirdly, it seems that tornado is realized as a special type of air flow, the form of which will be specified below. The combination of these three factors is uncommon, which explains the exclusivity of the tornado phenomenon.

There are dozens of works on the numerical and analytical modeling of processes in a tornado. Let us briefly dwell on some of the most important of these publications.

In [5], an analytical model is proposed for a vortex non-viscous flow with a vertical symmetry axis and a motionless core in the form of a funnel. It is assumed that the air density within the

Content from this work may be used under the terms of the Creative Commons Attribution 3.0 licence. Any further distribution of this work must maintain attribution to the author(s) and the title of the work, journal citation and DOI.
Published under licence by IOP Publishing Ltd

IOP Conf. Series: Journal of Physics: Conf. Series **1336** (2019) 012001 doi:10.1088/1742-6596/1336/1/012001

core is smaller than that of the moving air whose rotation velocity is growing in the direction of the funnel border and is maximal near the ground. In [6], a model is proposed for the formation of atmospheric tornado-type vortices due to instability caused by the growth of the vertical component of the air velocity in the direction of the earths surface or an increase of concentration of suspended particles. Such conditions are typical for the initial stage of the development of a tornado and are realized in thunderclouds in the atmosphere. A class of analytical solutions of the Navier-Stokes equations for viscous incompressible fluids is obtained in [7]. These solutions can be used to predict the characteristics of certain vortex flows, in particular, tornadoes. An analytical study of ascending swirling air flow is carried out in [1]. According to this investigation, the Coriolis force plays a key role in the formation of tornadoes. It should be observed that a number of assumptions made in [1] contradict the generally accepted views on the formation and stability of tornadoes.

Numerical simulation of a tornado faces the problem of setting correct initial and boundary conditions. Moreover, when using turbulent models of air flows in a tornado, the problem is the calculation of empirical turbulent transport coefficients. Nevertheless, some important results have been obtained in this direction. In [8], one considers the process of tornado formation due to convective instability near the earths ground. In [9–12], following [8], calculations of vortices are carried out and the determining role of the swirl ratio in the formation and stability of the vortex structure of a tornado is established. The process of tornado formation due to air rotation in a thundercloud is studied in [13], with the effects of turbulence of air currents, water or sand impurities, and compressibility taken into account. In [14,15], developing the approaches of [13], one calculates three-dimensional velocity fields and air pressure in a tornado. An analysis of the dynamics of tornado-like flows on the basis of the large vortex method (LES method) was carried out in [16–20], where is was shown, in particular, that the compressibility of air has almost no effect on the vortex dynamics in a tornado. A two-fluid model of tornado is examined in [21,22], the first (main) liquid being water vapor, which condenses upon a sharp change in pressure, the second liquid consisting of solid particles involved in the air flow by the vortex tornado flow.

Here, we propose a simple mathematical model of tornado stated in terms of continuum mechanics and based on a generalization of the exact von Karman solution [23,24] that describes the motion of a viscous incompressible fluid over a uniformly rotating infinite horizontal disk. The von Karman solution shows that the disk rotation results in a "vacuum cleaner" effect: the fluid is drawn from infinity with a constant velocity in the direction of the disk and spreads over it, the fluid pressure may vary in height, but remains constant on each horizontal plane parallel to the disk. If the last condition is dropped, i.e., if the pressure on each horizontal plane is allowed to vary, being greater on the periphery than at the center, then, under certain conditions, the behavior of the flow above the disk drastically changes. The fluid moves along the disk to the vertical axis, while rotating around it, and is then ejected to infinity along that axis. A similar behavior is exhibited by the air masses in a tornado. Thus, we come to a simple mathematical model of tornado.

2. Basic Equations

Our starting point is the dynamic system of equations describing the motion of a viscous incompressible fluid in a constant field of gravity,

$$\begin{aligned} &\operatorname{div}\mathbf{U} = 0, \quad \rho = \text{const}, \\ &\frac{\partial \rho \mathbf{U}}{\partial t} + \operatorname{Div}(\rho \mathbf{U}\mathbf{U} + p\mathrm{I}_3) = 2\mu \operatorname{Div}\operatorname{def}\mathbf{U} + \rho \mathbf{g}, \end{aligned} \tag{1}$$

where ρ is the density of the fluid, p is pressure, $\mathbf{U}$ is its hydrodynamic velocity, μ is the dynamic viscosity coefficient, $\mathbf{g}$ is the constant acceleration of gravity, I_3 is the three-dimensional identity tensor, def $\mathbf{U}$ is the strain-rate tensor, which in the Cartesian coordinates has the form

$$\operatorname{def}\mathbf{U} = \left[\frac{1}{2}\left(\frac{\partial U^k}{\partial x^s} + \frac{\partial U^s}{\partial x^k}\right)\right]_{1\le k,s\le 3}.$$

It is assumed that $\mu = \text{const}$, $\rho = \text{const}$ are given, and we seek four unknown functions, p and three components of $\mathbf{U}$, satisfying four scalar equations (1).

In the axisymmetric case ($\partial/\partial\varphi = 0$), system (1) has a particular solution of the form

$$U_r = rA(t,z), \quad U_\varphi = rB(t,z), \quad U_z = C(t,z), \quad p = Q(t)r^2 + R(t,z), \tag{2}$$

where the axis z has the same direction as $\mathbf{g} = (0,0,-g)$, the function $Q(t)$ is given and A, B, C, R are sought by substituting the expressions (2) into system (1). Taking into account the condition of axial symmetry $\partial/\partial\varphi = 0$ and carrying out some simple transformations, we come to the following result.

Theorem 1. *In the axisymmetric case,* (2) *yields a solution of system* (1) *if and only if* A, B, C *satisfy the equations*

$$\begin{aligned} &\frac{\partial A}{\partial t} + A^2 - B^2 + C\frac{\partial A}{\partial z} - \frac{\mu}{\rho}\frac{\partial^2 A}{\partial z^2} + \frac{2Q}{\rho} = 0, \\ &\frac{\partial B}{\partial t} + 2AB + C\frac{\partial B}{\partial z} - \frac{\mu}{\rho}\frac{\partial^2 B}{\partial z^2} = 0, \quad \frac{\partial C}{\partial z} = -2A. \end{aligned} \tag{3}$$

If A, B, C *satisfy equations* (3)*, then the derivative* $\partial R/\partial z$ *is expressed from the equation*

$$\frac{\partial C}{\partial t} + \frac{\partial}{\partial z}\left(\frac{C^2}{2} + \frac{2\mu}{\rho}A + \frac{R}{\rho} + gz\right) = 0, \tag{4}$$

which yields R *to within an arbitrary function of time.*

Introducing a new unknown complex-valued function $u \underset{\text{def}}{=} A + \mathrm{i}B$ and taking into account the relations $A = \operatorname{Re} u$, $u^2 = (A^2 - B^2) + 2\mathrm{i}AB$, we can rewrite system (3) in the form

$$\frac{\partial u}{\partial t} + C\frac{\partial u}{\partial z} - \frac{\mu}{\rho}\frac{\partial^2 u}{\partial z^2} + u^2 + \frac{2Q(t)}{\rho} = 0, \quad \frac{\partial C}{\partial z} = -2\operatorname{Re} u, \tag{5}$$

where $C(t,z) \in \mathbb{R}$, $u(t,z) \in \mathbb{C}$ are unknown real- and complex-valued functions, respectively. System (5) can be written as a single integro-differential equation. Indeed, taking an arbitrary real $C_0(t) \underset{\text{def}}{=} -C(t,0)$, we see that system (5) is equivalent to the following integro-differential equation:

$$\frac{\partial u}{\partial t} - \left(C_0(t) + 2\int_0^z \operatorname{Re} u \,\mathrm{d}z\right)\frac{\partial u}{\partial z} - \frac{\mu}{\rho}\frac{\partial^2 u}{\partial z^2} + u^2 + \frac{2Q(t)}{\rho} = 0, \tag{6}$$

where $C_0(t)$, $Q(t)$ are arbitrary given real functions.

Some stationary ($\partial/\partial t = 0$) solutions of system (3) with $Q(t) \equiv 0$ were studied in [23] (see also [24]). In this case, (3) reduces to an autonomous nonlinear system of fifth-order ordinary differential equations. In [23,24], one considers numerical solutions of a boundary value problem for the said nonlinear fifth-order system on the semi-axis $0 \le z < +\infty$ with certain (see below) boundary conditions at the points $z = 0$ and $z = +\infty$.

Let us consider some initial boundary value problems for system (5).

3. Excitation of a Viscous Incompressible Fluid by a Rotating Disk

Consider the half-space $z \geq 0$ occupied by a viscous incompressible fluid. Its boundary $z = 0$ coincides with an infinite rigid disk rotating around the axis z with constant angular velocity Ω. This rotation causes, in the domain $z > 0$, stationary motion of the fluid described by a solution of the form (2) with A, B, C, Q, R independent of t. This stationary solution can be obtained from a stabilizing solution of system (5) on the half-axis $0 \leq z < +\infty$ with certain boundary conditions at $z = 0$ and $z = +\infty$. To obtain these boundary conditions, one should take into account the conditions of *adhesion* and *nonpenetration* on the surface of the rigid disk at each time instant. For $z = 0$, the adhesion condition for a solution of the form (2) is $B(t, 0) = \Omega$, $A(t, 0) = 0$, while the nonpenetration condition has the form $C(t, 0) = 0$. In other words, at each time instant, the identities $u(t, 0) = \mathrm{i}\Omega$, $C(t, 0) = 0$ are required to hold. For $z = +\infty$, no separate condition is imposed on C, while the boundary conditions on u depend on our assumptions on the behavior of u and C at infinity, i.e., as $z \to +\infty$. For instance, suppose that for $z \to +\infty$ and any t, there exist finite limits

$$u \to u(\infty), \quad C\frac{\partial u}{\partial z} \to 0, \quad \frac{\partial^2 u}{\partial z^2} \to 0, \tag{7}$$

where $u(\infty)$ does not depend on t and the first convergence in (7) can be differentiated in t (i.e., $\partial u/\partial t \to \partial u(\infty)/\partial t = 0$ as $z \to +\infty$). Note that the second and the third conditions in (7) follow from the first if, for each t, the function $C(t, z)$ is bounded at infinity with respect to z and the real and the imaginary parts of $\partial u/\partial z$ and $\partial^2 u/\partial z^2$ are monotone in z. From (5), (7), we obtain the following boundary condition for $u(\infty)$:

$$u(\infty)^2 + \gamma_0 = 0 \Leftrightarrow u(\infty) = \sqrt{-\gamma_0}, \quad \gamma_0 = 2Q/\rho.$$

Thus, system (5) is supplemented with the following boundary conditions for any fixed t:

$$u|_{z=0} = \mathrm{i}\Omega, \quad C|_{z=0} = 0, \quad u|_{z=+\infty} = \sqrt{-\gamma_0}, \quad \gamma_0 = 2Q/\rho. \tag{8}$$

In order to find a stationary solution of the form (2) in the domain $z > 0$, we need to find a stabilizing solution of system (5), (8) with an arbitrary initial condition $u|_{t=0} = u_0(z)$, $0 \leq z < +\infty$, whose choice should not affect the solution. The case $\gamma_0 = 0$ was considered in [23,24]. For $\gamma_0 > 0$, condition (8) ensures that at infinity, $z = +\infty$, the incompressible fluid is rotating as a rigid body with constant angular velocity $\sqrt{\gamma_0}$. For $\gamma_0 < 0$, the fluid spreads along the radii at infinity with radial velocity linearly depending on the radial coordinate, the coefficient $\sqrt{|\gamma_0|}$ being the same for all radial directions. Keeping in mind that our theory is aimed at studying tornado, we assume that $\gamma_0 > 0$ in what follows.

Note that the condition that $u(t, z)$ is bounded in t and z does not guarantee that the vertical velocity $C(t, z)$ is bounded in z. The numerical analysis described below shows that for the stationary solution of problem (5), (8), the function $C(z)$ is not only bounded, but has a finite limit $\lim_{z \to +\infty} C(z)$.

Let us pass to dimensionless quantities in problem (5), (8) by choosing the following characteristic scales: $t_0 = 1/\Omega$ (time), $U_0 = \sqrt{\mu\Omega/\rho}$ (velocity), $L_0 = U_0 t_0 = \sqrt{\mu/(\rho\Omega)}$ (length), $p_0 = \mu\Omega/2$ (pressure), assuming that $u_0 = 1/t_0 = \Omega$, $C_0 = U_0$. Then system (5), (8) in dimensionless variables takes the form

$$\begin{aligned} &\frac{\partial u}{\partial t} + C\frac{\partial u}{\partial z} - \frac{\partial^2 u}{\partial z^2} + u^2 + \Gamma = 0, \quad \frac{\partial C}{\partial z} = -2\mathrm{Re}\, u, \\ &u|_{z=0} = \mathrm{i}, \quad C|_{z=0} = 0, \quad u|_{z=+\infty} = \sqrt{-\Gamma}, \\ &0 \leq z < +\infty, \quad t \geq 0, \quad \Gamma = 2Q/(\rho\Omega^2). \end{aligned} \tag{9}$$

IOP Conf. Series: Journal of Physics: Conf. Series **1336** (2019) 012001 doi:10.1088/1742-6596/1336/1/012001

Problem (9) can be reduced to the following boundary value problem for a single integro-differential equation:

$$
\begin{aligned}
&\frac{\partial u}{\partial t} - 2\int_0^z \operatorname{Re} u \, \mathrm{d}z \cdot \frac{\partial u}{\partial z} - \frac{\partial^2 u}{\partial z^2} + u^2 + \Gamma = 0, \\
&u\big|_{z=0} = \mathrm{i}, \quad u\big|_{z=+\infty} = \sqrt{-\Gamma}, \\
&0 \le z < +\infty, \quad t \ge 0.
\end{aligned} \tag{10}
$$

As the initial value $u_0(z)$ in (9), (10), one can take any smooth function on $[0, +\infty]$ satisfying the boundary conditions, for instance,

$$
u_0(z) = \frac{\mathrm{i} + z\sqrt{-\Gamma}}{1+z}, \quad 0 \le z < +\infty.
$$

4. A Simple Tornado Model

The phenomenon of tornado is caused by physical processes occurring in the atmospheric air. To describe these processes we use the mathematical tools of continuum mechanics. It is necessary to take into account the rotation of the Earth with a constant angular velocity Ω around the axis passing through the North (N) and South (S) Poles.

Consider two coordinate systems: system $Oxyz$ is fixed, its origin O coincides with the Earth's center and its axis Oz is directed along the rotation axis of the Earth; and system $O'x'y'z'$ is moving, being rigidly fixed to the surface of the Earth, with its axis $O'x'$ directed to the East along the parallel passing through O' and its axis $O'y'$ directed to the North along the meridian passing through O' (Figure 1).

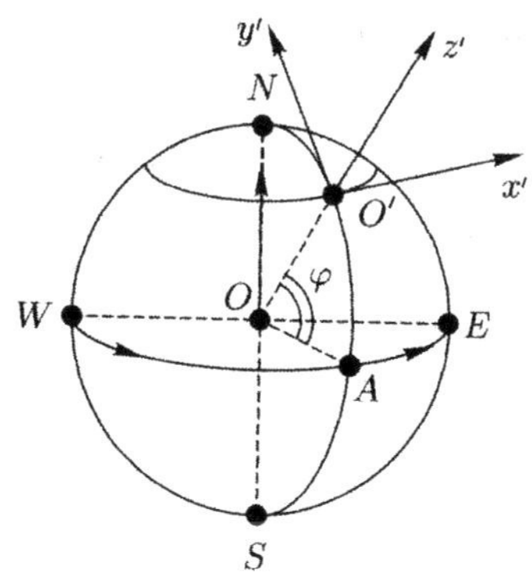

Figure 1. Coordinate systems.

We want to write the equation of atmospheric air dynamics in the moving coordinate system. Assume that the fixed coordinate system is inertial. Then, according to Newton's second law, the equation of motion of a material point of mass m in the gravity field near the surface of the Earth and the origin O' has the form

$$
m\mathbf{w}_a = m\mathbf{g}, \quad \mathbf{g} = -g\frac{\mathbf{r}}{r} \cong -g\frac{\mathbf{r}_0}{r_0}, \tag{11}
$$

where $\mathbf{r}_0$ is the radius vector of the point O', $\mathbf{r}$ is the radius vector of a moving point in the fixed coordinate system, $g = 980.665\,\mathrm{cm/s^2}$ is the acceleration of gravity near the surface, $\mathbf{w}_a$ is the absolute acceleration of a point. Assuming that the speed of rotation of the Earth is constant, $\Omega = \mathrm{const}$, we have (see [25,26])

$$
\mathbf{w}_a = \mathbf{w}_r + 2[\boldsymbol{\Omega}, \mathbf{v}_r] + [\boldsymbol{\Omega}, [\boldsymbol{\Omega}, \mathbf{r}']] + \frac{\mathrm{d}\mathbf{v}_0}{\mathrm{d}t}, \tag{12}
$$

where $\mathbf{v}_r, \mathbf{w}_r$ are, respectively, the velocity and the acceleration of a moving point, $\mathbf{r}'$ is the radius vector of a point in the moving coordinate system, $\mathbf{\Omega} = (0, 0, \Omega)$ in the fixed system, and $\mathbf{\Omega} = (0, \Omega\cos\varphi, \sin\varphi)$ in the moving system, φ is the latitude of O'. In the moving coordinate system, $\mathbf{g} = (0, 0, -g)$ and $\mathbf{r}_0(t) = r_0(\cos\varphi\cos\Omega t, \cos\varphi\sin\Omega t, \sin\varphi)$. From (11), (12), we obtain the equation

$$m\mathbf{w}_r = m\mathbf{g} - m\frac{\mathrm{d}\mathbf{v}_0}{\mathrm{d}t} - 2m[\mathbf{\Omega}, \mathbf{v}_r] - m[\mathbf{\Omega}, [\mathbf{\Omega}, \mathbf{r}']], \quad \mathbf{v}_0 = \frac{\mathrm{d}\mathbf{r}_0}{\mathrm{d}t}.$$

From the explicit formula for $\mathbf{r}_0$, we get

$$\frac{\mathrm{d}\mathbf{v}_0}{\mathrm{d}t} = \frac{\mathrm{d}^2\mathbf{r}_0}{\mathrm{d}t^2} = \frac{\mathrm{d}}{\mathrm{d}t}[\mathbf{\Omega}, \mathbf{r}_0] = [\mathbf{\Omega}, [\mathbf{\Omega}, \mathbf{r}_0]].$$

Substituting this into the preceding relation and taking into account the obvious identity $\mathbf{r} = \mathbf{r}_0 + \mathbf{r}'$, we finally obtain

$$m\mathbf{w}_r = \underbrace{m\mathbf{g}}_{\text{gravity force}} \underbrace{- 2m[\mathbf{\Omega}, \mathbf{v}_r]}_{\text{Coriolis force}} - \underbrace{m[\mathbf{\Omega}, [\mathbf{\Omega}, \mathbf{r}]]}_{\text{centrifugal force}} = \mathbf{F}_{\mathrm{g}} + \mathbf{F}_{\mathrm{Cor}} + \mathbf{F}_{\mathrm{cen}}. \tag{13}$$

Assuming air to be a continuous medium, making the replacements $m \to \rho$, $\mathbf{w}_r \to \mathrm{d}\mathbf{U}/\mathrm{d}t$, $\mathbf{v}_r = \mathbf{U}$ (i.e., applying equation (13) to the unit air volume), and taking into account viscous stresses and hydrodynamic air pressure, we obtain the following system from (13):

$$\begin{gathered}\frac{\mathrm{d}\rho}{\mathrm{d}t} + \rho\,\mathrm{div}\mathbf{U} = 0, \quad \frac{\mathrm{d}\mathbf{U}}{\mathrm{d}t} = \mathbf{g} - 2[\mathbf{\Omega}, \mathbf{U}] - [\mathbf{\Omega}, [\mathbf{\Omega}, \mathbf{r}]] + \frac{\mu}{\rho}\Delta\mathbf{U} - \frac{\nabla p}{\rho}, \\ \frac{\mathrm{d}}{\mathrm{d}t} = \frac{\partial}{\partial t} + \mathbf{U}\cdot\nabla, \quad \mathbf{r} = \mathbf{r}_0(t) + \mathbf{r}',\end{gathered} \tag{14}$$

where all spatial derivatives are in the variables $\mathbf{r}'$. System (14) should be supplemented by an equation for pressure p, but for the sake of simplicity, we consider only incompressible air flows in tornado, and therefore, assume that $\rho = \mathrm{const}$, $\mathrm{div}\mathbf{U} = 0$. Let us compare the orders of magnitude of the terms in the right-hand side of (14). Since $r = |\mathbf{r}| \cong R_{\mathrm{E}} = 6371\,\mathrm{km} = 6.371 \cdot 10^8\,\mathrm{cm}$, $\Omega = 7.3 \cdot 10^{-5}\,\mathrm{s}^{-1}$, $\mu/\rho = 0.15\,\mathrm{cm}^2/\mathrm{s}$ (for air temperature 20°C [24]), we have: $|\mathbf{F}_{\mathrm{g}}|/|\mathbf{F}_{\mathrm{cen}}| \geq g/(\Omega^2 R_{\mathrm{E}}) \cong 3\cdot 10^2$, and thus, the centrifugal force is much smaller than the gravity force. On the other hand, we have $|\mathbf{F}_{\mathrm{g}}|/|\mathbf{F}_{\mathrm{Cor}}| \geq g/(2\Omega|\mathbf{U}|) = 671.7 \cdot 10^4/|\mathbf{U}|$. Assuming that the air speed in a tornado is about $|\mathbf{U}| \simeq 100\,\mathrm{m/s} = 10^4\,\mathrm{cm/s}$ [3,6], we conclude that the Coriolis force is hundreds of times smaller than the gravity force. Now, let us compare the Coriolis force with the viscous stresses: $|\mathbf{F}_{\mathrm{Cor}}|/(\mu\rho^{-1}|\Delta^2\mathbf{U}|) \simeq 2\Omega L_0^2/(\mu\rho^{-1}) = 1.82 \cdot 10^{-4} \cdot L_0^2$, where L_0 is the characteristic length. It follows that when studying small-scale processes ($L_0 \sim 1\mathrm{cm}$), we can neglect the Coriolis force, as compared with viscous stresses, while for large-scale processes ($L_0 \sim 1\,\mathrm{m}$), the Coriolis force and viscous stresses are of the same order of magnitude. Below, we take into account viscous effects (which are essentially small-scale) and show that they are responsible for swirling and vertical acceleration of air masses typical for tornado. It should be emphasized (and this is one of the main results of this work) that, in our model, the said phenomena are caused exclusively by the viscous effects and not by the Coriolis force or convective processes. Therefore, in what follows, we neglect the Coriolis and the centrifugal forces and our final system of equations takes the form (1).

Note that the incompressible fluid approximation is justified if $U^2 \ll c_s^2$, where c_s is the speed of sound in air. According the the Fujita scale [2], the maximal value of U in a powerful tornado is about $U \simeq 100\,\mathrm{m/s}$, while the average speed of sound in air is $c_s \simeq 330\,\mathrm{m/s}$; thus, $(U/c_s)^2 < 0.1$. For a tornado of medium power, the Fujita scale [2] yields $(U/c_s)^2 \simeq 10^{-2}$. Therefore, it can be assumed that the relation $(U/c_s)^2 \ll 1$ is approximately satisfied for air flows in tornado, and this justifies the above incompressible fluid approximation for this phenomenon.

Our numerical analysis of some particular solutions of system (1) in the form (2) shows that they describe the motion of air masses typical for tornadoes.

IOP Conf. Series: Journal of Physics: Conf. Series **1336** (2019) 012001 doi:10.1088/1742-6596/1336/1/012001

5. Numerical Solution of the Boundary Value Problem (9)

Consider an interval $[0, L]$ with a sufficiently large L (the results should not substantially change with the growth of L) and a uniform grid $x_k = kh$ with integer $0 \le k \le N$, $N = L/h$, and step $h > 0$. We use the difference scheme

$$\begin{gathered}\frac{u_k^1 - u_k^0}{\tau} + C_k^0 \frac{u_{k+1}^1 - u_{k-1}^1}{2h} - \frac{u_{k+1}^1 - 2u_k^1 + u_{k-1}^1}{h^2} + u_k^1 u_k^0 + \mathit{\Gamma} = 0, \quad 0 < k < N, \\ u_0^1 = \mathrm{i}, \quad u_N^1 = \sqrt{-\mathit{\Gamma}},\end{gathered} \tag{15}$$

and the sweep method to find a solution u_k^1, $0 < k < N$, of the following system of linear equations:

$$\begin{gathered}u_{k+1}^1 \left[\frac{\tau}{2h} C_k^0 - \frac{\tau}{h^2}\right] + u_k^1 \left[1 + \tau u_k^0 + \frac{2\tau}{h^2}\right] + u_{k-1}^1 \left[-\frac{\tau}{2h} C_k^0 - \frac{\tau}{h^2}\right] = u_k^0 - \tau \mathit{\Gamma},\ 0 < k < N, \\ u_0^1 = \mathrm{i}, \quad u_N^1 = \sqrt{-\mathit{\Gamma}}.\end{gathered} \tag{16}$$

We seek C on each upper layer as a solution of the following Cauchy problem on the interval $[0, L]$:

$$\frac{\mathrm{d}C^1}{\mathrm{d}z} = -2\mathrm{Re}\, u^1, \quad C^1(0) = 0, \tag{17}$$

taking into account that the right-hand side of (17) is given at the nodes x_k. Since the approximation order of system (15) is $O(\tau + h^2)$, we can use a method of a relatively small order for finding a solution of problem (17). For instance, we can use the Runge-Kutta second order method, which reduces to the quadrature trapezoid formula

$$\begin{gathered}C_k^1 = C_{k-1}^1 - h\left(A_k^1 + A_{k-1}^1\right), \quad k = 1, \ldots, N, \quad C_0^1 = 0, \\ A_k^1 = \mathrm{Re}\, u_k^1.\end{gathered} \tag{18}$$

The sweep stability condition is satisfied if the diagonal coefficients are predominant in the matrix of system (16),

$$\left|\frac{\tau}{2h} C_k^0 - \frac{\tau}{h^2}\right| + \left|\frac{\tau}{2h} C_k^0 + \frac{\tau}{h^2}\right| < \left|1 + \tau u_k^0 + \frac{2\tau}{h^2}\right|, \quad 1 \le k \le N-1.$$

This condition holds a fortiori if

$$\tau < \frac{h}{2 \max\limits_{0 \le k \le N} \left|C_k^0\right|}, \qquad \tau < \frac{1}{2 \max\limits_{0 \le k \le N} \left|u_k^0\right|}. \tag{19}$$

The inequalities (19) yield a constraint on the time step τ.

The simplest condition for stopping the iterations for the stabilization problem has the form

$$\max_{0 \le k \le N} \max\left\{\left|u_k^1 - u_k^0\right|, \left|C_k^1 - C_k^0\right|\right\} < \varepsilon,$$

where ε is a given small positive parameter, its typical value being equal to $\varepsilon = 10^{-9}$. As soon as the last inequality is satisfied, the calculations stop and u_k^1, C_k^1, $0 \le k \le N$, are taken as the grid approximations of the desired stationary flow. The above algorithm for finding stationary solutions of the boundary value problem (9) is much simpler than direct methods, such as the shooting method [27] or the Bellman quasilinearization method [28], used for finding a solution of the boundary value problem for the nonlinear 5th order system of ODE's obtained from (9) by setting $\partial/\partial t = 0$. This can be explained by the fact that in the latter algorithms, the key role belongs to the Newton method for solving systems of nonlinear equations and the convergence of this method is determined by the initial approximation of the unknown solution, which makes calculations problematic.

Prior to describing our calculation results, we discuss the role of viscosity with regard to the existence of solutions corresponding to tornadoes.

6. The Role of Viscosity

In the case of zero viscosity, $\mu = 0$, the boundary value problem (9) does not make sense, since the boundary conditions have no physical meaning. One can try to reformulate the boundary value problem in the case of $\mu = 0$, but most likely with no result, since (as it is shown below) among the stationary solutions of system (9) with $\mu = 0$, there are no flows whose the vertical velocity grows and stabilizes as $z \to +\infty$. There is a possibility that such flows exist among nonstationary solutions of the form (2). This is an important problem, since real tornadoes are nonstationary and the funnel moves over the surface of the Earth. However, in order to analyze these processes, the original setting of the problem should take into account additional physical factors [29,30], for instance, winds over the surface. Moreover, for $\mu = 0$, the Coriolis force might take precedence [31,32], but then functions of the form (2) would not yield particular solutions of system (1) with the additional Coriolis term and the above approach to finding tornado solutions of the form (2) would become inconsistent. Note also that the role of viscosity in the generation of vortex flows is well known: the layers of a viscous fluid adhere to each other, transferring their motion to the neighboring layers, which contributes to the formation of vortices. It should also be taken into account that air in a real tornado contains a lot of water and sand dust (depending on whether it is formed above the surface of water or land), which substantially increases the hydrodynamic viscosity of the air mixture. Viscosity is also responsible for the transformation of azimuthal and radial movements of air in a tornado into a vertical movement.

Thus, consider system (9) with $\partial/\partial t = 0$:

$$C\frac{\mathrm{d}u}{\mathrm{d}z} + u^2 + \Gamma = 0, \qquad \frac{\mathrm{d}C}{\mathrm{d}z} = -2\mathrm{Re}\, u. \tag{20}$$

It turns out that one can find all solutions of system (20) in explicit form.

Theorem 2. 1) *For $\Gamma = 0$, the general solution of system (20) has the form $u = A + \mathrm{i}B$* with

$$A(z) = -\frac{Rk}{2}\sin 2kz, \quad B(z) = Rk\sin^2 kz, \quad C(z) = R\sin^2 kz, \tag{21}$$

where $R, k \in \mathbb{R}$ are arbitrary real constants. Moreover, there is a particular solution of the form

$$A(z) = -Rz, \quad B(z) = 0, \quad C(z) = Rz^2, \tag{22}$$

where $R \in \mathbb{R}$ is an arbitrary constant.

2) *For $\Gamma > 0$, the general solution of system (20) has the form*

$$A(z) = -\sqrt{\Gamma}\frac{\sin 2pz}{\mathrm{sh}2\sqrt{\Gamma}R}, \quad B(z) = -\sqrt{\Gamma}\left(\mathrm{th}\sqrt{\Gamma}R\cos^2 pz + \mathrm{cth}\sqrt{\Gamma}R\sin^2 pz\right),$$
$$C(z) = \frac{1}{2D}\left(\frac{\cos^2 pz}{\mathrm{ch}^2\sqrt{\Gamma}R} + \frac{\sin^2 pz}{\mathrm{sh}^2\sqrt{\Gamma}R}\right), \tag{23}$$

where $p = D\sqrt{\Gamma}\,\mathrm{sh}\, 2\sqrt{\Gamma}R$, $R \neq 0$, $D \neq 0$ are arbitrary constants. Moreover, there is a particular solution of the form

$$A(z) = -\Gamma Dz, \quad B(z) = 0, \quad C(z) = \frac{1}{D} + \Gamma Dz^2, \tag{24}$$

IOP Conf. Series: Journal of Physics: Conf. Series **1336** (2019) 012001 doi:10.1088/1742-6596/1336/1/012001

where $D \neq 0$ is an arbitrary constant.

Theorem 2 implies that in the absence of viscosity, $\mu = 0$, the vertical speed of the stationary flow (2) is a periodic function of the height or, in exceptional cases, it has quadratic dependence on the height, which does not agree with the vertical speed variation in a tornado.

7. Some Calculation Results

Our numerical analysis of stationary solutions of the boundary value problem (9) is based on approximate solutions of this problem obtained by the stabilization method described above. These calculation results do not depend on the initial data. For instance, one can take $u(0,z) = (\mathrm{i} + z\sqrt{-\Gamma})/(1+z)$, $0 \leq z < +\infty$. The algorithm was tested on known solutions: for $\Gamma = 0$, we obtain the von Karman solution [23,24] and for $\Gamma = 1$, the constant solution $u(z) \equiv \mathrm{i}$, $C(z) \equiv 0$. For $0 \leq \Gamma < 1$, the vertical speed $C(z)$ obtained in our calculations is always negative ("vacuum cleaner") and for $\Gamma > 1$, it is positive ("tornado"). Thus, $\Gamma = 1$ is the bifurcational value of the parameter Γ equal to the squared angular vortex velocity in upper atmosphere. Typical profiles of $A(z)$, $B(z)$, $C(z)$ for stationary flows for various values of Γ are represented in Figure 2. Thus, the vertical velocity $C(z)$ for $\Gamma > 1$ is positive and tends to the limit value $C(\infty) > 0$ as $z \to +\infty$. The dependence of $C(\infty)$ on Γ, obtained numerically, is represented in Figure 3 , which shows that with the growth of the angular vortex velocity in upper atmosphere, $C(\infty)$ grows monotonically as $\sim \sqrt{\Gamma}$ for large Γ. Another important result, which has been obtained for the left boundary condition $u(0) = 0$ and the other boundary data unchanged, shows (see Figure 4) that the presence of vortex motion near the surface due to, say, the Coriolis force, cannot be regarded as a necessary condition for the birth of tornado. Moreover, comparing the graphs of the vertical velocity in Figures 2 and 4, we see that without any vortex motion at $z = 0$, the limit vertical velocity $C(\infty)$ has increased $\cong 2.5$ times. Thus, air rotation near the surface reduces the air flow energy at infinity, while air rotation in the upper atmosphere increases that energy (see Figure 2). Another important consequence of our calculations is that if the rotation of air at infinity and that near the origin have opposite directions (say, $u(0) = \mathrm{i}$, $u(\infty) = -\mathrm{i}\sqrt{\Gamma}$), then no stabilization of the solution is observed, which means that in this case the problem has no stationary solutions.

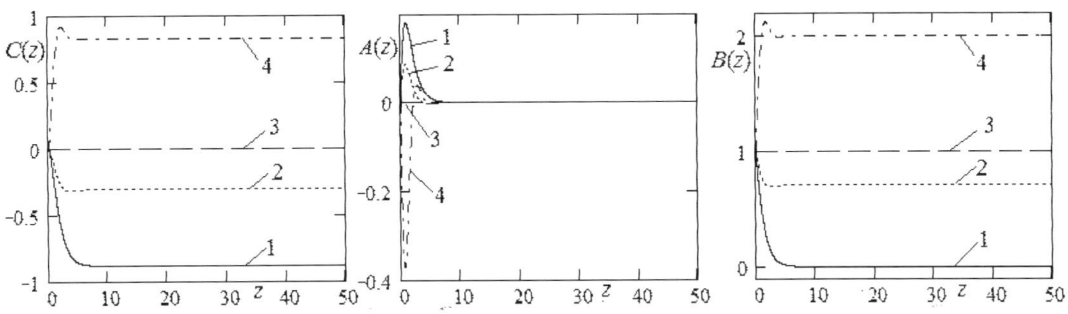

Figure 2. Steady profiles of $C(z)$, $A(z)$, $B(z)$ for Γ: (1) $\Gamma = 0$; (2) $\Gamma = 0.5$; (3) $\Gamma = 1$; (4) $\Gamma = 4$.

IOP Conf. Series: Journal of Physics: Conf. Series **1336** (2019) 012001 doi:10.1088/1742-6596/1336/1/012001

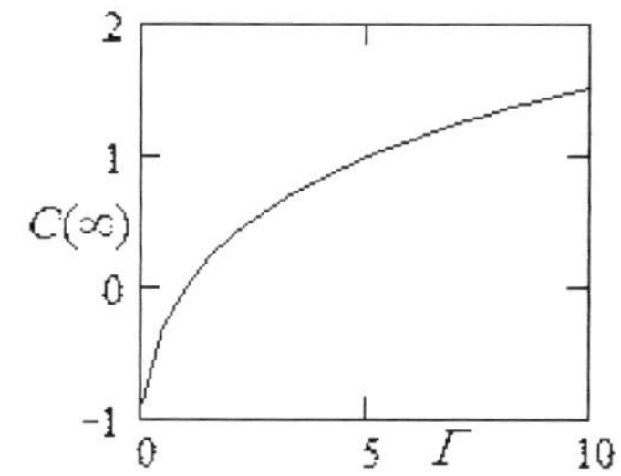

Figure 3. Vertical speed C at infinity versus Γ.

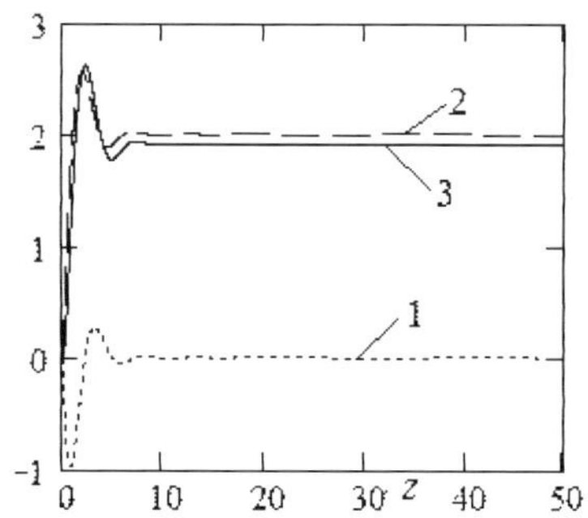

Figure 4. Steady profiles for $u(0) = 0$ and $\Gamma = 4$: (1) $A(z)$; (2) $B(z)$; (3) $C(z)$.

8. Conclusion

The numerical analysis described in this article can be used for some practical recommendations with regard to the tornado problem. First, it is necessary to monitor upper layers of the atmosphere for the presence of vortices and make a map of air pressure near the surface of the Earth. Secondly, tornadoes can occur only in those areas of land or water surface which are located under the corresponding vortices in upper atmosphere. A necessary condition for the emergence of a tornado is

$$\Omega^2 = \frac{1}{\rho}\left\langle \frac{\partial^2 p}{\partial r^2} \right\rangle, \tag{25}$$

where Ω is the angular velocity of a vortex, ρ is the air density, $\langle \partial^2 p/\partial r^2 \rangle$ is the spatial average of the second radial derivative of air pressure, the radius being measured from the center of a possible tornado, i.e., the intersection of the Earth's surface with the straight line passing through the center of the Earth and that of a vortex in upper atmosphere. (Note that the constant Q in (2) is equal to $Q = 1/2 \cdot \partial^2 p/\partial r^2$). Thirdly, to prevent a tornado, one should either create an anti-vortex, at least of low intensity, in the center of the supposed tornado near the Earth's surface, or change the pressure on the periphery of the expected tornado in order to violate condition (25). The above considerations allow us to conclude that the role of the Coriolis force is opposite to the role usually attributed to it. The Coriolis force, which causes spinning of air currents near the surface of the Earth, does not lead to the formation of a tornado [1,2], but, on the contrary, prevents the emergence of a tornado if the direction of air rotation at the surface due to the Coriolis force is opposite to the direction of rotation of air masses in the upper layers of the atmosphere.

References

[1] Bautin S P 2008 *Tornado and the Coriolis Force* (Novosibirsk: Nauka)
[2] Varaksin A Yu, Roman M E and Kopeytsev V N 2011 *Tornado* (Moscow: Fizmatlit)

[3] Nalovkin D V 1969 *Hurricanes, Storms and Tornadoes. Geographical Features and Geological Activity* (Leningrad: Nauka)
[4] Bautin S P 1997 *Mathematical Theory of Shock-Free Strong Compression of Ideal Gas* (Novosibirsk: Nauka)
[5] Yih C-S 2007 *Physics of Fluids* **19** pp 076601-1–076601-6
[6] Arsen'yev S A, Nikolaevskii V N and Shelkovnikov V N 2000 *Moscow University Physics Bulletin* **55** 1 P 57
[7] Shtern V, Borissov A and Hussain F 1997 *Physics of Fluids* textbf9 pp 2941–2959
[8] Rotunno R 1977 *J. Atmospheric Sciences* **34** 12 pp 1942–1956
[9] Rotunno R 1979 *J. Atmospheric Sciences* **36** 1 pp 140–155
[10] Gall R L 1982 *J. Atmospheric Sciences* **39** 12 pp 2721–2736
[11] Walko R 1986 and Gall R L *J. Atmospheric Sciences* **43** 20 pp 2137–2148
[12] Fiedler B H and Rotunno R 1986 *J. Atmospheric Sciences* **43** 21 pp 2328–2340
[13] Kiemp J B and Wilhelmson R B 1978 *J. Atmospheric Sciences* **35** 6 pp 1070–1096
[14] Wicker L J and Wilhelmson R B 1995 *J. Atmospheric Sciences* **52** 15 pp 2675–2703
[15] Grasso L D and Cotton W R 1995 *J. Atmospheric Sciences* **52** 12 pp 1192–1203
[16] Lewellen W S, Lewellen D C and Sykes R I 1997 *J. Atmospheric Sciences* **54** 5 pp 581–605
[17] Lewellen D C, Lewellen W S and Xia J 2000 *J. Atmospheric Sciences* **57** 4 pp 527–544
[18] Xia J, Lewellen W S and Lewellen D C 2003 *J. Atmospheric Sciences* **60** 22 pp 2820–2825
[19] Lewellen D C and Lewellen W S 2007 *J. Atmospheric Sciences* **64** 7 pp 2176–2194
[20] Lewellen D C and Lewellen W S 2007 *J. Atmospheric Sciences* **64** 7 pp 2195–2209
[21] Liu S, Wang Z, Gang Z, Chen F and Peng Q 2006 *J. Zhejiang University Science* A **7** 7 pp 1099–1106
[22] Liu S, Wang Z, Gang Z and Peng Q 2007 *The Visual Computer* **23** pp 559–567
[23] Karman T 1921 *ZAMM* **1** pp 244–247
[24] Landau L D and Lifshitz E M 1987 *Fluid Mechanics* vol 6 (2nd ed Butterworth-Heinemann)
[25] Loitsyansky L G and Lurie A I 1982 *The course of theoretical mechanics* vol 1 (Moscow: Nauka)
[26] Arnold V I 1989 *Mathematical methods of classical mechanics* (Moscow: Nauka)
[27] Hall G and Watt J M 1976 *Modern Numerical Methods for Ordinary Differential Equations* (Oxford: Oxford University Press)
[28] Bellman R and Kalaba R 1965 *Quasilinearization and nonlinear boundary-value problems* (New York: Elsevier) 218 p
[29] Bautin S P 2007 *Mathematical Modeling of Strong Compression of Gases* (Novosibirsk: Nauka)
[30] Varaksin A Yu, Romash M E, Kopeytsev V N and Taekin S I 2010 *High Temperature* **48** 2 pp 251–255
[31] Alekseenko S V, Kuybin P A and Okulov V L 2005 *An Introduction to the Theory of Concentrated Vortices* (Moscow-Izhevsk: Institute for Computer research)
[32] Varaksin A Yu, Roman M E and Kopeytsev V N 2010 *High Temperature* **48** 4 pp 617–622

Simulation of forest fires based on a two-dimensional three-phase model

A A Kuleshov[1,2,4], E E Myshetskaya[1,5] and S E Yakush[3,6]

[1] Keldysh Institute of Applied Mathematics, Russian Academy of Sciences, Miusskaya sq. 4, Moscow, 125047, Russia
[2] Author to whom any correspondence should be addressed.
[3] Ishlinskii Institute for Problems in Mechanics, Russian Academy of Sciences, Prospekt Vernadskogo, 101-1, Moscow, 119526, Russia

[4]andrew_kuleshov@mail.ru
[5]kuleshov@imamod.ru
[6]yakush@ipmnet.ru

Abstract. A two-dimensional three-phase mathematical model of forest fires constructed by the method of averaging over the height of the forest fuel layer is considered. The gas phase in the model is described by gas dynamics equations with the $k-\varepsilon$ turbulence model and the eddy breakup model to describe the rate of turbulent combustion and the rate of heat generation; the solid phase is described by a system of evolution equations for the change in the volume fractions of forest fuel and solid products of their pyrolysis with kinetic formulas for the rates of chemical reactions; and the disperse phase is described by the transfer equations for the mass and energy of solid burning particles in the wind field above the forest fuel layer. For the numerical implementation of the proposed model, a parallel software package is created. The results of calculations of the forest fire propagation with consideration of the basic physical-chemical processes and mechanisms typical of the real forest wildfire propagation are presented.

1. Introduction

Forest fires are a dangerous natural phenomenon that causes annually a significant economic and ecological damage. An important element for a successful forest-fire fighting is the prediction of the process of their propagation; the most efficient tool for such predictions is a mathematical modeling method that does not require considerable computational resources.

Mathematical modeling of forest fires has intensively developed over the past twenty years; many various models of forest fires are created up to the present time; a brief review of these models is presented in [1]. One of the directions of modeling of forest fires which was developed by the authors of [1-3] is the prediction of the fire-front propagation based on two-dimensional mathematical models. In the early studies of the authors of [2, 3], on the basis of the general approach to the description of forest fires as a multiphase reactive medium [4], a two-dimensional two-phase model of forest fires in which the medium consists of an air-gas (gas) phase and a solid phase was developed. This model was

Content from this work may be used under the terms of the Creative Commons Attribution 3.0 licence. Any further distribution of this work must maintain attribution to the author(s) and the title of the work, journal citation and DOI.
Published under licence by IOP Publishing Ltd

based on the fundamental mass, momentum, and energy conservation laws; the equations for these laws were obtained by averaging the initial three-dimensional equations over the height of the forest fuel (FF) layer. The model considers all basic physical processes that pass in the fire zone and may be applied on modern multiprocessor computer complexes to perform practically real-time modeling of the forest wildfire propagation [2, 3]. However, this model used a simplified description of turbulent mixing; the heat generation in the gas phase (the rate of fire-front propagation depends on this parameter) was described by kinetic formulas for the rates of chemical reactions with constants assigned with a significant uncertainty. Recently, the authors have proposed a modified two-dimensional model [1] which describes turbulence by means of the known $k-\varepsilon$ model with additional terms for the turbulence generation and dissipation in the woodland [5, 6]. In this model, the rate of turbulent combustion in the gas phase is described by the eddy breakup model, according to which the rate of combustion at high temperatures does not depend on the kinetics of chemical reactions and is determined only by the rate of turbulent mixing of the components [7-10].

In models [1-3], one of the components of the gas phase is highly dispersed soot which is transported with the gas phase. In our study, we consider a three-phase model as a development of the modified two-dimensional model [1]; in this three-phase model, the third phase consists of large burning disperse particles transported in the wind field above the FF layer. Considering the disperse particles as a separate phase moving with a velocity different from that of the gas phase, it is possible to describe the fire propagation mechanism observed in real forest fires with the help of burning particles, which are transported by the wind above the forest canopy and, precipitating, generate new sources of fire ahead of the leading edge of the fire front. The distances to which the large burning particles are transported in the horizontal direction depend on the wind velocity, burning time, mass of particles and their geometric shape that influences on their aerodynamics.

2. Mathematical model

Let us consider a one-stratum model of a forest wildfire in one FF layer of a constant thickness, assuming that the physical-chemical parameters (real density, moisture content, calorific power, etc.) are constant in this FF layer. This model considers a three-phase medium which consists of the gas, solid, and disperse phases. The gas phase consists of the following six components: combustible gas CO (with the mass concentration C_1), oxidizer O_2 (C_2), carbon dioxide CO_2 (C_3), water vapor H_2O (C_4), nitrogen N_2 (C_5), and highly-dispersed soot (C_s). It is assumed that the particles of highly-dispersed soot move together with the gas phase; in the process of soot burning, the heat exchange passes rapidly and, thus, we can consider that the gas phase temperature is unified. The solid phase is also multicomponent; it consists of the FF (with the volume fraction ϕ_1) and the FF pyrolysis products – coke breeze (ϕ_2) and ash (ϕ_3). Coke breeze and fine-dispersed soot consist almost entirely of carbon and, thus, no solid remainder is left after the combustion. The disperse phase (with the volume fraction ϕ_d) consists of burning carbon particles of the same shape and dimension (monodisperse mixture) which are transported by the wind above the FF layer and settle down due to the action of the gravity force.

The system of equations that describe the two-dimensional three-phase model was obtained by integrating the original three-dimensional equations over the FF layer (the forest stratum); it consists of the subsystem of the Favre-average gas dynamics equations for the multicomponent gas phase, the subsystem of equations for the mass of the components of the solid phase and its energy, and the subsystem of the mass and energy transfer equations for the large disperse particles.

The model equations for the gas phase have the following form:

$$\frac{\partial \rho}{\partial t}+\nabla\cdot(\rho\mathbf{V})=Q-J_{\rho},$$

$$\frac{\partial \rho\mathbf{V}}{\partial t}+\nabla\cdot(\rho\mathbf{V}\mathbf{V})+\nabla(\phi p)=\nabla\cdot(\mu_{eff}\nabla\mathbf{V})+\mathbf{F}+\mathbf{F}_w-\mathbf{J}_{\mathbf{V}},$$

$$\frac{\partial \rho E}{\partial t}+\nabla\cdot\left(\rho \mathbf{V}E+\phi p\mathbf{V}\right)=\nabla\cdot\left(k_{eff}\nabla T\right)+\alpha\left(T_1-T\right)+\sigma\left(æ_1T_1^4-æT^4\right)+$$
$$+\alpha_1(T_d-T)+\sigma\left(æ_dT_d^4-æT^4\right)-\nabla\cdot\mathbf{W}_r+f_\Gamma+f_{\Gamma R}+q_1W+q_sW_s-J_E,$$
$$\frac{\partial \rho C_i}{\partial t}+\nabla\cdot\left(\rho\mathbf{V}C_i\right)=\nabla\cdot\left(\rho D_{eff}\nabla C_i\right)+Q_i+W_i+W_i^s-J_{C_i},\quad i=\overline{1,5},\qquad (1)$$
$$\frac{\partial \rho C_s}{\partial t}+\nabla\cdot\left(\rho\mathbf{V}C_s\right)=Q_s-W_s-J_s,$$
$$\frac{\partial \rho k}{\partial t}+\nabla\cdot\left(\rho\mathbf{V}k\right)=\nabla\cdot\left(\frac{\mu_{eff}}{\sigma_k}\nabla k\right)+G+P_k-\rho\varepsilon-L_k,$$
$$\frac{\partial \rho \varepsilon}{\partial t}+\nabla\cdot\left(\rho\mathbf{V}\varepsilon\right)=\nabla\cdot\left(\frac{\mu_{eff}}{\sigma_\varepsilon}\nabla \varepsilon\right)+c_1G\frac{\varepsilon}{k}+P_\varepsilon-c_2\frac{\rho\varepsilon^2}{k}-L_\varepsilon,$$
$$p=\rho^0RT\sum_{i=1}^{5}\frac{C_i}{M_i};$$

for the solid phase:

$$\rho_j^0\frac{\partial\phi_j}{\partial t}=R_{\phi_j},\quad j=\overline{1,3};$$
$$\frac{\partial}{\partial t}\sum_{j=1}^{3}\rho_j^0\phi_jc_{pj}T_1=-\alpha\left(T_1-T\right)-\sigma\left(æ_1T_1^4-æT^4\right)-\alpha_2(T_1-T_d)-\sigma\left(æT_1^4-æ_dT_d^4\right)+Q_{T_1};\qquad (2)$$

for the disperse phase:

$$\frac{\partial\mathbf{V}_d}{\partial t}=-c_f(\mathbf{V}_d-\mathbf{V}_w)\left|\mathbf{V}_d-\mathbf{V}_w\right|,$$
$$\frac{\partial\rho_d}{\partial t}+\nabla\cdot\left(\rho_d\mathbf{V}_d\right)=R_d-J_d,$$
$$\frac{\partial\rho_dc_{pd}T_d}{\partial t}+\nabla\cdot\left(\rho_d\mathbf{V}_dc_{pd}T_d\right)=-\alpha_1(T_d-T)-\alpha_2(T_d-T_1)-\qquad (3)$$
$$-\sigma(æ_dT_d^4-æT^4)-\sigma(æ_dT_d^4-æ_1T_1^4)-J_{T_d}+Q_{T_d};$$

the normalizing and balance relations are as follows:

$$\phi+\sum_{j=1}^{3}\phi_j+\phi_d=1,\quad \sum_{i=1}^{5}C_i+C_s=1,\quad \sum_{i=1}^{5}(Q_i+W_i+W_i^s)+Q_s-W_s=Q,\quad \sum_{j=1}^{3}R_{\phi_j}=-Q.\qquad (4)$$

The system of equations (1)–(4) is considered in the Cartesian coordinates in the rectangular domain $\Omega=\{0\le x\le l_1, 0\le y\le l_2\}$ on the horizontal plane *XY*; ρ is the partial density of the gas phase; *p* is the total pressure of the multiphase medium; $\mathbf{V}=(u,v)$ is the velocity vector of the gas phase; *T* is the temperature of the gas phase; *E* is the total energy of the gas phase, $E=0.5(u^2+v^2)+c_vT$; c_v is the specific heat capacity of the gas phase; ϕ is its volume fraction; $\rho^0=\rho/\phi$ is its real density; *R* is the universal gas constant; M_i is the molecular mass of component i of the gas phase; $\mathbf{F}=-\rho c_d s\mathbf{V}|\mathbf{V}|$ is the volume force associated with the interphase momentum exchange (the interphase friction force); c_d is the empirical vegetation resistance coefficient; s is the FF specific surface; $\mathbf{F}_w=-\rho\xi(\mathbf{V}-\mathbf{V}_w)|\mathbf{V}-\mathbf{V}_w|$ is the friction force conditioned by the wind with the velocity $\mathbf{V}_w$ at the upper boundary of the FF layer; ξ is the dimensionless empirical friction coefficient; Q is the rate of the gas inflow into the gas medium due to chemical processes in the solid phase; ρ_j^0, $j=\overline{1,3}$ is the real density of component j of the solid phase, $\rho_j^0=\text{const}$; R_{ϕ_j} is the rate of the mass variation for component j of the solid phase due to chemical reactions; T_1 is the temperature of the solid phase; Q_{T_1} is the heat generation in the solid phase in the combustion process; c_{pj} is the specific heat capacity of the solid phase; W is the mass rate of the CO combustion reaction; W_s is the mass rate of the soot combustion; W_i, W_i^s, $i=\overline{1,5}$ are the production and consumption rates of the gas phase components in the corresponding reactions, $\rho_d, T_d, \mathbf{V}_d=(u_d, v_d), \phi_d$ are the partial density, temperature, velocity, and volume fraction, respectively; the particle resistance coefficient c_f depends on the shape and cross section of the particles which are assumed to be identical (we consider a monodisperse mixture), and the Reynolds number; $\mathbf{V}_w=(u_w, v_w)$ is the wind velocity above the woodland; $\phi_d=\rho_d/\rho_d^0$, where ρ_d^0 and c_{pd} are the real density and specific heat capacity of the material of disperse particles, respectively, and, since the disperse particles and coke breeze consist of carbon, $\rho_d^0=\rho_2^0$, $c_{pd}=c_{p2}$ and they are equal to the corresponding parameters of carbon.

The turbulent closure of the system of equations for the gas phase, is the modified model of turbulence in which the right-hand sides of equations for the turbulent kinetic energy k and its dissipation rate ε contain, in addition to the usual expression for the generation of the turbulence energy G, extra terms P_k, P_ε and L_k, L_ε, which describe, respectively, the turbulence generation and dissipation due to the interaction between the flow and FF [5, 6]. The effective viscosity coefficient is equal to the sum $\mu_{eff}=\mu_l+\mu_t$ of the laminar and turbulent viscosities, where the laminar viscosity is calculated by using the Sutherland formula $\mu_l=\dfrac{1.458\,T^{3/2}}{T+110.4}10^{-6}$, while the turbulence viscosity is calculated according to the formula $\mu_t=c_\mu\rho k^2/\varepsilon$;

$$G=\mu_t\left\{2\left[\left(\frac{\partial u}{\partial x}\right)^2+\left(\frac{\partial v}{\partial y}\right)^2\right]+\left(\frac{\partial u}{\partial y}+\frac{\partial v}{\partial x}\right)^2\right\};$$

$$P_k=\frac{1}{2}c_d s\rho|\mathbf{V}|^3,\quad P_\varepsilon=c_{\varepsilon1}c_d s\rho|\mathbf{V}|^3\frac{\varepsilon^0}{k^0},\quad L_k=2c_d s\rho|\mathbf{V}|k,\quad L_\varepsilon=4c_{\varepsilon2}c_d s\rho|\mathbf{V}|\varepsilon,$$

where the following standard set of constants of the $k-\varepsilon$ model of turbulence is used: $c_\mu=0.09$, $c_1=1.44$, $c_2=1.92$, $\sigma_k=1.0$, $\sigma_\varepsilon=1.3$, while for the terms $P_\varepsilon, L_\varepsilon$ the following constants obtained with the help of the in-situ experiments [5, 6] are used: $c_{\varepsilon1}=1.8, c_{\varepsilon2}=0.6$; k^0,ε^0 are the typical values of turbulence parameters in a woodland which are also used as the initial values for k and ε.

The effective heat conductivity coefficient $k_{eff}=k_l+k_t+k_r$ is the sum of the molecular, turbulent, and radiation heat conductivity coefficients, respectively, where $k_l=c_p\mu_l/\text{Pr}$, $k_t=c_p\mu_t/\text{Pr}_t$, $\text{Pr}=\text{Pr}_t=0.7$ are the laminar and turbulent Prandtl numbers, respectively. The radiation heat conductivity model was considered as a model of radiation transport. In this model, the radiation energy flux density is defined by the formula $\mathbf{W}_r = -k_r \nabla T$, $k_r = 16\sigma l_s T^3/3$, l_s is the radiation path length, σ is the Stefan–Boltzmann constant.

The effective diffusion coefficient is defined by the formula $D_{eff} = \mu_l/\rho\mathrm{Sc} + \mu_t/\rho\mathrm{Sc}_t$, where $\mathrm{Sc} = \mathrm{Sc}_t = 0.7$ are the laminar and turbulent Schmidt numbers, respectively.

The combustion model for the gas phase consists of the two following single-stage irreversible CO-combustion brutto-reactions for the dispersed soot:

$$1\text{ kg CO} + \nu_{O_2}\text{ kg O}_2 \Rightarrow \nu_{CO_2}\text{ kg CO}_2,\quad 1\text{ kg C} + \nu^s_{O_2}\text{ kg O}_2 \Rightarrow \nu^s_{CO_2}\text{ kg CO}_2,$$

where ν_i, ν_i^s, $i = \overline{1,5}$ are the mass stoichiometric coefficients of the gas-phase and dispersed-soot combustion reactions, respectively. The turbulent combustion rate is described by the eddy breakup model [7-10]. The mass combustion rates for CO and the soot are expressed by the formulas (the chemical formulas of the components are used instead of index *i*)

$$W = \rho A \frac{\varepsilon}{k} \min\left(C_{CO}, \frac{C_{O_2}}{\nu_{O_2}}, \frac{BC_{CO_2}}{\nu_{CO_2}}\right),\qquad W_s = \rho A \frac{\varepsilon}{k} \min\left(C_s, \frac{C_{O_2}}{\nu^s_{O_2}}, \frac{C_s \nu^s_{O_2}}{C_s \nu^s_{O_2} + C_{CO}\nu_{O_2}}\right),$$

respectively, were $A = 4$, $B = 0.5$. The heat generation in the gas phase occurs due to the combustion of CO and dispersed soot; it is equal to $q_1 W + q_s W_s$, where q_1, q_s are the combustion heat for the gas and soot, respectively; *W* is the rate of the chemical reaction of gas combustion; W_s is the soot burn-out rate; W_i, W_i^s, $i = \overline{1,5}$ are the formation or consumption rates for the gas-phase components in the corresponding reactions, which are expressed via the rates *W*, W_s and the mass stoichiometric coefficients by the formulas $W_i = \pm\nu_i W$, $W_i^s = \pm\nu_i^s W_s$, where the "plus" sign is valid for the reaction products and the "minus" sign is valid for the fuel and oxidizer; $\nu_{CO} = 1$, according to the definition of *W*.

Let us consider the coke breeze pyrolysis and combustion reactions as the basic chemical reactions in the solid phase. Let us describe the rates of these reactions according to the Arrhenius law. The rate of the FF pyrolysis reaction is described by the formula $R_0 = \rho_1 \phi_1 k_0 \exp(-E_0/(RT_1))$ [4]. Let us introduce the following designations: $\alpha_i, \alpha_s, \alpha_c, \alpha_{as}$ are the portions of gas *i*, dispersed soot, coke breeze, and ash, respectively, which are produced due to the pyrolysis; for oxygen and nitrogen, $\alpha_2 = \alpha_5 = 0$. Then, the generation of gases, soot, coke breeze, and ash due to the pyrolysis is described by the terms $Q_1 = \alpha_1 R_0$, $\alpha_3 R_0$, $Q_4 = \alpha_4 R_0$, $Q_s = \alpha_s R_0$, $\alpha_c R_0$, $R_{\phi_3} = \alpha_{as} R_0$, respectively; the FF burn-out rate is $R_{\phi_1} = -R_0$.

The coke breeze combustion rate is described by the formula $R_c = \rho C_2 k_c \exp(-E_c/(RT))$ [4]. The coke breeze combustion with the combustion rate R_c is accompanied with the oxygen burn-out in the gas phase and generation of CO_2 which inflows into the gas phase. Then, $Q_2 = -R_c$, $Q_3 = \alpha_3 R_0 + R_c$, $R_{\phi_2} = \alpha_c R_0 - R_c$.

The change in the mass of the disperse phase is described by the formula $R_d = \alpha_d R_0 - R_c$, where the first term describes the generation of the disperse particles due to the pyrolysis; the second term, their burn-out.

The heat generation due to the breeze coke combustion is described by the formula $Q_{T_1} = q_c \phi_2 R_c$, where q_c is the heat combustion for carbon. The heat generation due to the disperse particle combustion is described by the formula $Q_{T_d} = q_c \phi_d R_c$.

The matter, momentum, and energy flows through the upper and lower boundaries of the FF layer

$J_\rho, \mathbf{J}_V, J_E, \mathbf{J}_C, J_s, J_d$, respectively are described by the formula

$$\mathbf{J}_\Phi = \frac{\Phi\, w|_{z=h_2}}{h} - \frac{\Phi\, w|_{z=h_1}}{h}, \quad \Phi = (\rho, \rho u, \rho v, \rho E, \rho\mathbf{C}, \rho C_s)^{\mathrm{T}},$$

where h_1, h_2 are the heights of the upper and lower boundaries of the FF layer, respectively; h is the FF layer thickness, $h = h_2 - h_1$; the vertical velocity w at the upper and lower boundaries of the FF layer is defined by the semi empirical expression [4] $w|_{z=h_2} = (g(T - T_2)h/T_2)^{1/2}$, T_2 is the temperature above the upper boundary of the FF layer (for the top forest stratum, it is the temperature of the unperturbed atmosphere). In the considered one-stratum forest model, $w|_{z=h_1} = 0$. For the multistratum model, the value of $w|_{z=h_1}$ has to be assigned equal to the value of $w|_{z=h_2}$ for the lower stratum.

The term $\alpha(T_1 - T)$ describes the interphase heat exchange, where α is the heat-exchange coefficient; $\sigma(æ_1 T_1^4 - æT^4)$ is the interphase radiant energy exchange, where æ is the gas-phase radiation absorption coefficient, which takes into account the dispersed soot; $æ_1$ is the solid-phase radiation absorption coefficient.

The fallout of burning disperse particles from the surface atmosphere layer is modeled with the help of a random number generator; the terms $\alpha_1(T_d - T)$, $\alpha_2(T_d - T_1)$, $\sigma(æ_d T_d^4 - æT^4)$, and $\sigma(æ_d T_d^4 - æ_1 T_1^4)$ are the point energy sources in the gas and solid phases that act at the sites of fallout deposition of burning particles up to their complete burn-out; $æ_d$ is the absorption coefficient for the disperse particle radiation.

The laws for the heat exchange f_Γ and radiant energy exchange $f_{\Gamma R}$ at the upper and lower boundaries of the FF layer are the same as those for the interphase heat exchange and radiant energy exchange, respectively.

Boundary and initial conditions

The boundary conditions for the system of equations (1)–(4) are to be set at the open lateral boundary Γ of the computational domain. For the parts of the boundary where the flow enters the computational domain, the unperturbed environment parameters (with index a) are set as follows:

$$\Phi|_\Gamma = \Phi_a, \quad \Phi = (\rho, u, v, T)^{\mathrm{T}}, \quad (\mathbf{V}, \mathbf{n}) \le 0;$$

for the parts of the boundary where the gas flows out of the computational domain, the following conditions are set:

$$\left.\frac{\partial \Phi}{\partial \mathbf{n}}\right|_\Gamma = 0, \quad \Phi = (\rho, u, v, E, \mathbf{C}, \rho_d, T_d)^{\mathrm{T}}, \quad (\mathbf{V}, \mathbf{n}) > 0.$$

The initial values of the parameters outside the fire source are the environment parameters; inside the fire source, we set the values of the parameters characteristic for the considered kind of fire and FF type $(T^0, T_1^0, T_d^0, \mathbf{C}^0, C_s^0, \rho_d^0)$; the initial velocities of the gas and disperse phases are $\mathbf{V}^0 = 0, \mathbf{V}_d^0 = 0$; the pressure at the initial instant of time is set to be identical to the atmospheric pressure, $p^0 = p_a$; the unknown value of the initial gas-phase density can be derived from the state equation.

3. Results of numerical simulation

The algorithm and software for the numerical implementation of model (1)–(4) were based on the algorithm and software developed for the two-dimensional two-phase model described in [1], with inclusion of additional stages to the scheme of splitting with respect to physical processes to calculate the velocity, mass, and energy of the disperse phase. In the calculations of the interphase heat exchange and radiant energy exchange, we have considered additionally the exchange between the gas and disperse phases and between the solid and disperse phases.

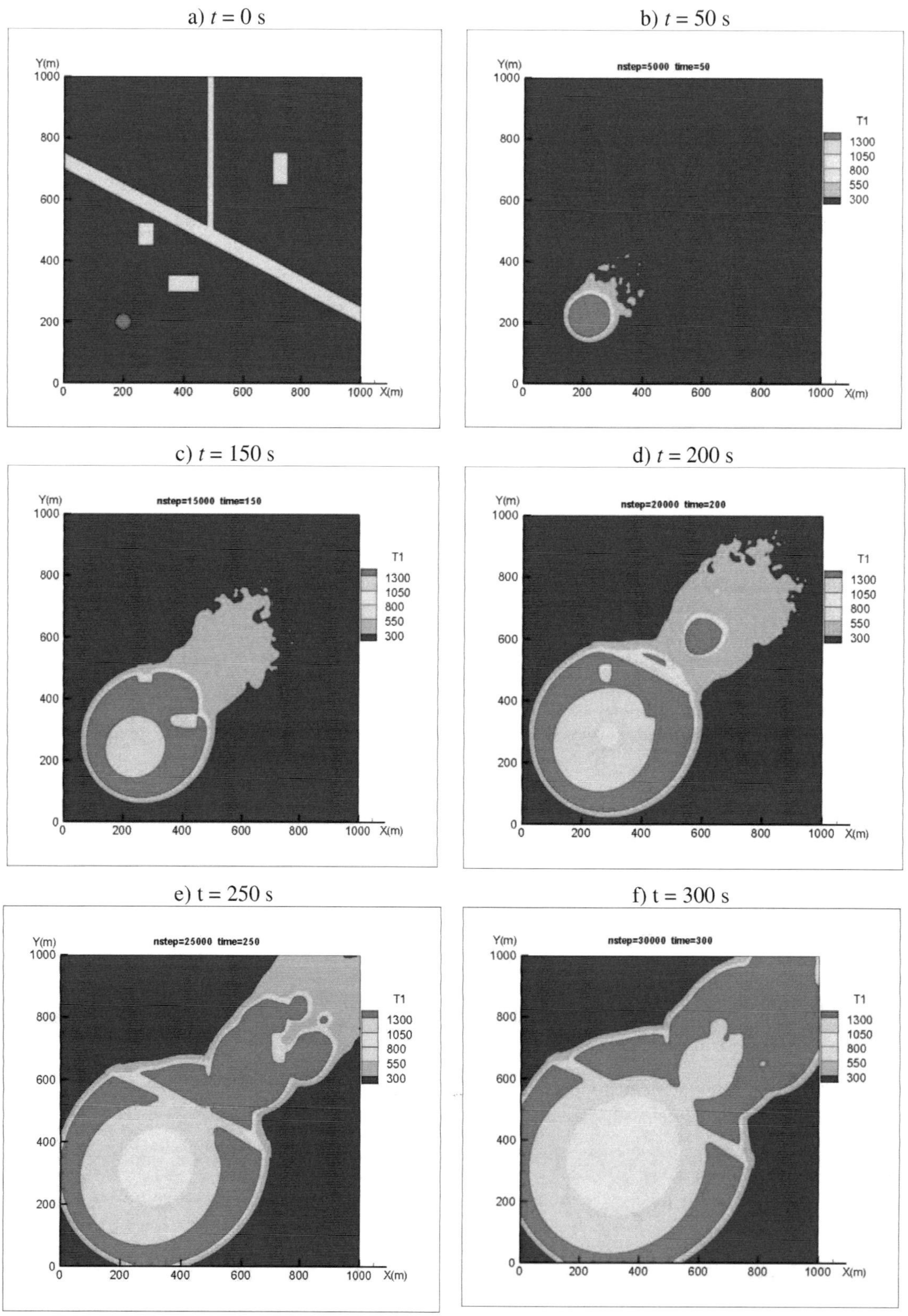

Figure 1. Propagation of the forest fire front.

Numerical simulation of forest fires was carried out on the HPC MBC-15000BM in Russian Academy of Sciences, Moscow.

Figure 1 shows the results of computations of the forest fire propagation by using the two-dimensional three-phase model (1)–(4) under the conditions of inhomogeneous FF distribution and the presence of the wind.

Figure 1a corresponds to the initial time; it shows schematically three rectangular glades and two roads with no vegetation. The fire source has a circular shape. The wind velocity is $\mathbf{V}_w = (4,4)$ m/s. Figures 1b–1f show the propagation of the forest fire temperature front in time. The fire front propagates in the wind direction. The wind transports the burning particles which are brought by the wind ahead of the main fire front, precipitate and form new fire sources which join the main fire front. Such a mechanism of fire propagation is observed in the real forest fires. It is seen in Fig. 1 how the fire front bypasses the glades and propagates over the roads. As the FF burns out, the combustion process stops, the temperature in this region gradually decreases and the back edge of the fire front is formed.

4. Conclusions

A two-dimensional three-phase model of forest fire propagation is developed and implemented; this model describes the fire-front propagation with consideration of the main physicochemical processes typical of the real forest fire propagation, including a local inhomogeneous FF distribution, the presence of obstacles for the fire propagation, the presence of a wind, as well as with consideration of the mechanism of fire propagation due to the burning particles transported by the wind over the FF layer.

References

[1] Kuleshov A A, Myshetskaya E E and Yakush S E 2017 *Math. Model. Comput. Simul.* **9** (4) 437–447
[2] Kuleshov A A and Myshetskaya E E 2009 *Math. Model. Comput. Simul.* **1** (5) 629–634
[3] Kuleshov A A, Chetverushkin B N and Myshetskaya E E 2013 *Computers&Fluids* **80** 202–206
[4] Grishin A M 1992 Mathematical Modeling of Forest Fires and New Methods of Fighting Them (Novosibirsk: Nauka) 404 (in Russian)
[5] Liang Li, Xiaofeng Li, Borong L and Yinghin Z 2006 *Atmos. Environ.* **40** 762–770
[6] Hiraoka H and Ohashi M 2008 *J. Wind Eng. Ind. Aerodyn.* **96** 2139–2149
[7] Magnussen B F and Hjertager B H 1976 *Proc. 16th Int. Symp. on Combustion (Pittsburgh)* (The Comb. Institute., Pittsburg, PA) 711–729
[8] Galant S, Grouset D, Martinez G, Micheau P and Allemand J B 1984 *Proc. 20th Int. Symp. on Combustion (Pittsburgh)* (The Comb. Institute., Pittsburg, PA) 531–540
[9] Makhviladze G M, Roberts J P and Yakush S E 1999 *Combust. and Flame* **118** 583–605
[10] Makhviladze G M and Yakush S E 2004 *IChemE Symp. Series* **150** 270–282

Optimization of the radiation transport calculation for quasi-one-dimensional model of the ionizing gas flows

A N Kozlov [1, 2] **and V S Konovalov** [1]

[1] Keldysh Institute of Applied Mathematics, Russian Academy of Sciences, Miusskaya sq. 4, Moscow, 125047, Russian Federation
[2] Lomonosov Moscow State University, Faculty of Mechanics and Mathematics, Leninskiye Gory 1, Moscow, 119991, Russian Federation

andrey-n-kozlov@mail.ru & v.s.konovalov@yandex.ru

Abstract. Algorithms were developed to solve the problem of radiation transport based on the method of long characteristics, taking into account its optimization in the framework of the quasi-one-dimensional approximation used to study the axisymmetric flows of the ionizing gas in coaxial channels of plasma accelerators. The optimization of calculations of the integral radiation parameters allowed to significantly reduce the calculation time in comparison with three-dimensional model of radiation transport. Calculations of the radiation energy density and the radiation energy flux density, based on various methods, have shown that new algorithms provide high-quality solution of the radiation transport problem.

1. Introduction

The study of radiation transport is one of the actual problems of modern computational plasma dynamics. The radiation leads to the redistribution of energy in the plasma volume, and also affects the ionization process in the channel of the quasi-stationary plasma accelerator (QSPA) [1–9], schematically shown in Fig. 1a. The plasma in the presence of an azimuthal magnetic field and a predominantly radial plasma current is accelerated in the QSPA due to the Ampere force $\mathbf{j}\times\mathbf{H}/c$, where $\mathbf{j}$ is the current in the plasma.

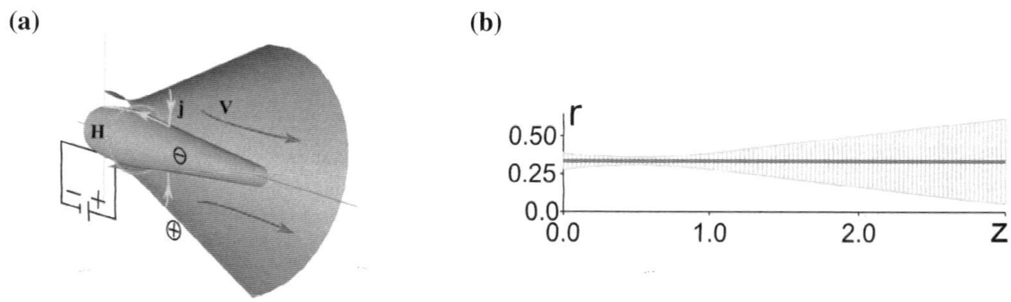

Figure 1. (a) Scheme of the QSPA. (b) Channel section

Different models of radiation transport in the channel of the QSPA have been developed earlier, including the model of the radiant heat conductivity and the diffusion approach (see e.g. [10]). Recently, the 3D model of radiation transport in the QSPA has been developed based on the method of long characteristics (see e.g. [11-13]), which allows to take into account the details of the geometry of the accelerator channel and the formation of shadow regions [14]. The solution of the spectral

Content from this work may be used under the terms of the Creative Commons Attribution 3.0 licence. Any further distribution of this work must maintain attribution to the author(s) and the title of the work, journal citation and DOI.
Published under licence by IOP Publishing Ltd

radiation transport equation in three-dimensional formulation is time-consuming task in terms of computer time. In this connection, it became necessary to construct the simplified model that would allow qualitatively and quickly to find the solution of the radiation transport problem based to study the ionizing gas flow in the quasi-one-dimensional approximation, taking into account the channel geometry, the emission spectrum and the possibility of parallel implementation of computational code.

The quasi-one-dimensional approximation is used to numerically simulate the axisymmetric flows in the narrow channel [15–17] and allows to effectively reduce the calculation time. The study of the ionizing gas flow with the formation of an ionization front in the QSPA channel is carried out on the basis of the MHD models of various levels of complexity, using the local thermodynamic equilibrium (LTE) approximation (see e.g. [14, 15]) and the modified diffusion approximation (MDA) for the equation of the ionization and recombination kinetics [16-17]. The most complete MHD model includes the set of equations of the level kinetics for populations of atomic levels. In this paper, we propose the appropriate optimization method to calculate the radiation transport in quasi-one-dimensional model of observable axisymmetric flows. Calculations based on the developed algorithm are compared with the results obtained by means of the 3D model of radiation transport. The proposed method of the simplified calculation was used to study the effect of radiation transport on the ionization process in the QSPA channel.

2. Radiation transport equation

Stationary radiation transport equation contains absorption coefficient $\kappa_\nu(\mathbf{r})$ and emissivity $\eta_\nu(\mathbf{r})$:

$$\mathbf{\Omega}\cdot\nabla I_\nu(\mathbf{r},\mathbf{\Omega})=\eta_\nu(\mathbf{r})-\kappa_\nu(\mathbf{r})\cdot I_\nu(\mathbf{r},\mathbf{\Omega}) \tag{1}$$

Here, the radiation intensity $I_\nu(\mathbf{r},\mathbf{\Omega})$ in the direction of the spatial angle $\mathbf{\Omega}$ is a function of coordinate $\mathbf{r}$ and photon frequency ν. The corresponding integrals of the radiation intensity determine the spectral density of the radiation energy $U_\nu(\mathbf{r})$ and the spectral flux of the radiation energy $\mathbf{W}_\nu(\mathbf{r})$:

$$U_\nu(\mathbf{r})=\frac{1}{c}\int_0^{4\pi} I_\nu(\mathbf{r},\mathbf{\Omega})\,d\Omega\,, \qquad \mathbf{W}_\nu(\mathbf{r})=\int_0^{4\pi} I_\nu(\mathbf{r},\mathbf{\Omega})\,\mathbf{\Omega}\,d\Omega\,. \tag{2}$$

The equations of radiation magnetic gas dynamics contain the integral values of the radiation energy density $U(\mathbf{r})=\int_0^\infty U_\nu(\mathbf{r})\,d\nu$ and the radiation energy flux density $\mathbf{W}(\mathbf{r})=\int_0^\infty \mathbf{W}_\nu(\mathbf{r})\,d\nu$.

Further, we consider one-group approximation and will not indicate the index ν in (1) and (2). Figure 1(b) shows the geometry of plasma accelerator channel and the grid in quasi-one-dimensional approximation. The line marked in red in the figure corresponds to the middle radius in the channel.

The integral radiation parameters will calculate at points located on the middle line between the electrodes. If parameter N_z is the number of grid nodes along the z axis, equal to $N_z=201$, for example, then the number of layers or cells is equal to $N_l=200$, and the number of elements of boundary surfaces will be equal to $N_b=402$, including the ring boundaries for each layer on the outer and inner surfaces of the coaxial electrodes, as well as the inlet and outlet boundaries. For each node of the coordinate grid, we construct an angular grid with the number of rays N_Ω, which can vary.

The absorption coefficient $\kappa_\nu(\mathbf{r})$ and emissivity $\eta_\nu(\mathbf{r})$ in equation (1) depend on the state of the medium, its density and temperature, as well as the photon energy $h\nu$. These values are determined by means of the known relations (see e.g. [14, 18-22]).

Graphs of typical dependences κ_ν and η_ν on photon energy are presented in figure 2 for hydrogen plasma taking into account the first 20 energy levels at characteristic temperature $T=1\,eV$ and particle density $n_o=10^{16}\,cm^{-3}$. The populations of atomic states are related by the Saha relation and the Boltzmann formula under conditions of local thermodynamic equilibrium. In the calculations, the whole spectrum is divided into 521 groups. Value inverse of the absorption coefficient corresponds to the characteristic free path of photons $l_\nu=1/\kappa_\nu$ with a frequency ν.

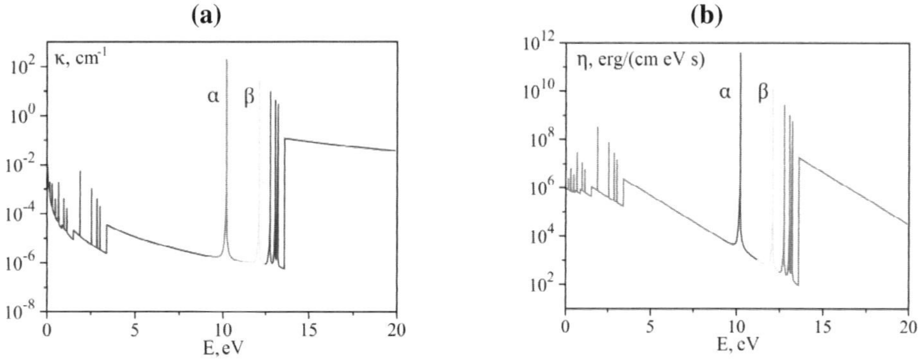

Figure 2. Dependences of (a) absorption coefficient and (b) emissivity on radiation energy

3. Calculation of radiation field based on the method of long characteristics

The radiation transport equation (1) is the first order equation for which any ray or direction of the radiation propagation can be considered as a characteristic. The phenomenon of refraction of rays is not taken into account.

We consider a solution on the characteristics (see e.g. [11-14]). Figure 3 shows one-dimensional spatial grid, where points correspond to the nodes at which the intensity is determined in order to then find the integral radiation parameters. Within each cell, the absorption coefficient and emissivity are considered constant. Solution on the characteristic with a direction $\mathbf{\Omega}_i$ at node n is constructed from the boundary of the computational domain at $\mathbf{r} = \mathbf{r}_b$ with the known intensity value $I(\mathbf{r}_b, \mathbf{\Omega}_i)$:

$$I(\mathbf{r}_n, \mathbf{\Omega}_i) = I(\mathbf{r}_b, \mathbf{\Omega}_i) \cdot e^{-\tau_{nk}^i} + \sum_{j=n+1}^{k} \frac{\eta_j}{\kappa_j}\left(1 - e^{-\tau_{jj}^i}\right) \cdot e^{-\tau_{nj}^i} + \frac{\eta_n}{\kappa_n}\left(1 - e^{-\tau_{nn}^i}\right), \tag{3}$$

where $\tau_{nk}^i = \sum\limits_{j=n,\dots,k} \kappa_j \cdot \Delta l_j^i$ is the photon optical path along the given characteristic with the direction $\mathbf{\Omega}_i$ from cell n to the cell k, value Δl_j^i is length of the path inside the cell j, the index b corresponds to the boundary element, the index i indicates the characteristic direction marked in the figure 3. In this case, we have $\Delta l_j^i = \Delta z_j / \cos\theta_i$. Further, we use the notation $I_{ni} \equiv I(\mathbf{r}_n, \mathbf{\Omega}_i)$.

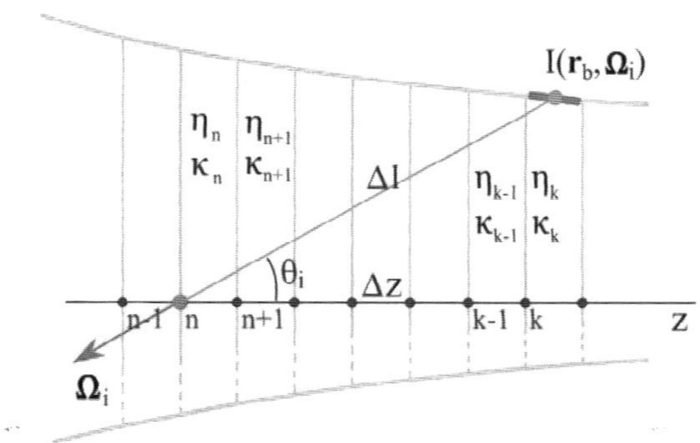

Figure 3. Ray tracing through cells

The number of terms in the sum (3) is equal to the number of layers or cells intersected by the characteristic. As a result, radiation intensities on the given spatial and angular grids are calculated for all rays and nodes $\left\{I\left(\mathbf{r}_n, \Omega_i\right)\right\}_{n=1,\dots,N_z}^{i=1,\dots,N_\Omega}$, where N_Ω is the number of rays of the angular grid. Search for a solution is assumed for all spectral groups taking into account the dependence on the photon frequency ν. Further, the radiation energy density $U_n = U(\mathbf{r}_n)$ and the radiation energy flux density

$\mathbf{W}_n = \mathbf{W}(\mathbf{r}_n)$ are calculated at the nodes of the spatial grid for the found intensities $\{I_{ni}\}_{n=1,...,N_z}^{i=1,...,N_\Omega}$, using the relations (2). In this case, the integrals correspond to the sums, in which the frequency index is not specified for convenience:

$$U_n = \frac{1}{c} \sum_{i=1}^{N_\Omega} I_{ni} \, \Delta\Omega_i \,, \qquad \mathbf{W}_n = \sum_{i=1}^{N_\Omega} I_{ni} \, \mathbf{\Omega}_i \, \Delta\Omega_i \,. \qquad (4)$$

The results of solving the problem of radiation transport in the 3D formulation will be presented for the points of the middle coordinate line between the electrodes, shown in figure 1(b). The calculation of the ionizing gas flow with the formation of an ionization front was carried out on the basis of the MHD model taking into account the ionization and recombination kinetics in the framework of the MDA approximation [16, 17, 22]. The distributions of density and temperature in the channel are presented in figure 4 for the following characteristic values of the gas parameters at the channel inlet: $n_o = 8 \cdot 10^{16} cm^{-3}$ and $T_o = 500\ ^oK$. In this case, the discharge current in the accelerator is equal $J_d = 60\ kA$, and the unit of the channel length corresponds to value $L = 20\ cm$.

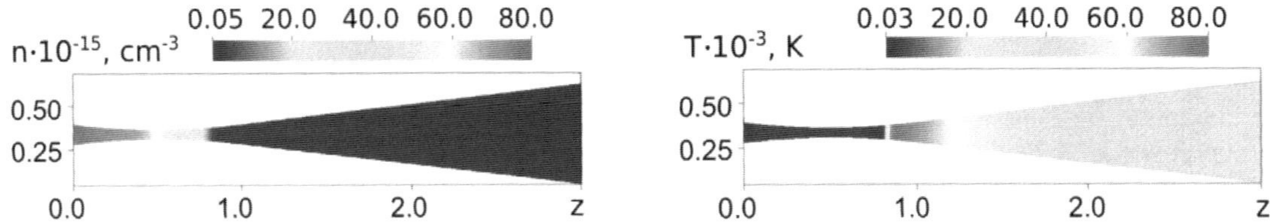

Figure 4. Distributions of density and temperature in the accelerator channel

The solid curves in figure 5 describe the corresponding dependences of the integral values of the radiation energy density $U(z)$ and the longitudinal component of the radiation energy flux density $W_z(z)$ calculated in the 3D model by means of the method of long characteristics for the angular grid with the number of rays $N_\Omega = 3280$. Calculation time in seconds is indicated in brackets. The number of nodes of the grid is 201, and the number of spectral groups is 521.

Relatively close results are obtained for various values N_Ω. The calculation of radiation transport problem takes 43 seconds for $N_\Omega = 220$. It needs 175 seconds for $N_\Omega = 840$, and calculation takes 670 seconds for $N_\Omega = 3280$ on the personal computer. This does not allow to use the method of long characteristics at each time step of the MHD problem, since the calculation of the quasi-stationary flow requires several thousand time steps. At the same time, the calculation of one time step in the numerical solution of the MHD problem takes approximately 0.1 seconds on the personal computer.

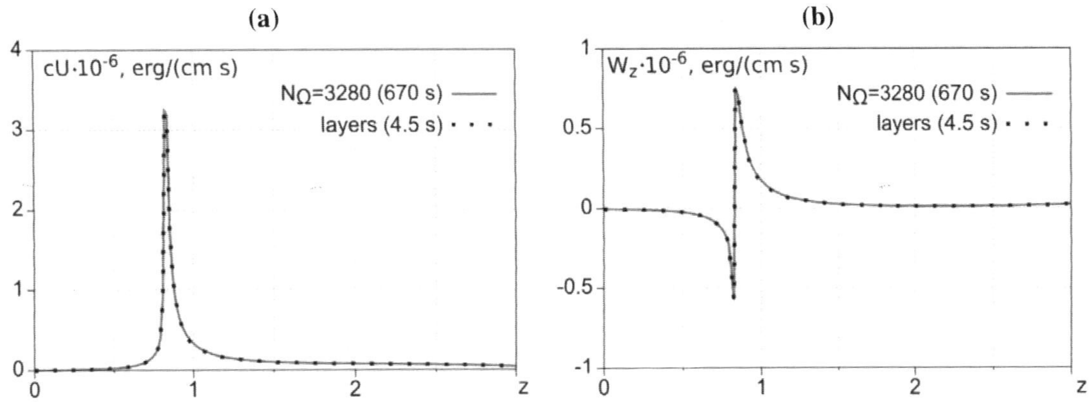

Figure 5. Changes in (a) radiation energy density and (b) radiation energy flux density, calculated by the method of long characteristics and the method of summation "by layers" of the region

In order to solve the radiation transport equation, a calculation of the large number of exponents is required, which is approximately equal to twice the number of terms in formula (3) and is proportional to the number of layers or segments of the computational domain (see figure 3). On average, each characteristic passes through $N_s \approx 50$ layers. Then the total number of calculated exponents for one spectral group will be approximately equal $2 \cdot N_s \cdot N_z \cdot N_{\boldsymbol{\Omega}} = 2 \cdot 50 \cdot 201 \cdot 60 \approx 1.2 \cdot 10^6$. This explains the long time to calculate the radiation field. Ways to optimize the calculation of radiation transport are associated with a decrease in the number of the calculated exponents.

4. Optimization of radiation transport calculation

The number of operations in the method of long characteristics and, accordingly, the calculation time depend on the number of rays $N_{\boldsymbol{\Omega}}$ of the angular grid. The time needed to solve the problem of radiation transport can be significantly reduced for the quasi-one-dimensional distributions of temperature and density, taking into account the layered structure of the distribution of thermodynamic parameters in the channel.

4.1. Optimization by mean of the rays coming from boundary

Summation over the directions of the angular grid can be replaced by summation over the elements of the boundary. In this case, the sum of the exponents in formulas (3) and (4) in the directions of the rays corresponding to a certain element of the boundary can be replaced by one exponential function of some tabulated function. This allows you to significantly reduce the number of the calculated exponential functions $\sum_{j=1}^{N} \alpha_j e^{-\tau\beta_j} \approx a e^{-b(\tau)}$, as illustrated in figure 6.

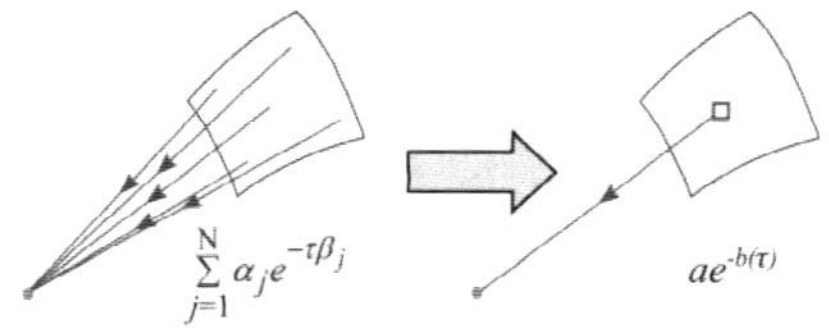

Figure 6. Illustration of replacement of sum calculation

We make the detailed tracing and consider the group of rays O_{nb} coming out of the ring element of the boundary b and coming into the node of the coordinate grid with the index n in accordance with the figure 7(a). Then relations (4) can be rewritten as follows:

$$U_n = \frac{1}{c} \sum_{b=1}^{N_b} \sum_{i \in \mathrm{O}_{nb}} I_{ni} \, \Delta\Omega_i \,, \qquad \mathbf{W}_n = \sum_{b=1}^{N_b} \sum_{i \in \mathrm{O}_{nb}} I_{ni} \, \boldsymbol{\Omega}_i \Delta\Omega_i \,,$$

where N_b is the number of ring elements of the boundary. We consider in more detail the sums over the group of rays O_{nb} in these relations and introduce the notation:

$$U_{nb} = \frac{1}{c} \sum_{i \in \mathrm{O}_{nb}} I_{ni} \, \Delta\Omega_i \,, \qquad \mathbf{W}_{nb} = \sum_{i \in \mathrm{O}_{nb}} I_{ni} \, \boldsymbol{\Omega}_i \Delta\Omega_i \,. \tag{5}$$

(a) (b)

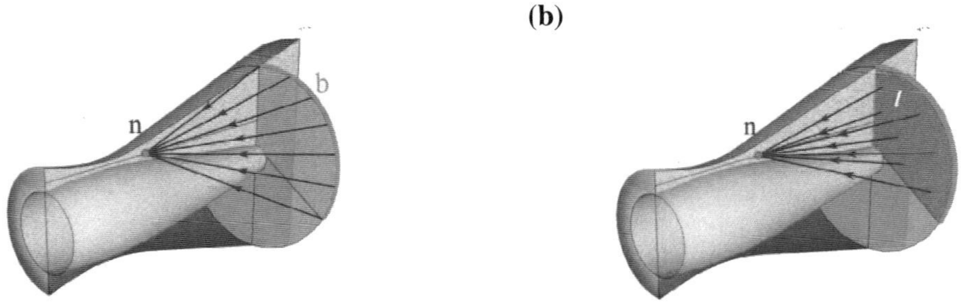

Figure 7. Tracing of rays coming from (a) boundary section and (b) plasma layer

IOP Conf. Series: Journal of Physics: Conf. Series **1336** (2019) 012003 doi:10.1088/1742-6596/1336/1/012003

We take into account that only the flow in the direction of the z axis is of interest in the quasi-one-dimensional formulation of the problem. Sums (5) can be represented in more detail:

$$U_{nb} = \frac{1}{c}\frac{\eta_{l_b}}{\kappa_{l_b}} \sum_{i\in O_{nb}} F_i \cdot \Delta\Omega_i\,, \quad W_{z\,nb} = \frac{\eta_{l_b}}{\kappa_{l_b}} \sum_{i\in O_{nb}} F_i \cdot \cos\theta_i\, \Delta\Omega_i\,,$$

$$F_i = \left(1-\exp(-\tau^i_{l_b l_b})\right)\exp(-\tau^i_{n l_b -1}) \tag{6}$$

where l_b is the index of the layer or cell related to the element of the boundary b, $\tau^i_{l_b l_b}$ is the length of the ray path in the layer or cell corresponding to the element of the boundary b, $\tau^i_{nl_b-1}$ is the length of the ray path from the node n to the cell related to the element of the boundary b. The length of the optical path of the ray from the layer m to the node n is calculated as the sum of:

$$\tau^i_{nm} = \sum_{j=n}^{j=m} \kappa_j\, \Delta l^i_j = \sum_{j=n}^{j=m} \kappa_j\, \Delta z_j / \cos\theta_i = \tau_{nm} / \cos\theta_i\,, \qquad \tau_{nm} = \sum_{j=n}^{j=m} \kappa_j\, \Delta z_j \tag{7}$$

Here, the summation is carried out over the cells through which the characteristic passes from the node with the index n to the cell with the number m. As a result, the sum is allocated that does not depend on the direction of the ray and $\cos\theta_i$.

Substituting formulas (7) into relations (6), we obtain the following dependences:

$$U_{nb} = \frac{1}{c}\frac{\eta_{l_b}}{\kappa_{l_b}} \sum_{i\in O_{nb}} G_i \cdot \Delta\Omega_i\,, \qquad W_{z\,nb} = \frac{\eta_{l_b}}{\kappa_{l_b}} \sum_{i\in O_{nb}} G_i \cdot \cos\theta_i\, \Delta\Omega_i\,,$$

$$G_i = \exp(-\tau_{nl_b-1}/\cos\theta_i) - \exp(-\tau_{nl_b}/\cos\theta_i)$$

in which there are typical sums:

$$P_{nb}(\tau) = \sum_{i\in O_{nb}} e^{-\tau/\cos\theta_i}\, \Delta\Omega_i\,, \qquad R_{nb}(\tau) = \sum_{i\in O_{nb}} e^{-\tau/\cos\theta_i} \cos\theta_i\, \Delta\Omega_i\,.$$

Formulas for calculating values U_n and W_{zn} can be rewritten taking into account new designations $P_{nb}(\tau)$ and $R_{nb}(\tau)$:

$$U_n = \frac{1}{c}\sum_{b=1}^{N_b} \frac{\eta_b}{\kappa_b} \cdot \left(P_{nb\text{-}1}(\tau_{nb\text{-}1}) - P_{nb}(\tau_{nb})\right),$$

$$W_{zn} = \sum_{b=1}^{N_b} \frac{\eta_b}{\kappa_b} \cdot \left(R_{nb\text{-}1}(\tau_{nb\text{-}1}) - R_{nb}(\tau_{nb})\right). \tag{8}$$

Functions $P_{nb}(\tau)$ and $R_{nb}(\tau)$ in relations (8) can be represented as follows:

$$P_{nb}(\tau) \approx \Omega_{nb} \cdot e^{-\mathrm{p}_{nb}(\tau)}\,, \qquad R_{nb}(\tau) \approx \Omega_{znb} \cdot e^{-\mathrm{r}_{nb}(\tau)}\,, \tag{9}$$

where $\mathrm{p}_{nb}(\tau)$ and $\mathrm{r}_{nb}(\tau)$ are the function parameters calculated for the given geometry of the accelerator channel. The value $\Omega_{nb} = P_{nb}(0) = \sum_{i\in O_{nb}} \Delta\Omega_i$ corresponds to the spatial angle of visibility of the layer boundary b from the point n. The value $\Omega_{znb} = R_{nb}(0) = \sum_{i\in O_{nb}} \cos\theta_i\, \Delta\Omega_i$ can be interpreted as a projection onto the z axis of the vector of the spatial angle of visibility of the layer boundary b from the point n. Accordingly, we rewrite the relations for U_n and W_{zn}:

$$U_n = \frac{1}{c}\sum_{b=1}^{N_b} \frac{\eta_b}{\kappa_b} \cdot \Omega_{nb}\left(e^{-\mathrm{p}_{nb}(\tau_{nb\text{-}1})} - e^{-\mathrm{p}_{nb}(\tau_{nb})}\right),$$

$$W_{zn} = \sum_{b=1}^{N_b} \frac{\eta_b}{\kappa_b} \cdot \Omega_{znb}\left(e^{-\mathrm{r}_{nb}(\tau_{nb\text{-}1})} - e^{-\mathrm{r}_{nb}(\tau_{nb})}\right). \tag{10}$$

Summation now does not occur to all the rays N_Ω of the angular grid, but along the elements of the boundary with the number of terms equal to N_b, each of which contains only two exponential functions. In this case, the calculation time does not depend on the number of rays of the angular grid, since the tracing is performed once at the preparatory stage to determine the function parameters $p_{ik}(\tau)$ and $r_{ik}(\tau)$. The preparatory procedure based on the 3D model can be performed in sufficient detail for the given channel geometry. As a result, data tables are formed to calculate the function parameters with different indices i and k.

4.2. Optimization by mean of the rays coming from plasma layers

The technique presented in the previous section for the boundary of the layer can significantly reduce the calculation time. We apply it to the whole layer, introducing the designation O_{nl} as a group of rays coming from the layer l and coming to the node n of the computational grid in accordance with figure 7(b). For the radiation energy density and the radiation energy flux density, we obtain the following relations by analogy with (10):

$$U_n = \frac{1}{c}\sum_{l=1}^{N_l} \frac{\eta_l}{\kappa_l}\cdot\Omega_{nl}\left(e^{-\mathrm{p}_{nl}(\tau_{n\,l\text{-}1})} - e^{-\mathrm{p}_{nl}(\tau_{n\,l})}\right),$$

$$W_{zn} = \sum_{l=1}^{N_l} \frac{\eta_l}{\kappa_l}\cdot\Omega_{znl}\left(e^{-\mathrm{r}_{nl}(\tau_{n\,l\text{-}1})} - e^{-\mathrm{r}_{nl}(\tau_{n\,l})}\right), \quad (11)$$

where N_l is the number of layers.

We assume that the number of layers or volume elements corresponds to the number of nodes in the coordinate grid $N_l = N_z = 200$. This number is 2 times less than the number of boundary ring elements $N_b = 400$, taking into account the surfaces of the inner and outer electrodes of plasma accelerator. Accordingly, the calculation time for the radiation transport problem is reduced by about two times compared with optimization based on the rays coming from the boundary.

It should be noted that the optimization by means of the boundary elements involves the setting of the boundary conditions on the surface of the coaxial electrodes. At the same time, the optimization method based on integration over the layers or the corresponding volume elements using relations (11) excludes the specification of boundary conditions on the surface of accelerator electrodes and involves the specification of boundary conditions at the channel inlet and outlet, which also provides the qualitative solution of the radiation transport problem.

4.3. Specification of function parameters to optimize calculations

In this section, we consider the function-parameters $\mathrm{p}_{ik}(\tau)$ and $\mathrm{r}_{ik}(\tau)$, which include relations (10) and (11). The characteristic dependences of the function parameters on the optical length τ are presented in figure 8.

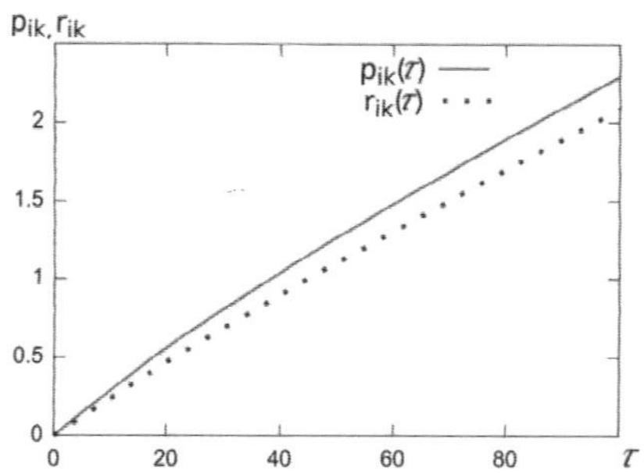

Figure 8. Typical dependencies of function-parameters on optical length

Functions $\mathrm{p}_{ik}(\tau)$ and $\mathrm{r}_{ik}(\tau)$ are equal to zero at zero values τ. They asymptotically tend to a linear function for large values τ and can be quite accurately approximated by a function of the form:

$$\mathrm{f}(\tau)=\frac{h\tau}{a\tau+1}+b\tau \qquad (12)$$

The parameters h, a and b in (12) are easily determined if the values $\mathrm{f}(\tau)$ are known at three arbitrary points τ_1, τ_2 and τ_3 other than zero. Values $\mathrm{f}_1=\mathrm{f}(\tau_1)$, $\mathrm{f}_2=\mathrm{f}(\tau_2)$ and $\mathrm{f}_3=\mathrm{f}(\tau_3)$ can be easily calculated after the tracing. To determine the parameters h, a and b in (12), we introduce additional values:

$$D_{10}=\frac{\mathrm{f}_1}{\tau_1}=\frac{h}{a\tau_1+1}+b,\quad D_{21}\equiv\frac{\mathrm{f}_2-\mathrm{f}_1}{\tau_2-\tau_1}=\frac{h}{(a\tau_1+1)(a\tau_2+1)}+b,\quad D_{32}\equiv\frac{\mathrm{f}_3-\mathrm{f}_2}{\tau_3-\tau_2}=\frac{h}{(a\tau_2+1)(a\tau_3+1)}+b,$$

$$\chi\equiv\frac{D_{10}-D_{21}}{\tau_2}=\frac{ha}{(a\tau_1+1)(a\tau_2+1)},\qquad \varphi\equiv\frac{D_{32}-D_{21}}{\tau_3-\tau_1}=\frac{ha}{(a\tau_1+1)(a\tau_2+1)(a\tau_3+1)}.$$

The consequence of the last two relationships is the equation $\varphi=\chi/(a\tau_3+1)$. Then we can consistently calculate the values of a ($\varphi\neq 0$), h ($a\neq 0$) and b:

$$a=(\chi/\varphi-1)/\tau_3,\ h=\chi(\tau_1+1/a)(\tau_2+1/a),\ b=D_{10}-h/(a\tau_1+1).$$

The degeneracy case $\varphi=0$ and $a=0$ is possible only under the condition $D_{10}=D_{21}=D_{32}$ that corresponds to a linear dependence $\mathrm{f}(\tau)=b\tau$. In this case, we have $a=0$, $h=0$ and $b=\mathrm{f}_1/\tau_1$.

As a result, we determine function-parameters $P_{ik}(\tau)=\Omega_{ik}\cdot e^{-\mathrm{p}_{ik}(\tau)}$ and $R_{ik}(\tau)=\Omega_{zik}\cdot e^{-\mathrm{r}_{ik}(\tau)}$ in relations (10) and (11), where the functions $\mathrm{p}_{ik}(\tau)$ and $\mathrm{r}_{ik}(\tau)$ have the form:

$$\mathrm{p}_{ik}(\tau)=\frac{h_{ik}^{p}\cdot\tau}{a_{ik}^{p}\cdot\tau+1}+b_{ik}^{p}\cdot\tau,\qquad \mathrm{r}_{ik}(\tau)=\frac{h_{ik}^{r}\cdot\tau}{a_{ik}^{r}\cdot\tau+1}+b_{ik}^{r}\cdot\tau.$$

The sets of parameters $\{\Omega_{ik},h_{ik}^{p},a_{ik}^{p},b_{ik}^{p}\}$ and $\{\Omega_{zik},h_{ik}^{r},a_{ik}^{r},b_{ik}^{r}\}$ can be saved for the given geometry and used in the numerical solution of the evolutionary problem of radiation magnetic gas dynamics, excluding the necessity for re-tracing at each time step. As reference points, we can choose values $\tau_1=1$, $\tau_2=10$ and $\tau_3=100$, for example, as well as a different set of values.

5. Results of radiation transport calculations based on optimization methods

Calculations of the radiation energy density and the radiation energy flux density in the quasi-one-dimensional model of the ionizing gas flow are based on the methods of summation over the elements of the boundary and summation over layers or elements of the volume. Comparison of these calculations among themselves showed that both of these optimization methods give almost identical results. The calculation time of the integral characteristics of radiation based on the method of summation over layers was reduced to 4.5 seconds. This is two times less than the calculation using the summing over the elements of the boundary, for which the calculation time was 9 seconds. In both cases, the computation time with optimization is significantly less than the computation time of the integral radiation parameters using the characteristics method in the 3D model. Results of calculations based on the standard method of characteristics were represented by solid curves in Fig. 5. In this case, the calculation time was 670 seconds for the angular grid with the number of rays $N_{\Omega}=3280$.

The distributions of the radiation energy density and the radiation energy flux density calculated on the basis of the summing method over layers are represented by the dashed curves in figures 5(a) and 5(b). It can be seen that calculations based on the method of summation over layers and the standard method of characteristics give practically identical results. Thus, the use of optimization methods can significantly reduce the computation time, ensuring the quality and accuracy of solving the problem of radiation transport.

The study of radiation transport in streams of ionizing gas showed that radiation, coming from the ionization front, in certain frequency range can penetrate deep into the region of the incoming neutral gas. Therefore, along with the redistribution of the internal energy of the medium and temperature, the radiation can lead to the effect of pre-ionization in the gas flow ahead of the ionization front.

The frequency in the radiation transport equation (1) plays the role of an independent parameter, which allowed to develop parallel program codes that were created using of the OpenMP technology.

6. Conclusion

Numerical algorithms have been developed for solving the problem of radiation transport in the axisymmetric ionizing gas flows with the formation of an ionization front in the coaxial channels of plasma accelerators. The proposed optimization methods to calculate of radiation transport in the framework of the quasi-one-dimensional MHD model of flows can significantly reduce the computation time of the integral radiation parameters in comparison with the 3D model of radiation transport based on the method of long characteristics. Comparison of the calculation results for the radiation energy density and the radiation energy flux density showed that optimization of calculations based on the summation of radiation from the ring elements of the boundary and the summation over the plasma layers allows to obtain the high-quality solutions of the problem.

References

[1] Morozov A I 2008 *Introduction to plasmadynamics (in Russian)* (Moscow: Fizmatlit)
[2] Morozov A I 1990 *Sov. J. Plasma Physics (in Russian)* **16** 69-78
[3] Brushlinskii K V, Zaborov A M, Kozlov A N, Morozov A I and Savelyev V V 1990 *Sov. J. Plasma Physics (in Russian)* **16** 79-87
[4] Garkusha I E *et al* 2017 *Nuclear Fusion* **57** 116011
[5] Klimov N *et al* 2009 *J. Nuclear Materials* **390-91** 721-26
[6] Astashynski V M *et al* 2004 *J. Surface and Coating Technology* **180-81** 392-95
[7] Kozlov A N, Drukarenko S P, Klimov N S, Moskacheva A A and Podkovyrov V L 2009 *Problems of Atomic Science and Technology. Series: Plasma Physics* **1** 92-4
[8] Kozlov A N 2008 *J. Plasma Physics* **74** 261-86
[9] Kozlov A N 2017 *Plasma Physics and Controlled Fusion* **59** 115004
[10] Kozlov A N, Garkusha I E, Konovalov V S and Novikov V G 2013 *Problems of Atomic Science and Technology. Series: Plasma Physics* **1** 128-30
[11] Chetverushkin B N 1985 *Mathematical modeling of radiating gas dynamic problems (in Russian)* (Moscow: Nauka)
[12] Castor J I 2000 *Lectures on radiation hydrodynamics* (Livermore: Lawrence Livermore National Laboratory)
[13] Gasilov V A 2012 *Matematicheskoe modelirovanie (in Russian)* **24** 55–87
[14] Kozlov A N and Konovalov V S 2017 *Communications in Nonlinear Science and Numerical Simulation* **51** 169-79
[15] Brushlinskii K V 2009 *Mathematical and computational problems of magnetic fluid dynamics (in Russian)* (Moscow: BINOM, Laboratoriya znanii)
[16] Kozlov A N 2000 *Fluid Dynamics* **35** 784-90
[17] Brushlinskii K V, Kozlov A N and Konovalov V S 2015 *Computational Mathematics and Mathematical Physics* **55** 1370–80
[18] Mihalas D 1978 *Stellar atmospheres. Second edition* (San Francisco: W. H. Freeman)
[19] Veinstein L A, Sobel'man I I and Yukov E A 1995 *Excitation of atoms and broadening of spectral lines* (Springer Series on Atomic, Optical, and Plasma Physics)
[20] Fortov V E 2016 *Thermodynamics and equations of state for matter: from ideal gas to quark-gluon plasma* (Singapore: World Scientific Publishing)
[21] Nikiforov A F, Novikov V G and Uvarov V B 2005 *Quantum-statistical models of hot dense matter. Methods for computation opacity and equation of state* (Switzerland, Basel: Birkhauser Verlag)
[22] Biberman L M, Vorobev V S and Yakubov I T 1982 *Kinetics of nonequilibrium low-temperature plasma (in Russian)* (Moscow: Nauka)

IOP Conf. Series: Journal of Physics: Conf. Series **1336** (2019) 012004 doi:10.1088/1742-6596/1336/1/012004

The numerical modeling of moving of dwarf galaxy through the intracluster medium

Igor Kulikov[1], Igor Chernykh[1], Dmitry Karavaev[1], Viktor Protasov[2], Vladimir Prigarin[2], Ivan Ulyanichev[2], Eduard Vorobyov[3] and Alexander Tutukov[4]

[1]Institute of Computational Mathematics and Mathematical Geophysics SB RAS, 6 Lavrentyev av., Novosibirsk, Russia
[2]Novosibirsk State Technical University, 20 Prospekt Karla Marksa, Novosibirsk, Russia
[3]University of Vienna, 17 Tuerkenschanzstrasse, Vienna, Austria
[4] Institute of Astronomy RAS, 48-th Pyatnitskaya Street, Moscow, Russia

Abstract. In this paper, we present the new results of mathematical modeling of dwarf galaxy movement through the intracluster medium (ICM). Our numerical model includes self-gravity hydrodynamics equation for the gas component of the galaxy and collisionless Boltzmann equation for the stellar component. We also included important sub-grid physics: star-formation, supernova feedback, stellar wind, cooling, and heating function, and non-equilibrium chemistry to ion helium hydride. As a result of our simulation, we provide density, pressure, and temperature for destruction processes in the high-density intracluster medium.

1. Introduction

The subject of modern astrophysics is the study of physical processes in the universe, their influence on the self-organization and evolution of astronomical objects, as well as on their further dynamics and interaction. The description of astronomical objects is based on hydrodynamic processes. It is hydrodynamics that determines the character of astrophysical flows, which leads to the evolution of astrophysical objects. Mathematical modeling is the main and often the only way for theoretical investigation of astrophysical flows due to the impossibility of carrying out total experiments. The numerical model of interacting galaxies is based on gravitational gas dynamics equations to describe the gas component, and equations for the first moment of the collisionless Boltzmann equation with the full tensor of velocities dispersion to describe the star component [1].

2. The Physics Model

The numerical model of galaxies contains gas and stars components. To describe the gas components, we will use the system of single-speed component gravitational hydrodynamics equations. To describe the collisionless components, we will use the system of equations for the first moments of the Boltzmann collisionless equation. The details of initial profile construction for galaxies can be found in [2].

We use the supernovae feedback for supernovae core-collapse, supernovae Ia, and stellar wind of other medium mass stars. To describe the supernova feedback we are using the initial mass function from [3], stellar lifetimes function [4] and nucleosynthesis [5].

Content from this work may be used under the terms of the Creative Commons Attribution 3.0 licence. Any further distribution of this work must maintain attribution to the author(s) and the title of the work, journal citation and DOI.
Published under licence by IOP Publishing Ltd

We will consider cooling functions in two temperature modes:

(i) The low-temperature cooling. At low temperatures, the ionization of the elements H, O, C, N, Si, and Fe occurs due to collision. The collision frequency and the corresponding cooling function can be found in the work [6].

(ii) The high-temperature cooling. At high temperatures, the emission process for the elements H, He, C, N, O, Ne, Mg, Si, and Fe occurs. The cooling function can be found in paper [7].

To describe the heating function, we will consider the following two processes:

(i) Cosmic ray heating. The process of ionization of hydrogen and helium atoms. The heating function can be found in operation [8].

(ii) Photoelectric heating from small dust grains [9].

There are many important results have been obtained in the study of the physics of supernovae, the remnants of their explosion, and their feedback on more massive structures in recent years. One of these results is that dwarf galaxies couldn't be the place of birth of supermassive black holes [10]. This is why we don't use a special model for supermassive black holes.

3. The Chemistry

The chemical composition of the gas mixture of interstellar gas and clouds: H, He, e, H^+, He^+, HeH^+, H_2^+. The chemical reaction network to helium hydride ion [11] with the reaction rate k (in cm^3/s) and the temperature T (in Kelvins):

(i) Ionization of hydrogen by cosmic rays $H + c.r. \to H^+ + e$ [12],

$$k_1 = 10^{-17}, T_1 = 0.$$

(ii) Collisional ionization of hydrogen $H + e \to H^+ + 2e$ [13],

$$k_2 = \exp\big(-32.7 + 13.5 \times \ln T - 5.7 \times \ln^2 T + 1.5 \times \ln^3 T$$
$$-0.3 \times \ln^4 T + 3.5 \times 10^{-2} \times \ln^5 T - 2.6 \times 10^{-3} \times \ln^6 T$$
$$+1.1 \times 10^{-4} \times \ln^7 T - 2.1 \times 10^{-6} \times \ln^8 T\big), T_2 = 0.$$

(iii) Collision ionization of helium $H + e \to He^+ + 2e$ [14],

$$k_3 = \exp\big(-32.7 + 13.5 \times \ln T - 5.7 \times \ln^2 T + 1.5 \times \ln^3 T$$
$$-0.3 \times \ln^4 T + 3.5 \times 10^{-2} \times \ln^5 T - 2.6 \times 10^{-3} \times \ln^6 T$$
$$+1.1 \times 10^{-4} \times \ln^7 T - 2.1 \times 10^{-6} \times \ln^8 T\big), T_3 = 900$$

(iv) Dissociative recombination of atomic hydrogen $H^+ + e \to H + \gamma$ [15],

$$k_{4,low} = 3.92 \times 10^{-13} \times T^{-0.6353}, T_{4,low} = 0,$$

$$k_{4,high} = \exp\big(-28.6 - 0.7 \times \ln T - 0.02 \times \ln^2 T + 2.4 \times 10^{-3} \times \ln^3 T$$
$$-3.1 \times 10^{-4} \times \ln^4 T + 1.4 \times 10^{-5} \times \ln^5 T + 5.0 \times 10^{-6} \times \ln^6 T$$
$$+5.8 \times 10^{-7} \times \ln^7 T - 1.8 \times 10^{-8} \times \ln^8 T\big), T_{4,high} = 5500.$$

(v) Dissociative recombination of molecular hydrogen $H_2^+ + e \rightarrow 2H + \gamma$ [16],

$$k_{5,low} = 10^{-8}, T_{5,low} = 0,$$

$$k_{5,high} = 1.32 \times 10^{-6} \times T^{-0.76}, T_{5,high} = 617.$$

(vi) Radiative association of the helium ion and hydrogen $He^+ + H \rightarrow HeH^+ + h\nu$ [17],

$$k_6 = 1.4 \times 10^{-16}, T_6 = 5000.$$

(vii) Dissociative recombination of the helium hydride ion $HeH^+ + e \rightarrow He + H$ [18],

$$k_7 = 3.0 \times 10^{-10} \times (T/10^4 K)^{-0.47}, T_7 = 1000.$$

(viii) Collisional recombination of the helium hydride ion $HeH^+ + H \rightarrow He + H_2^+$ [19],

$$k_8 = 1.2 \times 10^{-9} \times (T/10^4 K)^{-0.11}, T_8 = 200.$$

Adiabatic index [20]:

$$\gamma = \frac{5n_H + 5n_{He} + 5n_e + 5n_{H^+} + 5n_{He^+} + 7n_{HeH^+} 7n_{H_2^+}}{3n_H + 3n_{He} + 3n_e + 3n_{H^+} + 3n_{He^+} + 5n_{HeH^+} 5n_{H_2^+}}$$

4. Numerical simulation of moving of dwarf galaxy through the intracluster medium

The moving of dwarf galaxy through the intracluster medium is considered in the model of the motion of a dense gas cloud through a rarefied medium. The Characteristics of the gas components of the galaxy $n_A = 10^{-1}$ cm^{-3} is density of interstellar gas, $T_A = 10^4$ K is interstellar gas temperature, $n_H : n_{He} = 9 : 1$ is chemical composition of the gas component. Cloud Characteristics: $n_C = 1$ cm^{-3} is density of cloud, $T_C = 10K$ is cloud temperature, $v = 3 \times 10^5$ m/sec is cloud speed,$n_C = n_H$ is chemical composition of the cloud. The results of computational experiments of dwarf galaxy movement through the intracluster medium are presented in Fig. 1. The figure shows, that in front of the cloud a shock wave is formed, resulting from the interaction of cloud with ICM. At the edges of the galaxy, a shell is formed, the development of instability in which leads to the formation of Jellyfish-like galaxies. In areas of sufficiently high temperature, the conditions are created for the reactions that lead to the formation of the helium hydride ion.

5. Conclusion

We achieved new results of mathematical modeling of the dwarf galaxy movement through the intracluster medium. Our numerical model includes self-gravity hydrodynamics equation for the gas component of the galaxy and collisionless Boltzmann equation for the stellar component. We also included important sub-grid physics: star-formation, supernova feedback, stellar wind, cooling, and heating function, and non-equilibrium chemistry to ion helium hydride. The numerical simulation of ion helium hydride formation through destruction of the galaxy in high-density ICM is described.

Acknowledgements

This research was supported by the RFBR grants 18-01-00166 (high order parallel numerical methods), 18-07-00757 (astrochemical kinetics simulation) and by the ICMMG SB RAS (project 0315-2019-0009).

IOP Conf. Series: Journal of Physics: Conf. Series **1336** (2019) 012004 doi:10.1088/1742-6596/1336/1/012004

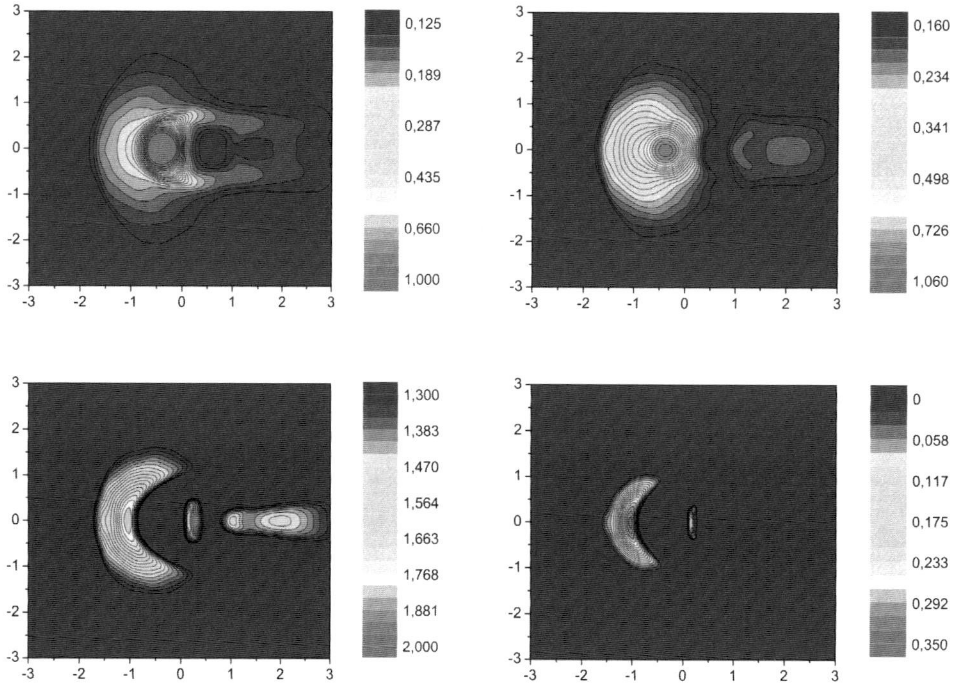

Figure 1. The density in cm^{-3} (top left), the relative pressure (top right), the temperature in T_A (bottom left) and density in 10^{-6} cm^{-3} of the helium hydride ion (bottom right).

References

[1] Vorobyov E., Recchi S., Hensler G. Stellar hydrodynamical modeling of dwarf galaxies: simulation methodology, tests, and first results // Astronomy & Astrophysics. – 2015. – Vol. 579. – Article Number A9.

[2] Springel V., Di Matteo T., Hernquist L. Modelling feedback from stars and black holes in galaxy mergers // Monthly Notices of the Royal Astronomical Society. – 2005. – V. 361. – P. 776-794.

[3] Kroupa P. On the variation of the initial mass function // Monthly Notices of the Royal Astronomical Society. – 2001. – V. 322, I. 2. – P. 231-246.

[4] Padovani P, Matteucci F. Stellar Mass Loss in Elliptical Galaxies and the Fueling of Active Galactic Nuclei // The Astrophysical Journal. – 1993. – V. 416. – P. 26-35.

[5] Van den Hoek L.B., Groenewegen M.A.T. New theoretical yields of intermediate mass stars // Astronomy & Astrophysics Supplement Series. – 1997. – V. 123, I. 2. – P. 305-328.

[6] Dalgarno A., McCray R.A. Heating and Ionization of HI Regions // Annual Review of Astronomy and Astrophysics. – 1972. – V. 10. – P. 375-426.

[7] Boehringer H., Hensler G. Metallicity-dependence of radiative cooling in optically thin, hot plasmas // Astronomy & Astrophysics. – 1989. – V. 215, I. 1. – P. 147-149.

[8] Van Dishoeck E.F., Black J.H. Comprehensive models of diffuse interstellar clouds - Physical conditions and molecular abundances // The Astrophysical Journal Supplement Series. – 1986. – V. 62. – P. 109-145.

[9] Bakes E.L.O., Tielens A.G.G.M. The photoelectric heating mechanism for very small graphitic grains and polycyclic aromatic hydrocarbons // The Astrophysical Journal. – 1994. – V. 427, I. 2. – P. 822-838.

[10] Mezcua M. Dwarf galaxies might not be the birth sites of supermassive black holes // Nature Astronomy. – 2019. – V. 3. – P. 6-7.

[11] Gusten R., et al. Astrophysical detection of the helium hydride ion HeH^+ // Nature. – 2019. – V. 568. – P. 357-359.

[12] Lepp S. The Cosmic-Ray Ionization Rate // Astrochemistry of Cosmic Phenomena: Proceedings of the 150th Symposium of the International Astronomical Union, held at Campos do Jordao, Sao Paulo, Brazil, August 5-9, 1991. – 1992. – P. 471-475.

[13] Abel T., Anninos P., Zhang Y., Norman M. Modeling primordial gas in numerical cosmology // New Astronomy. – 1997. – V. 2, I. 3. – P. 181-207.
[14] Janev R.K., Langer W.D., Evans K.J., Post D.E.J. Elementary Processes in Hydrogen-Helium Plasmas. Cross Sections and Reaction Rate Coefficients // Springer Series on Atomic, Optical, and Plasma Physics. – 1987. – V. 4. – 326 P.
[15] Ferland G.J., Peterson B.M., Horne K., Welsh W.F., Nahar S.N. Anisotropic line emission and the geometry of the broad-line region in active galactic nuclei // The Astrophysical Journal. – 1992. – V. 387. – P. 95-108.
[16] Schneider I.F., Dulieu O., Giusti-Suzor A., Roueff E. Dissociate recombination of H2(+) molecular ions in hydrogen plasmas between 20 K and 4000 K // The Astrophysical Journal. – 1994. – V. 424. – P. 983-987.
[17] Zygelman B., Dalgarno A. The radiative association of He+ and H // The Astrophysical Journal. - 1990. - V. 365. - P. 239-240.
[18] Stromholm C., et al. Dissociative recombination and dissociative excitation of 4HeH+: absolute cross sections and mechanisms // Physical Review A. – 1996. – V. 54. – P. 3086-3094.
[19] Bovino S., Tacconi M., Gianturco F.A., Galli D. Ion chemistry in the early universe. Revisiting the role of HeH+ with new quantum calculations // Astronomy & Astrophysics. – 2011. – V. 529. – Article Number A140.
[20] Glover S., Mac Low M. Simulating the Formation of Molecular Clouds. I. Slow Formation by Gravitational Collapse from Static Initial Conditions // The Astrophysical Journal Supplement Series. – 2007. – V. 169, I. 2. – P. 239-268.

IOP Conf. Series: Journal of Physics: Conf. Series **1336** (2019) 012005 doi:10.1088/1742-6596/1336/1/012005

High performance collisional PIC plasma simulation with modern GPUs

Alexey A Romanenko [1] and Alexey V Snytnikov [1,2,4] and Galina G Lazareva[1,2,3]

[1] Novosibirsk State University
[2] Institute of Computational Mathematics and Mathematical Geophysics SB RAS
[3] Peoples' Friendship University of Russia (RUDN University), Moscow, Russian Federation
[4] Institute of Automation and Electrometry SB RAS

E-mail: lazareva@ssd.sscc.ru,snytav@ssd.sscc.ru, arom@ccfit.nsu.ru

Abstract. GPU implementation of the collisional Particle-In-Cell plasma model is proposed. The goal of this model is the simulation of Plasma Enhanced Chemical Vapour Deposition (PECVD) reactors with glow discharge plasma and also for simulation of fusion plasma with sufficient role of collisions. Vlasov equation is being solved by the Particle-In-Cell (PIC) method. Collision are simulated with the null collision technique. The important difference from the collisionless approach is that major time is taken by collision evaluation and not by particle push.

GPU performance is increased by storing the particles in cells and also by the exclusion of synchronization of threads at the stage of current and density evaluation. Electromagnetic field is given by either the Maxwell equations or be the Poisson equation for electrostatic case. The Maxwell equations are solved by the FDTD method. The Poisson equation is solved by the hybrid method. Performance obtained at the moment is 0.5 TFLOPS for particle push with the Nvidia Tesla V100, and the parallels efficiency is 92% for a cluster with 250 nodes with 2 GPUs each resulting in 500 GPUs.

Acknowledgments
The publication has been prepared with the support of the "RUDN University Program 5-100", code development was supported by RFBR grant 18-07-00364, the numerical model was designed by RFBR grant 19-07-0085, computational experiments were performed under support of RFBR grant 19-07-00446, development of the collision model was supported by RFBR grant 19-07-00446.

1. Introduction
The main idea of this work is to achieve high performance with supercomputers and GPUs in order to conduct high resolution 3D simulations of the fusion plasma in different experimental facilities and also in astrophysical applications like supernovae expansions and solar bursts of II and III type. Although fusion plasma is geneally treated as being collisionless, there are cases when collisions might play a role. Moreover, we intend to use the same Particle-In-Cell engine for low temperature plasma problems, too. An important example of such a problem is the Plasma Enhanced Chemical Vapour Deposition (PECVD) reactor.

Content from this work may be used under the terms of the Creative Commons Attribution 3.0 licence. Any further distribution of this work must maintain attribution to the author(s) and the title of the work, journal citation and DOI.
Published under licence by IOP Publishing Ltd

IOP Conf. Series: Journal of Physics: Conf. Series **1336** (2019) 012005 doi:10.1088/1742-6596/1336/1/012005

Since two problems: fusion reactor and PECVD reactor are treated with the same code, there are two filed solvers, one for Maxwell equations and the other one for Poisson equation. Exactly the same particle pusher is used for both cases. Like other plasma physics problems with non-Maxwellian velocity distribution this problem requires kinetic treatment.

It means either direct finite-difference solution of Vlasov equation or Particle-In-Cell[1] method. While PIC method is noisy and time-consuming, still it is in many cases the only method that is able to treat the problem with no non-physical simplifications.

There are a lot of PIC packages, like OSIRIS[2], LCODE[3] or VLPL[4], including those adopted for GPU, e.g. PIConGPU[5], ALaDyn[6], and also some works aimed to particular questions of GPU implementation of PIC method, like [7] and [8], but since they are all problem-specific, we make an implementation on our own.

PIC method consists of two stages: field evaluation and particle pushing. Computational experiments with general purpose computers (e.g. the Lomonosov supercomputer at MSU) show that for the collisionless case most time is spent in particle pushing (from 60% to 90%). Since the total time is very big even for a small test case: about a week with 128 Intel Xeon cores for a 3D problem with just $100 \times 4 \times 4$ mesh nodes and 10000 particles for each cell [9], it requires hiring GPUs to solve real-size problem.

2. Basic equations

The basic equations are Vlasov equations for ion and electron components of the plasma and also of the Maxwell equation system. These equations in the usual notation have the following form:

$$\frac{\partial f_{i,e}}{\partial t} + \vec{v}\frac{\partial f_{i,e}}{\partial \vec{r}} + \vec{F}_{i,e}\frac{\partial f_{i,e}}{\partial \vec{p}} = St\{f\}, \qquad \vec{F}_{i,e} = q_{i,e}\left(\vec{E} + \frac{1}{c}[\vec{v}, \vec{B}]\right)$$

$$rot\vec{B} = \frac{4\pi}{c}\vec{j} + \frac{1}{c}\frac{\partial \vec{E}}{\partial t}$$

$$rot\vec{E} = -\frac{1}{c}\frac{\partial \vec{B}}{\partial t} \tag{1}$$

$$\nabla^2\varphi = 4\pi\rho$$

$$div\vec{B} = 0$$

Here $f_{i,e}$ is the distribution function for ions and electrons, respectively, t is time, $\vec{v}$ is the velocity vector, $\vec{r}$ is the coordinate vector, $\vec{F}$ is the Lorentz force, $\vec{p}$ is the impulse, $\vec{E}$ is the electric field, $\vec{B}$ is the magnetic field, φ is the electric potential, ρ is the charge density. Finally, $St\{f\}$ is the collision term.

All the equations below are given is non-dimensional units. The following quantities are used for transition to non-dimensional form:

- velocity of light $c = 3 \times 10^{10}$ cm/s
- mass of electron $m_e = 9.1 \times 10^{-28}$ g
- plasma density $n_0 = 10^{14}$
- time $t = \omega_{pe}^{-1}$, where plasma electron frequency $\omega_{pe} = 5.6 \times 10^{11}$ s^{-1}.

3. Computational domain

The 3D computational domain has the shape of a cube with the following dimensions:

$$0 \le x \le L_X, \qquad 0 \le y \le L_Y, \qquad 0 \le z \le L_Z$$

Within this domain there is the model plasma that consists from electrons and ions. The model plasma particles are distributed uniformly within the domain. The density of plasma is set by the user as well as the electron temperature. The temperature of ions is considered to be zero. Initial distribution of particles by velocities is Maxwellian:

$$f(v) = \frac{1}{\Delta v\sqrt{2\pi}} \exp\left(-\frac{(v - v_0)^2}{2\Delta v^2}\right)$$

here Δv - electron velocity dispersion, v_0 - average velocity ($v_0 = 0$ for plasma particles. Up to this moment it is the same for fusion plasma and for PECVD reactor plasma. But there are also differences. It is about the composition of plasma and the boundary conditions.

3.1. Fusion reactor plasma simulation

It this case the boundary conditions are periodical, and plasma consists of three sorts of particles:

- Hydrogen ions,
- plasma electrons,
- beam electrons.

Beam electrons are also distributed uniformly, but they have the same impulse directed along X axis and low temperature along Y and Z.

3.2. PECVD reactor simulation

In this case there is a periodic electric potential applied at $Z = L_Z$ and $\varphi = 0$ at all the other boundaries. At $z = 0$ there is substrate with amorphous silicon hydrid film deposed from plasma. In this case plasma consists of

- Hydrogen ions
- electrons
- SiH_2^+, SiH_3^- ions

4. Fusion reactor plasma simulation

The physical statement of the problem exactly follows the simulations in [9], the difference is in simulation time which in the present case is by orders of magnitude smaller. As it is stated in [9] beam relaxation may proceed in three modes, depending on the instability increment γ and beam velocity dispersion Δv: kinetic ($k\Delta v \gg \gamma$), hydrodynamical ($k\Delta v \ll \gamma$) and transit mode ($k\Delta v \sim \gamma$). Here k is the wavenumber.

The goal of the computations is not to simulate the facility as a whole - since no supercomputer is able to do it now or in the future. Instead, we are going to reveal the underlying instabilities and reproduce the turbulence observed in experiments.

In such a way, the computational domain corresponds to a very small (characteristic size $\sim 10^{-3}$ cm) volume of plasma.

5. Description of the GPU implementation

The Vlasov equation is solved by the Particle-In-Cell method [1]. In order to understand the ideas and bottlenecks of GPU implementation of PIC method let us first consider the underlying numerical techniques.

IOP Conf. Series: Journal of Physics: Conf. Series **1336** (2019) 012005 doi:10.1088/1742-6596/1336/1/012005

5.1. Numerical method

The essential of the PIC method is the substitution of the Vlasov equation by the set of characteristic equations. The characteristic equations exactly resemble the Newton motion equations for plasma particles, and due to this reason they are called the motion equations for model particles.

One must note that these model particles are just mathematical tool for solving the Vlasov equation. This is not a replacement of very big number ($N \sim 10^{23}$) of small particles in real plasma by a small number ($N \sim 10^{10}$) of big particles in model plasma.

5.1.1. Particle push In such a way, the equations of motion for model particles are the following:

$$\frac{\partial \vec{p}_{i,e}}{\partial t} = q_{i,e}\left(\vec{E} + \vec{v} \times \vec{B}\right), \qquad \frac{d\vec{r}_{i,e}}{dt} = \vec{v}_{i,e}, \qquad \vec{p}_{i,e} = \frac{\vec{v}_{i,e}}{\sqrt{1 - v_{i,e}^2}} \tag{2}$$

here i,e are the indices for ions and electrons, respectively, $\vec{p}$ is impulse, $\vec{r}_{i,e}$ is the coordinate vector, $\vec{E}$ - electric field, $\vec{v}$ is the velocity and $\vec{B}$ - magnetic field. Equations 2 are solved by the leapfrog scheme:

$$\frac{\vec{p}_\alpha^{\,m+1/2} - \vec{p}_\alpha^{\,m-1/2}}{\tau} = q_{i,e}\left(\vec{E}^m + \frac{\vec{v}_\alpha^{\,m+1/2} - \vec{v}_\alpha^{\,m-1/2}}{2} \times \vec{B}^m\right)$$

$$\frac{\vec{r}_\alpha^{\,m+1} - \vec{r}_\alpha^{\,m}}{\tau} = \vec{v}_{i,e}^{\,m+1/2}$$

here τ is the timestep.

5.1.2. Field solver

Maxwell equations For the evaluation of electromagnetic field the Langdon-Lasinski [1] method is employed:

$$\frac{\vec{B}^{m+1/2} - \vec{B}^{m-1/2}}{\tau} = rot_h \vec{E}^m$$

$$\frac{\vec{E}^{m+1} - \vec{E}^m}{\tau} = rot_h \vec{B}^{m+1/2} - \vec{j}^{m+1/2}$$

Poisson solver The method for solving the 3D Poisson equation is a mixture of FFT along Y, the Thomas elimination along X and Block Successive Over-Relaxation (BSOR) along Z. Is is shown that BSOR iteration works faster than even the FFT if the right hand side (charge density) varies not very much. One can see in figure 1 that when BSOR makes just one or two iterations, BSOR solution time is less than FFT solution time. The algebraic multigrid method was also applied in the Poisson solver [10] instead of BSOR.

5.1.3. Collision simulation The collisions of electrons with gas molecules are treated with the null collision technique [11]. First, for the given electron, the gas is defined (considering two gases with the densities n_1 and n_2) by means of the random number t, $0 < t < 1$. If the condition

$$t < \frac{n_1}{n_1 + n_2}$$

is valid, then the first gas is taken for collision, otherwise the second. In the same way, one of the possible collisional processes is selected (the processes are listed in Section 4). Since the

IOP Conf. Series: Journal of Physics: Conf. Series **1336** (2019) 012005 doi:10.1088/1742-6596/1336/1/012005

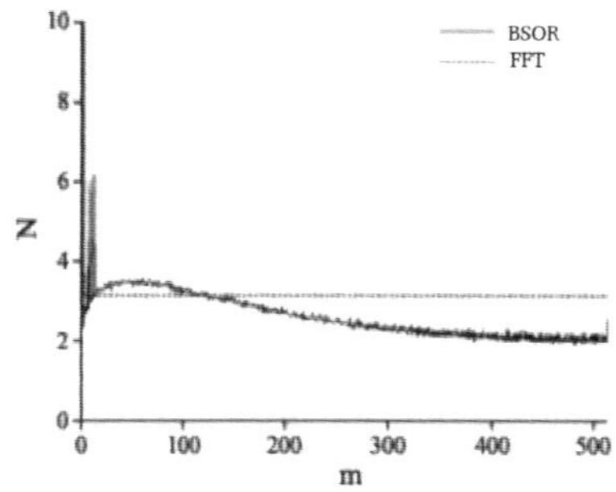

Figure 1. The number of iterations of the BSOR method depending on the simulation timestep number. The horizontal line shows the FFT solution time divided by a single iterations time.

type of the process is known, it is possible to find its cross-section σ. The cross-section is used to compute the collision probability

$$p = 1 - exp(\nu\tau\sigma n).$$

Here v is the electron velocity, τ is the timestep and n is the gas density. By means of a random number, the collision probability defines either this electron superparticles suffer a collision or not.

5.2. GPU implementaion principles

The following main principles are used to achieve high GPU performance:

- shared memory is used for computations with particles;
- particles are stored in arrays attached to cells in order to use the cache more efficiently;
- field and current values (8 of them for each field and current component) are computed and stored in cells;
- particle reordering is used to speedup particle pusher.

The computational algorithm consists of the following stages performed at each time step:

- particle push, that is the computation of new coordinates and impulses of each particle. At present, the evaluation of current is included in the push, [12, 13];
- electromagnetic field evaluation, [14];
- reordering of particle, which means the transfer of particles to the cells they belong to [15];
- copying of currents from cells to the global arrays [14];
- copying of fields from global arrays to cells [14].

For each of the stages a special kernel is devoted in GPU implementation. The implementation of each kernel is given in more detail in the above-listed papers.

Random values needed for PIC-MC simulation were evaluated with cuRand library.

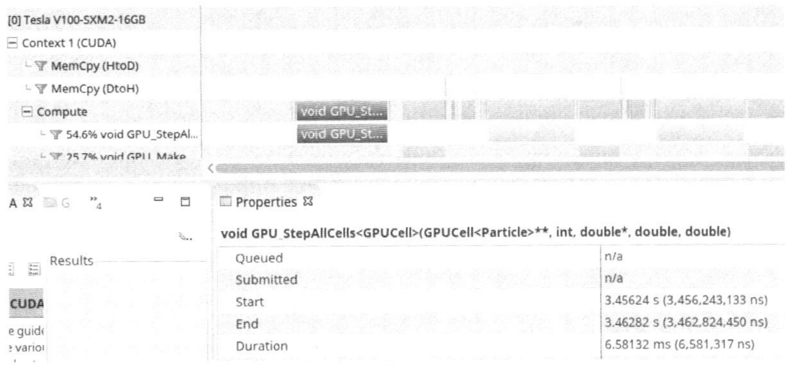

Figure 2. A part of the performance profile for V100 GPU. The focus is on the kernel GPU_StepAllCells that performs particle push.

6. Performance measurements with different GPUs

The results of profiling of the code execution is presented in the table 1.

Performance is evaluated in the following way: the computational domain has 6.4 million particles, and near 500 floating point operation are required to push a particle at a time step. The total number of operations is divided to the measured pushing time. As pushing takes the most time (from 60% to 95%), only push performance is given. Field solution time is given for both Poisson equation (solved by hybrid method) and for the Maxwell equations solved by FDTD. This does not mean that both solvers were used in one simulation. The Poisson equation was solved for PECVD simulation and The Maxwell equation – for fusion simulation, and the same pusher was used fot both of them.

Table 1. Worktime, measured by the Nvidia Visual Profiler and the resulting performance in flops for different GPUs, time in microseconds. Grid size $100 \times 6 \times 6$ with 1000 particles in cell.

Device name	Field evaluation time		Pushing time	Collision time	Push performance
	Poisson	Maxwell			
Geforce GTX 1050 Ti	166.2 ms	20 ms	225.4 ms	20.1 ms	14.1 GFLOPS
Tesla K40m	280 ms	36 ms	434.4 ms	38.3 ms	7.3 GFLOPS
Tesla V100	4 ms	0.5 ms	6.5 ms	2 ms	0.492 TFLOPS

A part of the profile for the NVidia Tesla V100 is shown in figure 2. One can see the device name (Tesla V100 SXM2) as well as kernel duration (6.58132 ms). It can be noticed also that there is no big pause between the kernels, so the device is busy most of the time.

7. Conclusion

The performance achievement of nearly 0.5 TFLOPS by a single NVidia Tesla V100 for 3D Particle-In-Cell simulation is presented. It means that now it is possible to perform the simulation of plasma of the size mentioned in Introduction in 4 hours with one V100 instead of a week with a ordinary cluster of 128 cores (the cluster at Siberian Supercomputer Center, based on rather old Intel Xeon 5540 4-core processors). It is also shown that the Plasma Enhanced Chemical Vapour Deposition reactor could be also simulated within reasonable time.

IOP Conf. Series: Journal of Physics: Conf. Series **1336** (2019) 012005 doi:10.1088/1742-6596/1336/1/012005

References

[1] Vshivkov V A, Grigoryev Y N and Fedoruk M P 2002 *Numerical Particle-in-Cell Methods. Theory and applications* (Utrecht-Boston: VSP)

[2] Fonseca R A, Silva L O, Tsung F S, Decyk V K, Lu W, Ren C, Mori W B, Deng S, Lee S, Katsouleas T and Adam J C 2002 *Computational Science — ICCS 2002* ed Sloot P M A, Hoekstra A G, Tan C J K and Dongarra J J (Berlin, Heidelberg: Springer Berlin Heidelberg) pp 342–351 ISBN 978-3-540-47789-1

[3] Lotov 2012 (accessed August 11, 2018) *LCODE main site* URL `http://www.inp.nsk.su/~lotov/lcode/`

[4] Tuckmantel T, Pukhov A, Liljo J and Hochbruck M 2010 *IEEE Transactions on Plasma Science* **38** 2383–2389

[5] Burau H, Widera R, Honig W, Juckeland G, Debus A, Kluge T, Schramm U, Cowan T E, Sauerbrey R and Bussmann M 2010 *IEEE Transactions on Plasma Science* **38** 2831–2839

[6] Rossi F, Londrillo P, Sgattoni A, Sinigardi S and Turchetti G 2012 *AIP Conference Proceedings* **1507**(1) 184–192 (*Preprint* `https://doi.org/10.1063/1.4773692`) URL `https://doi.org/10.1063/1.4773692`

[7] Kong X, C Huang M, Ren C and K Decyk V 2011 **230** 1676–1685

[8] Rieke M, Trost T and Grauer R 2015 *Journal of Computational Physics* **283** 436–452 (*Preprint* `https://doi.org/10.1016/j.jcp.2014.12.016`) URL `https://doi.org/10.1016/j.jcp.2014.12.016`

[9] Lotov K V, Timofeev I V, Mesyats E, Snytnikov A V and Vshivkov V A 2015 *Physics of Plasmas* **22**(2) (*Preprint* `https://doi.org/10.1063/1.4907223`) URL `https://doi.org/10.1063/1.4907223`

[10] Snytnikov A V and Vshivkov V A 2005 *Proceedings of the 8th International Conference on Parallel Computing Technologies* PaCT'05 (Berlin, Heidelberg: Springer-Verlag) pp 457–467 ISBN 3-540-28126-6, 978-3-540-28126-9

[11] Birdsall C K 1991 *IEEE Transactions on Plasma Science* **19** 65–85 ISSN 0093-3813

[12] Snytnikov A and Romanenko A 2014 *Computer Science* **36** 79–88

[13] Snytnikov A and Romanenko A 2013 *Numerical Analysis* **16** 81–92

[14] Snytnkov A V, Boronina M A, Mesyats E A and Romanenko A A 2015 *Russian Supercomputing Days 2015* (MSU)

[15] Romanenko A A and Snytnikov A V 2018 *Vestnik Udmurtskogo Universiteta*

IOP Conf. Series: Journal of Physics: Conf. Series **1336** (2019) 012006 doi:10.1088/1742-6596/1336/1/012006

Numerical simulation of an anisotropic heat transfer in magnetized neutron stars with 3D basic operators method

I A Kondratyev[1,2], S G Moiseenko[1], G S Bisnovatyi-Kogan[1,3], M V Glushikhina[1]

[1] Space Research Institute RAS, Profsoyuznaya 84-32, 117997, Moscow, Russia
[2] National Research University Higher School of Economics, Department of Physics, Moscow, Russia
[3] National Research Nuclear University MEPhI, Kashirskoe highway, 31, 115409, Moscow, Russia

E-mail: mrkondratyev95@gmail.com, moiseenko@iki.rssi.ru, gkogan@iki.rssi.ru, m.glushikhina@iki.rssi.ru

Abstract.
We have solved numerically a three dimensional boundary-value problem for a heat transfer equation in a magnetized neutron star crust with an updated tensorial heat conductivity coefficient. The temperature distribution in a neutron star crust in presence of a magnetic field was simulated. To calculate the surface temperature distribution, we have constructed a local one-dimensional plane-parallel model of a magnetized outer envelope of the neutron star and used it as an outer boundary condition for 3D problem to find a self-consistent solution.

This problem was solved with our extension of a basic (support) operators numerical method on a tetrahedral mesh. The idea of operator approach consists of inclusion the boundary conditions into difference form of the solving problem and its formulation as one operator equation. The finite difference operators are constructed in the way to fulfill corresponding relations between continuous operators (for example, div(curl)=0, div is conjugated to -grad so div(grad) is self-conjugated etc.). Such approach allows to obtain completely conservative implicit finite difference schemes. Efficient iterative methods can be used to find the solution because constructed matrixes have good properties, such as symmetry and positive definiteness.

1. Introduction

Neutron star (NS) surface magnetic fields can be $\sim 10^{12-13}$G. One of the possible ways to study the surface magnetic fields is observations of thermal radiation in a soft X-ray band [1, 2]. Periodic changes of spectra from some single NSs may indicate a nonuniform temperature distribution on their surfaces. Such heterogeneities are determined by anisotropic thermal conductivity of degenerate matter with presence of a strong magnetic field, because a heat flux is suppressed sufficiently across the magnetic field lines in the outer layers of magnetized neutron stars. Comparing results of simulations of the heat transfer processes with observational data can help to determine magnetic field structure on the NS surface.

Outer layers of the NS consist of a plasma of degenerate (and also ultrarelativistic at density $\rho \gg 10^6$ g/cm^3) electrons and almost nondegerate nonrelativistic ions. We separeate outer

Content from this work may be used under the terms of the Creative Commons Attribution 3.0 licence. Any further distribution of this work must maintain attribution to the author(s) and the title of the work, journal citation and DOI.
Published under licence by IOP Publishing Ltd

layers to the crust ($10^{10} < \rho < 2 \cdot 10^{14}$ g/cm^3) and an outer envelope ($\rho < 10^{10}$ g/cm^3). Pressure is detemined mostly by the electrons, and the matter can form a state of Coulomb crystal or liquid, and a heat also transfers mostly by the electrons in degenerate layers and by radiation in nondegenrate layer near the surface. A degree of the heat flux depression across the magnetic field is determined by a so called magnetization parameter $\omega\tau$, where τ is an average time between electron-nucleon collisions, and $\omega = eB/m_e^* c$ is a cyclotron plasma frequency, $m_e^* = m_e\sqrt{1 + {p_{fe}}^2/m_e c^2}$ is an effective electron mass, $p_{fe} = \hbar(3\pi^2 n_e)^{1/3}$ is an electron Fermi momentum, e is an electron charge, c is a speed of light, $\hbar$ is a reduced Planck constant, B is a magnetic induction. A thermal conductivity tensor for the plasma with degenerate electrons is derived as a solution of a Boltzmann equation with a Chapman-Enskog method in [3].

In this work we have found a stationary solution of a three-dimensional heat transfer equation with anisotropic thermal conductivity. Because of a sufficient nonlinearity of a radiative boundary condition ($F \sim T^4$, where F is a heat flux, T is a temperature), physical parameters also drastically change through the thin crustal and envelope regions. In such a case a full 3D approach can not be applied to the problem. Fortunately, the heat flux in the envelope is mostly radial, and the temperature distribution in an envelope region can be calculated separately from the crust but selfconsistently with it [4]. We have built a local one-dimensional model of a thermal structure of the outer envelope.

In 2D approach this problem was solved in many works with different magnetic fields and microphysical input (e.g. [4, 5], and [6] for a review).

2. Physical input

2.1. Heat transfer in a magnetic field

Temperature distribution is defined by the heat transfer equation

$$C\frac{\partial T}{\partial t} = \nabla \cdot \kappa \cdot \nabla T + f \tag{1}$$

where C is a heat capacity, κ is a thermal conductivity tensor, f is defined by heat sources and sinks (neutrino emission, Joule heating, etc.). We look for a stationary solution ($\frac{\partial T}{\partial t} = 0$) and also assume an absence of sources and sinks and consider $f = 0$.

Thermal conductivity tensor κ for strongly degenerate electrons in a magnetic field was obtained in [3, 7] using a Chapman-Enskog method for a Boltzmann equation. This tensor takes into account heat fluxes along and across the magnetic field, and also a Hall heat flux. In the Cartesian coordinates it is written as follows

$$\begin{cases} \kappa_{ij} = \frac{k_B^2 T n_e}{m_e^*}\tau\big(\kappa^{(1)}\delta_{ij} + \kappa^{(2)}\varepsilon_{ijk}\frac{B_k}{B} + \kappa^{(3)}\frac{B_i B_j}{B^2}\big) \\ \kappa^{(1)} = \frac{5\pi^2}{6}\big(\frac{1}{1+(\omega\tau)^2} - \frac{6}{5}\frac{(\omega\tau)^2}{(1+(\omega\tau)^2)^2}\big) \\ \kappa^{(2)} = -\frac{4\pi^2}{3}\omega\tau\big(\frac{1}{1+(\omega\tau)^2} - \frac{3}{4}\frac{(\omega\tau)^2}{(1+(\omega\tau)^2)^2}\big) \\ \kappa^{(3)} = \frac{5\pi^2}{6}(\omega\tau)^2\big(\frac{1}{1+(\omega\tau)^2} + \frac{6}{5}\frac{1}{(1+(\omega\tau)^2)^2}\big) \end{cases} \tag{2}$$

where $n_e = \frac{\rho Z}{A m_u}$ is an electron consentration, $\tau = \frac{3}{32\pi^2}\frac{h^3}{m_e^* Z e^4 \Lambda}$ is an average time between electron-nucleon collisions, k_B is a Boltzmann constant, h is a Planck constant, Λ is a Coulomb logarithm. Parameter $\omega\tau$ chages drastically in the crust and the envelope of the NS, for the density $\rho \sim 10^{10}$ g/cm^3, the value of the parameter $\omega\tau \sim 1$ when the value of a magnetic field induction $B \sim 10^{13}$G. Approximately $\omega\tau \sim B/\rho^{2/3}$in the curst, where $\rho = 10^{10} - 2 \cdot 10^{14}$g/cm^3. As it follows from (2), heat conductivity coefficients across and along the magnetic field can be written in the following form:

$$\begin{cases} \kappa_{e\perp} = \frac{k_B^2 T n_e}{m_e^*}\tau\kappa^{(1)}, \\ \kappa_{e\parallel} = \frac{k_B^2 T n_e}{m_e^*}\tau\big(\kappa^{(1)} + \kappa^{(3)}\big). \end{cases} \tag{3}$$

Through this work magnetic field is assumed to be dipolar

$$\mathbf{B} = \frac{B_p R_{NS}^3}{2} \frac{3(\mathbf{d}\cdot\mathbf{r})\,\mathbf{r} - \mathbf{d}\,r^2}{r^5}, \tag{4}$$

where R_{NS} is a NS radius, $\mathbf{d}$ is a unit vector aligned to magnetic dipole in the centre of a star, and B_p is a magnetic field induction on the pole.

The degree of the heat flux supression across the field ($\kappa^{(1)}$ in (2) and (3)) is stronger, than in previous works [8, 9], where a relation between heat fluxes along and across the magnetic field is $\frac{F_{\parallel}}{F_{\perp}} = 1 + (\omega\tau)^2$.

2.2. *Equation of state*

Density is explicitly appears in thermal conductivity (2). We have built a NS model solving the Tolman-Oppenheimer-Volkoff equations of hydrostatic equilibrium to get a density profile in the crust. For the NS interior we used moderately stiff equation of state SLy4 [10], which is based on the microscopic calculations with an effective nuclear potential [11]. We have chosen central density $\rho_c = 1 \cdot 10^{15}$ g/cm^3. NS mass is $M_{NS} = 1.42\,M_\odot$, where $M_\odot$ is a solar mass, and inner and outer radii of NS crust are $R_{in} = 10.594$ km at $\rho = \rho_{in} = 2 \cdot 10^{14}$ g/cm^3 and $R_{out} = R_{NS} = 11.618$ km at $\rho = 10^{10}$ g/cm^3 respectively.

For the outer envelope (see next subsection) we assumed nonquantizing ideal plasma with nondegerate nonrelativistic ions (nuclei) and degenerate relativistic electrons:

$$P = P_{id}^{(N)} + P_{id}^{(e)},$$

where $P_{id}^{(N)} = n_N k_B T$ is an ion pressure, and the pressure of the electrons of an arbitrary degree of degeneracy can be written in terms of Fermi-Dirac integrals [12, 13]:

$$P_{id}^{(e)} = \frac{(2m_e)^{3/2}}{3\pi^2\hbar^3\beta^{5/2}} \left(I_{3/2}(\chi,\tau) + \frac{\tau}{2} I_{5/2}(\chi,\tau)\right) \tag{5}$$

where $\beta = (k_B T)^{-1}$, $\chi = \beta\mu_{id}^{(e)}$ is an electron chemical potential, normalized on $k_B T$, $\tau = (\beta m_e c^2)^{-1}$, and the Fermi-Dirac integral is defined as follows:

$$I_\nu(\chi,\tau) = \int_0^\infty \frac{u^\nu \sqrt{1 + \tau u/2}}{\exp(u - \chi) + 1} du, \tag{6}$$

here $u = \beta m_e c^2 (\sqrt{1 + \frac{p^2c^2}{m_e^2 c^4}} - 1)$, and p is an electron momentum. In most of the envelope and crust electron gas is strongly degenerate, and pressure is written as (e.g. [14])

$$P_{sd}^{(e)} = \frac{m_e^4 c^5}{32\pi^2\hbar^3} \left(\frac{1}{3}\sinh\xi - \frac{8}{3}\sinh\frac{\xi}{2} + \xi\right), \tag{7}$$

where $\xi = 4\sinh^{-1}\frac{(3\pi^2 n_e)^{1/3}\hbar}{m_e^2 c^2}$, and in non-degenerate layer near the surface in ideal plasma approximation electron pressure (5) approaches the imit of ideal gas $P_{nd}^{(e)} = n_e k_B T$. We use analytical approximation for Fermi-Dirac integrals, which is considered in [12, 13].

IOP Conf. Series: Journal of Physics: Conf. Series **1336** (2019) 012006 doi:10.1088/1742-6596/1336/1/012006

2.3. Thermal structure of an outer envelope

The outer envelope of the NS is a thin layer ($\sim 10 - 100$ metres) of plasma with degenerate electrons and nondegenerate ions. The temperaure decreases by 2-3 orders of magnitude radially across this envelope, while surface temperature variations are in limits of one order of magnitude, so in the first approximation, heat flux is assumed to be only radial through the envelope, and it's local value is constant. In such approach the thermal structure equation for the envelope reads [15, 16]:

$$\frac{dT}{dP} = \frac{3K}{16g_s}\frac{T_s^4}{T^3}, \tag{8}$$

where T_s is a local surface temperature, $K = K(B, \theta_B, T, \rho)$ is an effective opacity, θ_B is a magnetic field inclination angle to a normal to the surface, $g_s = GM_{NS}/(R_{NS}^2\sqrt{1 - r_g/R_{NS}})$ is a surface gravity acceleration, where G is a gravitational constant, and r_g is a gravitational radius. For the effective opacity we have taken into account electon thermal conductivity (3), and radiative opacities for free-free and bound-free transitions and electron Thompson scattering. More detailed discussion can be found in [17]. Equation (8) can be solved as a Cauchy problem for the given values of surface temperature T_s and surface pressure P_s, which is calculated from Eddington approximation $P_s \approx \frac{2g_s}{3K(B,\theta_B,T_s,\rho_s)}$ [18], using equation of state $P_s = P(\rho_s, T_s)$ to obtain ρ_s, where ρ_s is a density at the NS surface. We solve this equation from the NS surface to the bottom of the envelope (crust-envelope boundary) at $\rho_b = 10^{10}$ g/cm^3 and obtain a temperature T_b at this density. Solving equation (8) numerically for different magnetic fields, inclination angles and surface temperatures leads to so-called $T_s - T_b$-relationship, tabulated or analytically approximated function, which relates the surface temperature T_s with the temperature on the crust-envelope boundary T_b. Results of several calculations are presented on Table 1. We used tabulated $T_s - T_b$-relationship as a boundary condition for the heat transfer equation in the crust (see next section).

$\lg B_p$	11	12	13
$T_s(\theta = 0)/10^6 K$	1.02	1.03	1.16
$T_s(\theta = \pi/2)/10^6 K$	0.71	0.35	0.18
$T_{s\parallel}/T_{s\perp}$	1.43	2.94	6.44

Table 1. Surface temperature on a NS magnetic pole ($\theta_B = 0$) and on an equator ($\theta_B = \pi/2$) and their relations for $T_b = 1 \cdot 10^8 K$ and dipolar magnetic field (4).

3. Boundary value problem

We assume, that heat flux radiates from the surface of the NS, and a temperature on the inner radius of the crust T_{core} is constant through the core because of sufficiently lagre value of a heat capacity. This temperature decreases with time as the core cools very slowly, after the fast neutrino cooling stage, so the thermal evolution of the NS can be considered as a sequence of cooling models with a statianory temperature distribution over the star.

In such consideration a value of the surface temperature T_s is determined by a solution of a stationary heat transfer equation in a spherical layer with a given temperature T_{core} on the inner boundary, and a radiative black-body boundary condition on the outer one, which is described by a Stephan-Boltzmann law $F_s = \sigma T_s^4$. In the outer envelope, because of smallness of its mass, the radial heat flux F_s is assumed to be local constant. For this reason, this heat flux on the outer boundary of the crust (inner boundary of the envelope), is equal to the radial part of heat flux calculated from the crust by the solution of the heat transfer equation. Boundary conditions read as

$$T|_{in} = T_{core}, \quad \kappa(\mathbf{B}, \rho, T)\nabla_r T + F_s|_{out} = 0, \tag{9}$$

where index "in" corresponds to the temperature value on the inner boundary of the crust with $r = R_{in}$, and index "out" corresponds outer one with $r = R_{out}$. Continuity of the temperature on the crust-envelope boundary is used with the $T_s - T_b$-relationship for different (B, θ_B), also $T_b = T_{out}$. As a result, we obtain inambiguous dependence $T_s(T_{out}, B, \theta_B)$, which determines surface temperature distribution in the magnetized NS. In spherical layer $R_{in} \leq r \leq R_{out}$ we solve boundary-value problem for the heat transfer equation

$$\nabla \cdot \kappa(\mathbf{B}, \rho, T) \cdot \nabla T = 0 \tag{10}$$

with boundary conditions (9). This problem is solved with our extension [19] of the Basic operators methods on unstructured tetrahedral mesh.

4. Numerical technique

An operator approach in numerical methods [20] was proposed by Soviet and Russian mathematician Alexander Andreyevich Samarskii. His idea consists of the inclusion of boundary conditions into a finite-difference form and formulation a finite-difference problem as an operator equation. It allows to study operators properties and use corresponding efficient methods for the solution of the numerical problem. The idea of the basic (support) operators method consists of the grid analogue construction of one of the continuous operator (e.g. div), which approximates corresponding continuous operator. The grid analogue of the relative conjugate operator (e.g. -grad) is constructed using a corresponding integral relation. In this way operator-difference equations will have the same properties as the differential problem. For example, self-conjugated continuous operator remains symmetrical in the grid form, and sparse and symmetrical matrix corresponds to it. Also basic operators method allows to obtain completely conservative finite-difference schemes. After obtaining the grid formulation of the solving problem, the sparse and symmetrical matrixes can be inversed efficiently by modern iterative or direct methods. Two dimensional operator-difference completely conservative Lagrangian method for the magnetohydrodynamical equations was built by Ardeljan et al. in the series of papers [21, 22, 23, 24] for triangular grid of a variable structure. 3D grid analogues for the differential operators in Cartesian coordinates were described in [19]. Following the operator formalism from [21, 22, 23, 24], the 3D grid analogues of basic differential operators were suggested and some test calculations were provided. In the next two subsections gradient and divergence operator are described.

4.1. Grid analogue of the gradient

We use the grid which consists of the tetrahedra. So-called cell-node approximation is used: node grid functions are defined in the grid nodes, and cell grid functions are defined to be constant in the grid cells and the boundary nodes. Following [21, 22, 23], let's introduce linear spaces of cell grid functions, node functions and boundary node functions. This formal approach allows to study the approximation and stability of the numerical schemes using efficient modern methods.

First we introduce the grid gradient operator, which converts the scalar node function into the cell vector function. Conjugated differential operators can be considered in pairs (for example, div and grad), so, in this approach we have to obtain the approximation of one operator from the pair naturally. In this case, we use the following invariant definition of the *grad* operator:

$$\nabla p = \lim_{V \to 0} \frac{1}{V} \int_S p * d\vec{S}. \tag{11}$$

Let p is a scalar node grid function. Inside each cell we additionally define it as a linear interpolation of node values of the cell's nodes. After integration of this linear interpolation, taking into account mean value theorem, we get the definition of operator *grad* grid analogue of:

$$(\nabla_{\triangle} p)_i = \frac{1}{V_i} \sum_{k=1}^{4} (\bar{p}_k S_k \vec{n}_k)_i. \tag{12}$$

Here $\bar{p}_k = \frac{(p_1+p_2+p_3)_k}{3}$ - is an average interpolated value of p in the k-th tetrahedron face, indexes 1,2 and 3 correspond to p node values on the k-th face; $\vec{n}_k$ is a unit external normal to the k-th face, S_k is a square of the k-th face and V_i is a volume of the cell with the index i. Expression (12) is the first order approximation for the differential operator *grad*. The grid analogues of other (nodes to cells) operators (such as $div(vector)$, $grad(vector)$, $curl(vector)$ etc.) can be derived in a similar way.

4.2. Corresponding divergence operator

Let us introduce the scalar product in the linear grid spaces $(p, g)_\alpha = \sum_s U_s p_s g_s$, α corresponds to node and cell spaces. For the cells $U_s = V_i$, for nodes $U_s = W_j$, here $W_j = \frac{1}{4}\sum_{k=1}^{K_j} V_k$ is a "node volume" (see [21] and references in it).

The grid analogue for divergence operator is constructed to be conjugated to the $-grad$ grid operator from (12). The Green formula and its grid analogue was used (for the case when all functions vanish on the domain boundary or the domain is infinite):

$$\int p \nabla \cdot \vec{v} dV + \int \vec{v} \cdot \nabla p dV = 0 \tag{13}$$

$$(\vec{v}, \nabla_{\triangle} p) + (\nabla_{\times} \cdot \vec{v}, p) = 0; \quad \sum_{l=1}^{N_l} \nabla_{\times} \cdot \vec{v}_l p_l W_l = -\sum_{k=1}^{K_j} \vec{v}_k \nabla_{\triangle} p_k V_k. \tag{14}$$

Grid analogue of the Green formula (14) is written in terms of a scalar product in grid spaces. After substitution the $\nabla_{\triangle}$ from (12) and rearrangement of terms for the j-th node the expression for the grid operator *div* can be written in the form:

$$(\nabla_{\times} \cdot \vec{v})_j = -\frac{1}{3W_j} \sum_{k=1}^{K_j} \tilde{\vec{v}}_k \cdot (\vec{n}_1 S_1 + \vec{n}_2 S_2 + \vec{n}_3 S_3)_k, \tag{15}$$

here K_j is a number of adjacent cells for j node, $\tilde{\vec{v}}_k$ - value of the cell function $\vec{v}$ in the k-th cell. Indexes $1, 2$ and 3 correspond only to those cell faces that contain the j-th node. Summation is done over all cells adjacent to the j-th node.

Difference analogues of differensial operators like (12) and (15) with corresponding boundary operators Φ (see e.g. [19, 21, 22]) allow to formulate grid analogues of boundary-value problems for various systems of partial differential equations.

4.3. Numerical impementation of the solving problem
To include boundary conditions in to the grid operator equation we follow the technique, suggested in [21, 22]. Operator-difference formulation of the problem (10) reads:

$$(I-\delta_1)\nabla_\times \cdot \kappa \cdot \nabla_\triangle (I-\delta_1)T + (I-\delta_1)\nabla_\times \cdot \kappa \cdot \nabla_\triangle T_{core} - \delta_2 \Phi \cdot \vec{n}\sigma T_s^4 = 0, \qquad (16)$$

where Φ is a boundary operator, I is a unit operator, and operator δ_k is defined as follows: it equals to 1 in the boundary nodes with the boundary sort k ($k = 1$ for inner boundary nodes and $k = 2$ for outer ones), and to 0 in other case.

This problem has to be solved self-consistently, because surface temperature T_s in the outer boundary condition is a function of a temperature in the crust itself. We implemented iterational procedure of relaxation: the problem is solved with boundary conditions of first type on the inner boundary and of the second type on the outer one n times until inequality $max|\frac{T_s^n - T_s^{n-1}}{T_s^{n-1}}| < \epsilon$ is not satisfied, where n is an iteration number, ϵ is a some small number. After each iteration the value of T_s is specified with the $T_s - T_b$-relationship with the surface temperature distribution, obtained from the previous iteration. The value T_s^0 for the boundary condition on the first iteration is obtained from the initial approach of the crust temperature. In some sense, this procedure is equivalent to a solving a time-dependent heat transfer equation with boundary conditions of the first and the third types, while the value of the heat flux density $\sigma T_s^4(T)$ in the outer boundary condition from (10) is taken from the previous "time"-step.

Operator-difference equation (16) is nonlinear at each "time"-step (tensor κ is a function of temperature) and it should be solved with Newton method of solving systems of nonlinear equations, and appeared system of linear algebraic equations on the each Newton method iteration is solved by the iterational Seidel method.

5. Results

In this work we have self-consistently obtained the temperature distribution inside the volume of the crust and on the surface of the NS with a three-dimensional code. On Fig.1 temperature distribution in the crust is shown for the densities $\rho = 10^{10} - 2\cdot 10^{14}$ g/cm^3 with the core temperature $T_{core} = 2\cdot 10^8 K$ for the dipolar magnetic field (4) with induction $B_p = 10^{13}$ G on the pole.

The temperature distribution at the outer boundary of the crust is inverted in comparison with the temperature distribution at the neutron star surface, the cause is as follows. The most crucial suppression of the heat flux is in the envelope, where the parameter $\omega\tau \gg 1$. In the envelope the suppressed heat flux on the magnetic equator causes decrease of a temperature gradient in the crust, so that a variation of the crust temperature on the magnetic pole is higher, than on the equator.

On Fig.2 surface temperature distribution is presented for the same parameters, as on Fig.1 with the semi-analytical temperature distribution from [25] with calculated pole and equator surface temperature values:

$$T_s(B,\theta) = [T_s^4(0)\cos^2\theta_B + T_s^4(\pi/2)\sin^2\theta_B]^{1/4} \qquad (17)$$

It is seen from Fig.2, that for a pure-dipolar magnetic field configurations an angular temperature dependence is in a good agreement with the formula (17) from [25]. Polar surface temperature exceeds equatorial one in a factor of ~ 5.

In this work we solved axially symmetric 2D heat transfer problem with an updated electron-ion thermal conductivity and an original three-dimensional extension of Basic operators method as a first step of our studies. Consideration of 3D magnetic field configurations and modelling of observational manifestations of arising 3D effects will be published elsewhere.

IOP Conf. Series: Journal of Physics: Conf. Series **1336** (2019) 012006 doi:10.1088/1742-6596/1336/1/012006

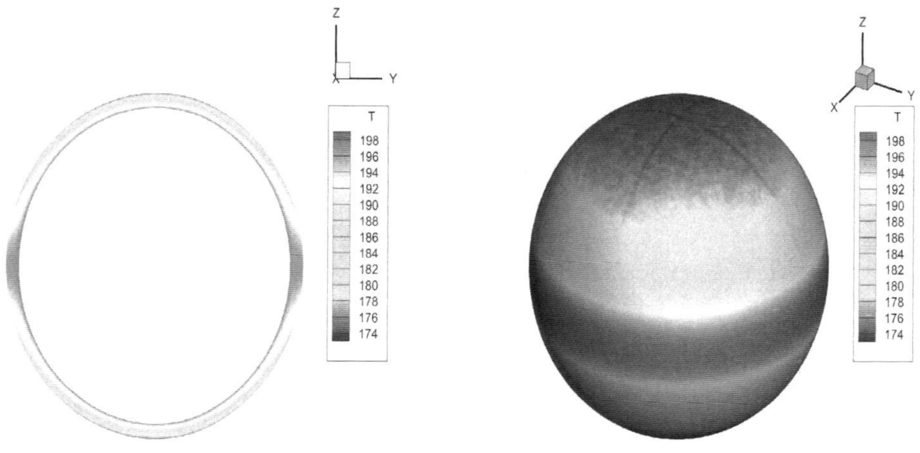

Figure 1. Crust temperature distribution for the dipolar magnetic field (4) with $B_p = 10^{13}$ G, $T_{core} = 2 \cdot 10^8 K$ (temperature T is in units of $10^6 K$) for the density values $\rho = 10^{10} - 2 \cdot 10^{14}$ g/cm^3. left - cross-section in Y-Z plane, right - crust-envelope surface at $\rho_{out} = 10^{10}$g/cm^3.

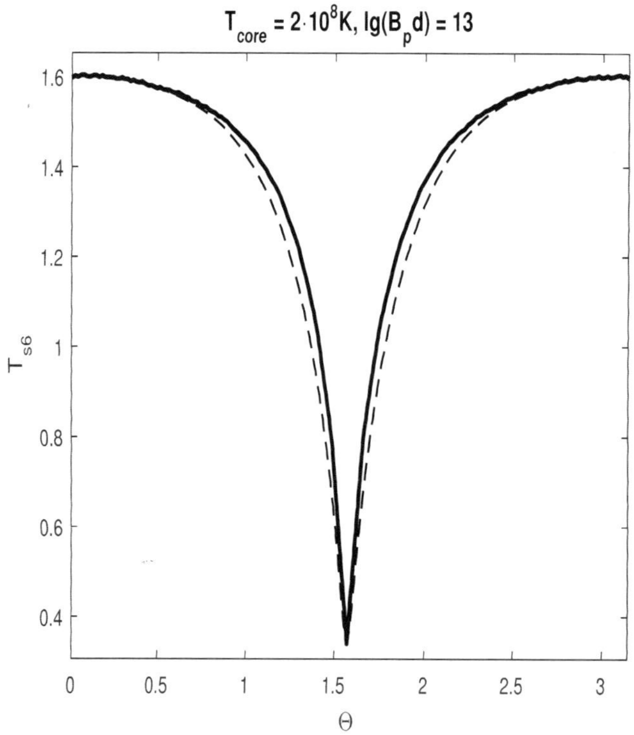

Figure 2. Surface temperature distribution (in units of $10^6 K$) for the same physical parameters as on Fig.1, Θ is a polar angle. Solid line - numerically calculated distribution, dashed line - calculated with the formula (17).

Acknowledgments
This work was partially supported by RFBR grants 17-02-00760 and 18-02-00619 and program of fundamental researches RAS No.12 "Problems of origin and evolution of the Universe".

References
[1] Page D, Geppert U and Weber F 2006 *Nuclear Phys. A* **777** 497
[2] Pons J A, Walter F M, Lattimer J M, Prakash M, Neuhauser R. and An P 2002 *Astrophys. J.*, **564** 981
[3] Bisnovatyi-Kogan G S and Glushikhina M V 2018 *Plasma Physics Reports* **44** 4 355
[4] Geppert U, Kuker D and Page D 2004 *Astron. Astrophys.* **426** 267
[5] Perez-Azorin J F, Miralles J A and Pons J A 2006 *Astron. Astrophys.* **451** 1009
[6] Potekhin A Yu, Pons J A and Page D 2015 *Space Science Rev.* **191** 239
[7] Bisnovatyi-Kogan G S and Glushikhina M V 2018 *Plasma Physics Reports* **44** 12 1114
[8] Flowers E and Itoh N 1976 *Astrophys. J.* **206** 218
[9] Yakovlev D G and Urpin V 1980 *Sov. Astron.* **24** 303
[10] Douchin F and Haensel P 2001 *Astron. Astrophys.* **380** 151
[11] E Chabanat, E Bonche, E Haensel, J Meyer, R Schaeffer 1997 *Nuclear Phys. A* **627** 710
[12] Blinnikov S I, Dunina-Barkovskaya N and Nadyozhin D K 1996 *Astrophys. J. Suppl. Ser.* **106** 171
[13] Chabrier G and Potekhin A Yu 1998 *Phys. Rev. E* **58** 4941
[14] Landau L D and Lifshitz E M 1980 *Statistical physics, part I* Oxford: Pergamon
[15] Gudmundsson E H, Pethick C J and Epstein R I 1983 *Astrophys. J.* **272** 286
[16] Potekhin A Yu and Yakovlev D G 2001 *Astron. Astrophys.* **374** 213
[17] Kondratyev I A, Moiseenko S G, Bisnovatyi-Kogan G S and Glushikhina M V 2019 *Astronomy Reports* Submitted
[18] Bisnovatyi-Kogan G S 2001 *Stellar Physics I: Fundamental Concepts and Stellar Equilibrium* Springer
[19] I A Kondratyev and S G Moiseenko 2019 *J. Phys.: Conf. Ser.* **1163** 012069
[20] Samarskii A A 1989 *Finite difference schemes theory* Moscow: Nauka
[21] Ardeljan N V and Kosmachevskii K V 1995 *Comput. Math. Modelling* **6** 209
[22] Ardeljan N V and Kosmachevskii K V 1987 *Problems of Construction and Research of Conservative Difference Schemes for Magneto-gas-dynamics* MSU, Moscow
[23] Ardelyan N V and Gushin I S 1982 *Vestnik MSU* **15**, 3 3
[24] Ardelyan N V 1983 *Comp. Math. Math. Phys.* **23** 1168
[25] Greenstein G and Hartke G J 1983 *Astrophys. J.* **271**, 283

IOP Conf. Series: Journal of Physics: Conf. Series **1336** (2019) 012007 doi:10.1088/1742-6596/1336/1/012007

A realization of the turbulent vortex dynamo in the atmosphere: based on the 21st century knowledge

G V Levina

Space Research Institute RAS, 84/32 Profsoyuznaya Street, Moscow, 117997, Russia

E-mail: levina@iki.rssi.ru

Abstract. A brief summary of the results of numerical modeling undertaken to substantiate the action of the turbulent vortex dynamo (TVD) in the atmosphere is presented. Two series of numerical experiments are discussed. The first of them was aimed at examining a mathematical model of TVD formulated as the mean-field equations, which describe the dynamo-effect in a non-uniformly heated medium. To simulate and demonstrate the main peculiarities of the new large-scale instability in turbulence, an ingenious statement of the problem for numerical analysis was produced, namely, within a simpler framework – the Rayleigh-Bénard problem on laminar thermal convection in an extended horizontal layer. Within the second series, our developed numerical approach, used and tested in the initial series as well as obtained results therein allowed applying the TVD theory to diagnosis of tropical cyclogenesis. To this purpose, the Russian-American collaboration was initiated and the most advanced to that date tools of atmospheric analysis were utilized. Tropical cyclone formation was examined through the analysis of the American data of near-cloud-resolving (2-3 km horizontal grid spacing) atmospheric numerical simulation.

1. Introduction

Modern astrophysical and geophysical researches show more and more examples of the emergence of organized structures in turbulence, the spatial and temporal scales of which can significantly exceed those of turbulent pulsations. Among them are the large-scale motions driven by non-uniform heating of the environment in the Sun and stars' convective zones, planets' atmospheres, and the Earth's atmosphere and oceans. Vivid illustrations for such phenomena are tropical cyclones, representing large-scale long-lived helical vortices of anomalous intensity, which arise in the atmosphere of the Earth dozens of times a year.

A way to understanding the phenomenon of tropical cyclone is our ability to find a reasonable explanation for the fact why such mode of flow organization becomes the most preferred in some specific states of the atmosphere. By other words, what are the conditions and physical mechanisms that initiate the process of self-organization of atmospheric turbulence and may provide the transformation of small-scale convective atmospheric cells, each with the horizontal dimension and lifetime typical for an ordinary cumulus cloud (several kilometers and several hours), to a rapidly rotating vortex with a special helical structure, of diameter of a few hundreds of kilometers and a lifetime of up to a number of days.

The problem of tropical cyclogenesis remains among the most complex and unresolved ones in meteorology. In order to puzzle out this phenomenon, one more idea may be tried, connected with searching for a solution based on the peculiarities of heat and mass transfer in pre-hurricane conditions

Content from this work may be used under the terms of the Creative Commons Attribution 3.0 licence. Any further distribution of this work must maintain attribution to the author(s) and the title of the work, journal citation and DOI.
Published under licence by IOP Publishing Ltd

of the tropical atmosphere. Indeed, it is well known, convective motions transporting latent and sensible heat from the underlying water surface are the main source of energy for these atmospheric vortices. A breakthrough hypothesis on this subject was put forward within the framework of the theory of turbulent vortex dynamo [1, 2, 3]. The authors [3] substantiated theoretically that the single large-scale vortex of a special spiral topology, characterized by the linkage of horizontal and vertical circulation on the vortex system scales, could be formed as a result of self-organization of small-scale helical convective turbulence. They hypothesized further that such new ordered large-scale flow configuration allows more efficient heat removing from the overheated ocean surface than numerous chaotic small convective cells. For instance, in tropical meteorology to evaluate a probability of cyclogenesis initiation, the following is applied for thermal conditions in most situations. Water temperatures of at least 26.5 °C are needed down to a depth of at least 50 m to provide the overlying atmosphere unstable enough for sustaining convection and thunderstorms [4].

However, as for any other theoretical findings, the final and reliable assessment can only be made on the basis of experimental evidence. In the case under discussion, numerical simulation was chosen as the primary tool to address the theory examination.

2. Numerical evidence of the turbulent vortex dynamo

We are going to briefly sum up results for two different series of numerical experiments. The first of them was undertaken to examine a mathematical model of the turbulent vortex dynamo (TVD) formulated as the mean-field equations, which describe the dynamo-effect in a non-uniformly heated medium. Within the second one, our developed numerical approach, used and tested in the initial series as well as obtained results therein allowed applying the TVD theory to diagnosis of tropical cyclogenesis. To this purpose, the Russian-American collaboration was initiated and the most advanced to that date tools of atmospheric analysis were utilized. Tropical cyclone formation was examined through the analysis of the American data of near-cloud-resolving (2-3 km horizontal grid spacing) atmospheric numerical simulation.

2.1. Helicity and helical turbulence: implication in self-organization and inverse energy cascade

According to the long-standing classical concept of turbulent flows, any large-scale formation, for instance a vortex of spontaneous or forced origin, should be destroyed by turbulence. It is quite another matter if the symmetry of turbulence is broken. The weakest of possible deviations of turbulence from completely disordered is a violation of reflection invariance (mirror-invariance breakdown). Turbulence showing this property is called helical. Apart from the mean energy it is also characterized by nonzero pseudoscalar $H=\int\langle \boldsymbol{V}\cdot\operatorname{curl}\boldsymbol{V}\rangle \mathrm{d}\boldsymbol{r}$ representing a mean helicity. Helicity is one of the main characteristics of the vector velocity field $\boldsymbol{V}\,(\boldsymbol{r},t)$ [5, 6, 7, 8]. This quantity is a topological invariant measuring the degree of linkage of the vortex lines of a flow. It is also a measure of the lack of mirror symmetry of the flow. Helicity is, like energy, an inviscid quadratic constant of motion in barotropic fluids. Existence of the second quadratic constant of motion (in addition to the energy) makes helical flows relatively more stable. Helical structures resist dissipation and survive longer [9]. However, unlike energy, helicity can be both positive and negative. Its sign determines the predominance of the left-handed or the right-handed spiral motions in the examined flow.

In nature, helical turbulence is generated by pseudovector physical fields, such as magnetic or Coriolis force field. In theoretical and numerical studies, to break the symmetry of turbulence an approach is often applied that employs a force of special kind called the forcing function or simply – the forcing.

Small-scale helical turbulence has a number of special features. Under certain conditions it can display a tendency towards self-organization and is capable of intensifying and sustaining large-scale vortex disturbances by means of energy transfer from small to large scales [6, 7, 8].

As an illustration for such process in the natural conditions of the Earth's tropical atmosphere may become a new scenario of tropical cyclogenesis advanced by Montgomery et al. [10] and known as "A vortical hot tower route to tropical cyclogenesis." The work demonstrated how a mesoscale tropical depression vortex of hundred kilometers in diameter could develop from rotating cumulonimbus convection cores with characteristic diameter of ten kilometers as a result of system-scale convergence and upscale vorticity growth. Though in seminal work [10] the helical features of atmospheric moist convective turbulence were not explicitly introduced into consideration, a process of upscale vorticity organization examined by the authors reveals lot in common with TVD's scenario [1, 2, 3].

Following this new atmospheric concept almost immediately, in the same 2006, collaborative Russian-American researches were started. They were based on near-cloud-resolving numerical data [10] and aimed at introducing the TVD theory into tropical cyclone investigations.

2.2. The turbulent vortex dynamo in a non-uniformly heated environment

The generating properties of small-scale helical turbulence leading to the large-scale structure formation were first discovered in magnetohydrodynamics (MHD) by Steenbeck et al. [11]. This phenomenon is known as the alpha-effect. The discovery of the alpha-effect paved the way towards a vigorous development of MHD-dynamo theory. The formal similarity of equations describing the magnetic field $\boldsymbol{B}$ in a moving electrically-conducting medium and vorticity $\boldsymbol{\omega} = \mathrm{curl}\,\boldsymbol{V}$ in non-conducting fluids [6, 7, 8]

$$\partial \boldsymbol{B} / \partial t = \nabla \times \left(\boldsymbol{V} \times \boldsymbol{B} \right) \tag{1}$$

$$\partial \boldsymbol{\omega} / \partial t = \nabla \times \left(\boldsymbol{V} \times \boldsymbol{\omega} \right) \tag{2}$$

gave an impetus to a search for analogs to this phenomenon in general hydrodynamics.

In non-MHD fluid dynamics, the first theoretical example of large-scale instability generated by specific properties of turbulence with a symmetry break was proposed by Moiseev et al. [1] for a compressible fluid in isothermal conditions. The finding gave the researchers a ground to hypothesize a mechanism for intensification of large-scale vortex disturbances in the atmosphere that was called the turbulent vortex dynamo [2].

A mathematical model of the vortex dynamo-mechanism operating in developed thermoconvective turbulence was advanced in [3] and was obtained by the same methods as the model in [1] without a temperature field, namely, by use of the mean-field theory [12].

Let us note that in both, MHD and non-MHD, cases the theory gives thresholds for the large-scale helical instability. Thus, theoretical ground [1] was applied in [2] to estimate the conditions in the Earth's atmosphere, which are corresponded to the formation of tropical cyclones. Later, the approach [1, 2] was extended by authors [13, 14] for the atmosphere of Jupiter to interpret the dimensions and lifetime of large-scale long-lived vortex disturbances caused by the fall of the fragments of comet Shoemaker-Levy 9 in 1994. Both applications to the atmospheres of planets, the Earth and Jupiter, were discussed in detail in review [15].

2.3. A generating term to provide the vortex dynamo in thermoconvective turbulence

By analogy with the model of the alpha effect in magnetohydrodynamics, the helical-vortex instability is ensured by a generating term obtained by averaging the equations for the turbulent velocity field. The term represents a force of some kind in the physical sense [16]

$$\boldsymbol{f} = C \left\{ \boldsymbol{e} \cdot \left(\mathrm{curl}\, \boldsymbol{V} \right)_z - \partial \left(\boldsymbol{e} \times \boldsymbol{V} \right) / \partial z \right\}, \qquad \boldsymbol{e} = \left\{ 0,0,1 \right\}. \tag{3}$$

The force parameterizes the influence of small-scale helical turbulence formed in a rotating non-uniformly heated medium. As it has been substantiated theoretically [3, 17] and illustrated by numerical simulations [18, 19], term (3) creates a positive feedback loop between the poloidal and toroidal component of the vector velocity field. This feedback provides mutual intensification of both components [18, 19]. The key role in the feedback generation and maintenance is held by the vertical component of vorticity, which appears in the equation for the vertical velocity as it follows from expression (3). In a more realistic sense, e.g. for tropical cyclones, such a hypothetical result would mean the formation of positive feedback between the tangential and transverse circulation in the atmospheric vortex through a rotating upward flow.

The dimensionless coefficient C characterizes the small-scale turbulence and is related in a rather complicated manner to its characteristics such as the scale of the energy-containing eddies of the turbulence λ and characteristic time τ of the turbulent velocity correlation [17, 18, 19], and is proportional to the angular velocity of a fluid layer rotation Ω, and the power of internal heat sources Λ. Two independent studies [17, 20] demonstrated that the necessary condition for reaching the vortex dynamo effect is volume heat release in the investigated region. Indeed, it follows from (3): if coefficient C turns to zero in the absence of volumetric heat release ($\Lambda = 0$), then generating term f also turns to zero, and the dynamo-effect is impossible. As applied to the atmosphere, this means that heating from below owing to solar radiation obtained by the underlying surface is insufficient for the initiation of large-scale instability.

It is important to note that atmospheric turbulence in pre-hurricane environment is formed precisely under these conditions [4, 10, 16].

However, in those years when the TVD was just discovered [1, 2, 3], in the last two decades of the 20th century, the immediate application of the new phenomenon for explaining the genesis of tropical cyclones caused serious difficulties. Indeed, unlike the magnetohydrodynamics, where the mean electromotive force acts as the generating term [6, 8, 11, 12] by linking the two different physical fields – the magnetic field and the velocity field – and ensuring the amplification and maintenance of large-scale magnetic field, the force described by formula (3) in the TVD theory could not get such transparent interpretation. In a non-conducting medium, the obtained hypothetical force should operate based on velocity field alone. Nothing similar to this was known to scientists at that time.

For further advancement in studying the new phenomenon, a numerical investigation of mathematical model [3] was undertaken.

2.4. Numerical simulation of helical-vortex effects in Rayleigh-Bénard convection [18, 19]

An ingenious statement of the problem for numerical analysis was chosen to simulate and demonstrate the main peculiarities of the helical-vortex instability by the simplest available way.

It is well known that the Rayleigh-Bénard convection produces a great number of similar convective cells. Authors [18, 19] suggested such cells in a stationary laminar convection regime to be considered as structures of an intermediate scale and used as an initial state for numerical modeling. Term (3) had to represent a forcing function included into equations of thermal convection to imitate the influence of small-scale helical turbulence generated in a rotating fluid with internal heat sources. Indeed, force (3) should pump energy into the system and being applied to the initial state, which includes a number of convective cells, may parameterize a transfer of the energy from turbulence to these relatively larger scale motions.

Thus, the strategy applied to numerical experiments included the study of Rayleigh-Bénard convection in an extended horizontal layer of incompressible fluid heated from below [18, 19]. The Boussinesq system with term (3) added as the forcing function to the momentum equation was used to this purpose. An essential difference of such approach from the conventional ones was in a way that the forcing acts. It does not generate the turbulence of prescribed properties but is applied to a developed natural convection flow and influences it relatively "softly": the forcing generates only a helical structure of flow and initiates a positive feedback between the poloidal and toroidal components of vector velocity field.

Further, within this framework, the researchers tried and implemented an intriguing idea: is it possible to simulate some signs of large-scale instability in these relatively simple conditions?

2.4.1. Numerical implementation. For numerical simulation the above problem was formulated for four physical variables (stream function, vorticity, azimuthal velocity and temperature) and was investigated by the finite differences method. One of the simplest numerical realizations, allowed transforming the three-dimensional problem to two dimensions but saving all three components of the velocity, was applied for a cylindrical computational domain with axial symmetry. Throughout the calculations, the cylinder had a radius much larger than its height – $R:h = 10:1$. All boundaries were considered as impenetrable solid surfaces. Thermal conditions included constant but different temperatures on the upper and lower surfaces what corresponded to heating from below, and the thermally insulated lateral surface.

The statement of the problem included four dimensionless parameters: Rayleigh, Prandtl and Taylor numbers – Ra, Pr, Ta, which are common to characterize the thermal convection in rotating fluids, and parameter S describing the intensity of the forcing [18].

2.4.2. The most striking manifestations of new large-scale instability. In numerical experiments, a statement of the problem with application of the forcing function was implemented. This allowed imitating the large-scale helical-vortex instability on the basis of laminar helical convection and demonstrating a number of qualitatively new effects in flow structure and energetics predicted within the theoretical model [3, 15].

It is necessary to briefly remind the main findings [18, 19] in order to emphasize those of them, which may have the most immediate relevance to recently highlighted processes in the Earth's tropical atmosphere [16, 21, 22, 23, 24, 25]:

- it has been shown how the nonzero mean-helicity of the velocity field $\langle H \rangle \neq 0$ is generated that implies the broken mirror symmetry;
- aimed at searching for large-scale instability and its expected excitation threshold, several modes of helical-vortex flow have been modeled at various combinations of dimensionless parameters;
- the excitation and action of positive helical feedback between the horizontal and vertical circulation, leading to a rapid increase in the kinetic energy of vortex system, has been investigated;
- how this feedback could be identified by examining integral kinetic energies of the horizontal and vertical circulation has been demonstrated;
- new effects in flow structure and energetics attributed to the large-scale helical-vortex instability have been found and examined;
- numerical experiments vividly illustrated a probable scenario for development of the new instability by merging of helical convective cells and consequent intensification of newly forming larger-scale helical vortices through the helical feedback linking the horizontal and vertical circulation.

The discovered effect of cell merging was observed in a wide range of governing parameters. Depending on the ratio between the intensity of convection, rotation and intensity of the helical forcing, several steady-state modes of mixed and forced convection with a different number of helical vortex cells were obtained and described in detail [18, 19]. As the most impressive illustration we present a formation of the stationary forced mode included the single helical vortex. The corresponding set of isoline maps for the stream function demonstrates a flow evolution under the action of forcing – figure 1. In cylindrical geometry using the axial symmetry condition (the axis of symmetry coincides with the left boundary of the domain in the figures), convective motions are realized as axially symmetrical annular rolls. Projected onto the computational domain lying in the (r, z) plane, these movements look like a system of cells with the direction of vertical circulation alternating from structure to structure.

To produce the presented flow evolution, the stationary convective regime at $Ra=3000$, $Pr=1$, $Ta=100$, $S=0$ was chosen as the initial state – figure 1(a). The number of cells that fit along the radius turns out to be nine, and their horizontal scale, as one would expect from classical knowledge on thermal convection, is comparable to the height of the cylindrical layer. The forcing of intensity $S=6.4$ was applied to this initial distribution, and the Rayleigh number was set $Ra=1100$, which is much lower than the critical number $Ra^* \approx 1724$ found for free convection in the applied conditions of these numerical experiments.

Under the action of forcing, the convective motions become spiral and the process of their merging starts – see, figure 1(b)-(e). As a result, one helical vortex structure with a horizontal scale of the order of the radius of computational domain is formed – figure 1(f).

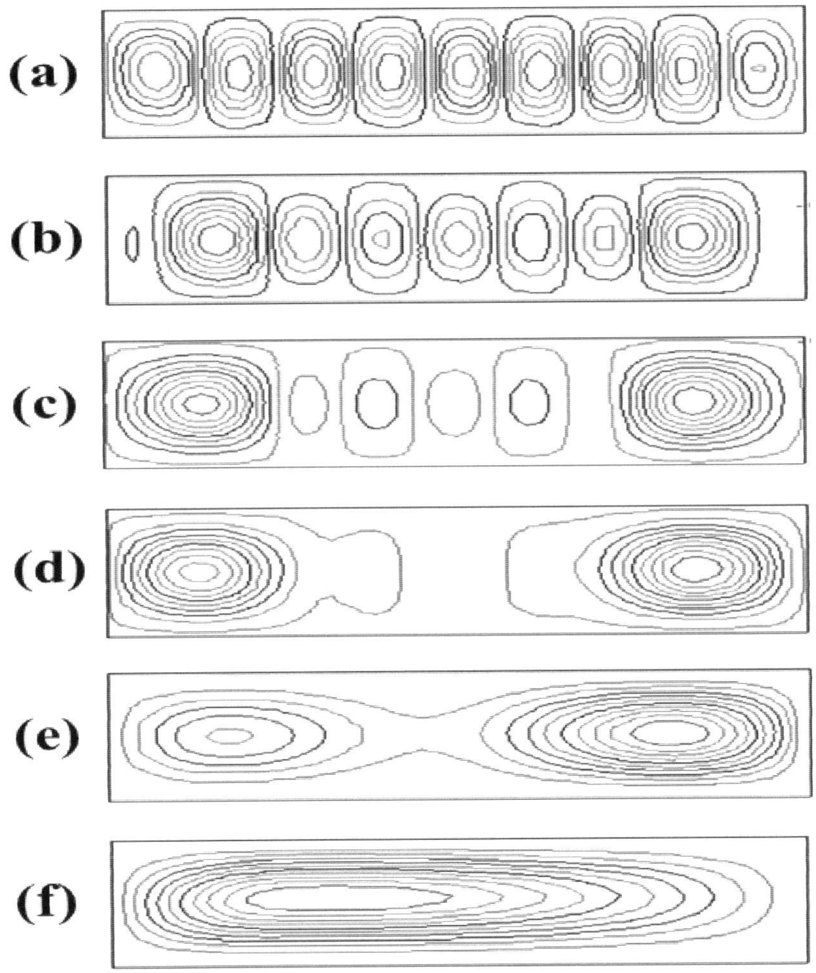

Figure 1. Merging of helical convective cells: (a) t=0; (b) 0.3; (c) 0.4; (d) 0.5; (e) 1.0; (f) 30.0.

The process of merging occurs quickly enough, covering only a span of approximately two units of dimensionless time (introduced as usually in Rayleigh-Bénard problem through relation of the layer height h and the coefficient of kinematic viscosity ν – h^2/ν [18]). At first, the intensity of observed flow decreases as compared with the initial mode, since there is a significant reduction in the heat flux due to the Rayleigh number change to its subcritical value for free convection equal to $Ra=1100$. According to [19], the evolution of helical instability in this time interval does not noticeably affect the intensity of movements, but manifests itself primarily in a qualitative change in the flow pattern, destroying the typical structure of the convective mode and forming instead one vortex cell by the time moment $t \approx 2$ (not shown). Yet the intensity of the formed large-scale flow is still very low. From this moment until the realization of the fully established regime at $t = 30$, shown in figure 1(f), fundamental changes in the flow structure are no longer observed, but a sharp increase in the kinetic energy of the vortex occurs. As a result, the kinetic energy of large-scale spiral vortex $E_f = 4335$ greatly exceeds the energy of the initial convective mode $E_a = 236$.

The flow mode that has been formed under the progressing helical-vortex instability is fundamentally different in its structure and energetics from the initial free convection regime. Under the action of helical forcing, initiating an intense tangential flow, the process of cells merging occurs and the dissipative losses associated with the existence of ascending and descending motions, which form the convective circulation within each cell, decrease. In such conditions, the nature of heat transfer in the layer changes radically. Heat transfer is carried out by forced convection due to intense helical-vortex motion in the large-scale cell. Consequently, the emerged toroidal field of tangential circulation on the one hand suppresses natural convective motions, and on the other hand, itself organizes a vortex cell.

Thus, the resulting large-scale motion represents a spiral vortex, in which the high-intensity horizontal circulation is linked with a weaker vertical circulation.

A number of mixed convection regimes leading up to different amount of helical vortex cells in established steady states were examined thoroughly in [19].

The undertaken numerical analysis of the mathematical model for TVD in a convective system culminated in three important conclusions:

- the existence of an excitation threshold for the helical-vortex instability has been confirmed;
- it has been shown that the increasing intensity of forcing, S, results in the threshold decrease and increase of horizontal dimensions of convective structures, and consequently, smaller number of cells in stationary states;
- increasing effectiveness of heat transfer (evaluated by the Nusselt number) within a helical and larger vortex flow configuration has been emphasized.

Thus, the presented results clearly show that the energy of additional helical source can be converted into the energy of intense large-scale vortex flows. A tendency was found to merging of convective cells that leads to the increase in the horizontal scale of vortex motions as well as to considerable intensification of fluid motion and heat transfer. The discovered new effects may have the immediate relevance to the large-scale instability in turbulent states.

These findings gave the authors of [18, 19] an impetus to search for possible application to tropical cyclone investigations.

3. Numerical analysis of self-organization of cloud convection in the tropical atmosphere

Bridging the turbulent vortex dynamo theory and tropical cyclone investigations was not possible, based on the knowledge of the 20th century, and it was awaiting "its time." Only recently it became possible to address the TVD theory examination by use of direct numerical simulation of atmospheric phenomena what implies a very high spatial resolution of 3-1 km and less horizontal grid spacing. This required the breakthrough of American scientists in tropical meteorology and cloud-resolving numerical modeling of atmospheric processes that happened in the 2000s.

A decisive opportunity for bringing both matters together became the discovery of rotating cloud convection in the tropics in 2004 [26], now known as vortical hot towers - VHTs. The new finding immediately got the observational evidence based on direct measurements in the tropical atmosphere in 2005 [27]. A new scenario of tropical cyclogenesis based on self-organization of rotating moist atmospheric convection followed that in 2006 [10].

3.1. Rotating moist convection – vortical hot towers (VHTs) are helical by definition.

By now, the VHTs and their role in tropical cyclone formation have become a subject of significant amount of studies and reliably confirmed in observations performed by researchers over the world, see e.g. [28, 29, 30], and references therein. These rotating convective structures were closely investigated in several field experiments in the Atlantic and Pacific oceans.

The cloud hot towers in the tropical atmosphere of the Earth were first described in 1958 [31] as horizontally small (~10-30 km wide) but intense cumulonimbus convection cores that reached the tropopause, that in the tropics typically lies at least 15 km above sea level. The term 'hot' comes not from the temperature of the air but because of the intense latent heat release due to phase transitions of moisture (vapor – water – ice) along the tower height.

The vortical tropical convection – VHTs – was first found nearly a half of the century later [26, 27]. Typically, VHTs exhibit convective lifetimes on the order of one hour. For example, a thorough observational evidence of VHTs [32] showed that the specific updraft was 10 km wide and had vertical velocities reaching 10-25 m/s above 6 km. The convective cell was extending up to a height of 17 km. The peak vertical velocity within this updraft exceeded 30 m/s. Maximum values of vertical relative vorticity reached up to $6\text{-}18 \cdot 10^{-3}$ 1/s, what by the one-two order of magnitude exceeds the planetary rotation.

The authors [33, 34] paid the closest attention to the helical features of convective towers. In [33], helicity was calculated in Hurricane Bonnie (1998) using tropospheric-deep dropsonde soundings carried out by reconnaissance aircrafts during the NASA Convection and Moisture Experiment (CAMEX) and the most extreme values of helicity, among the largest ever reported in the literature, were found in the vicinity of deep convective cells. These cells reached as high as 17.5 km. As it was noted in [34], in which helicity calculations were performed for eight tropical cyclones of 1998-2001 sampled during the CAMEX, VHTs are helical by definition because they contain coincident updrafts and vertical vorticity.

3.2. Self-organization of atmospheric moist convection in the tropics.

Near-cloud-resolving idealized simulations [10] showed how a mesoscale tropical depression vortex could develop from rotating deep cumulonimbus convection as a result of upscale vorticity growth. The simulations first generated a number of VHTs, each of 10-30 km horizontal scale, which eventually resulted in an intense mesoscale helical vortex. The enlargement of flow scales occurred by multiple mergers of small-scale convective structures. Such atmospheric scenario of self-organization had much in common with that observed in simulations of the helical-vortex Rayleigh-Bénard convection [18, 19]. This did become evident when both scenarios were brought together and discussed in seminars of Montgomery Research Group at Colorado State University in Fort Collins.

On the background of these findings, the ideas of vortex dynamo acquired a completely new meaning. It became evident that the vortical hot tower route to tropical cyclogenesis [10] represents the most appropriate basis to take into account helical features of convective atmospheric turbulence. This gave a start to collaborative Russian-American studies aimed at applying the TVD theory to examination of tropical cyclogenesis.

3.3. "Helical" post-processing of atmospheric simulation data.

To analyze the process of self-organization of moist atmospheric convection observed under conditions of tropical cyclogenesis as posed in [10], a set of helical characteristics was computed, as well as some other integral characteristics of the velocity field which were applied in [19].

The velocity fields used for post-processing in our studies [21, 22, 23, 24, 25] originated from [10] (in which a more detailed information can be found) and were obtained by use of three-dimensional non-hydrostatic Regional Atmospheric Modeling System (RAMS) comprising time-dependent equations for all three components of velocity, pressure, potential temperature, total water mixing ratio, and cloud microphysics and utilizing an interactive multiple nested grid scheme. For all numerical experiments [10] three nested grids were used. Our post-processing of the model data was carried out on the finest computational grid for subsequent times with a time increment of 10 minutes during 72 hours of numerical experiment. Characteristics were calculated in the computational domain of 276×276×20 km in Cartesian coordinates by use of uniform finite-difference grid. Throughout the post analysis the vertical increment was equal to 500 m; the horizontal increments were 2 and 3 km. We also applied an analysis of system-scale dynamics from a traditional vortex-centric perspective when the Cartesian model data were transformed into a local cylindrical coordinate system.

It is important to point out that no external assumptions were imposed on the fluid motions investigated and described in [21, 22, 23, 24, 25], i.e., no external forcing terms were imposed to mimic a "helical alpha effect". In other words, the presented results are the outcome of a direct numerical simulation subject to the usual caveats of a sub-grid scale closure that is used to remove small scale motions at the horizontal grid scales of the model (~ 3 km).

3.4. Helical tropical cyclogenesis.

For purposes of quantitative diagnosis, we have analyzed the evolution of energetics and structure of the forming vortex. The integral kinetic energy of primary tangential and secondary transverse circulation was applied to diagnose the onset of large-scale vortex instability. A pseudoscalar – helicity of the velocity field (helicity density, integral helicity as well as its horizontal and vertical contribution) was applied to quantitatively analyze the topology.

The first discovery of the undertaken collaborative Russian-American efforts [21] was the break of the mirror symmetry of atmospheric turbulence – non-zero mean helicity generation – during tropical cyclone formation that gave us the impetus to further search for the large-scale vortex instability.

The process of helicity generation was examined with a special focus on how this is realized by a single VHT generated by a local heating at low levels [16]. It can be summarized that the VHTs appear as a result of interaction between cloud moist convection and vertical wind shear. Each convective updraft generates the vertical vorticity by tilting of horizontal vortex filaments and amplifies it by stretching. This process provides a linkage of horizontal and vertical vortex lines and results in helicity generation on a local cloud scale whilst an evolving population of VHTs of different sizes and intensity during tropical cyclone evolution ensures the linkage of circulations on mesoscales.

For the first time in tropical cyclone research, a role of special topology of forming vortex provided by interaction of motions of different scales was emphasized [16]. It has been found that the newly forming mesoscale vortex becomes energy-self-sustaining when a helical structure of the system-scale circulation organizes. Such helical mesoscale organization is only possible due to the linkage of primary tangential and secondary transverse circulation, which is realized through rotating convective structures of cloud scales – VHTs. Helicity dynamics allows tracing such upscale flow organization.

The foregoing results allowed suggesting a key role of VHTs as a necessary element to provide the dynamo effect. Based on the mathematical model for the turbulent vortex dynamo, an analogy was traced [16] between the role of interaction 'moist convection – vertical wind shear' (see, hypothetical forcing (3)) in creating the vortex dynamo in the atmosphere and the role of the mean electromotive force providing the MHD dynamo in electrically conducting medium.

4. Diagnosis of tropical cyclogenesis

By adapting and using for analysis of atmospheric data our developed approach [18, 19], which was tested in the search for a large-scale vortex instability in the helical-vortex Rayleigh-Bénard convection, we have discovered similar instability in conditions of tropical cyclogenesis [23, 24, 25]. The onset of instability was identified as the start of mutual intensification of the primary and secondary circulation making a nascent hurricane vortex energy-self-sustaining. It must be stressed that the observed new instability begins several hours earlier than the vortex of tropical depression is formed. At present, it is the latter that is usually declared in meteorological observations as the formation of a tropical cyclone. If the discovered instability is reliably confirmed for tropical cyclones observed in real atmospheric conditions, it may be suggested to consider the emergence of this instability as the commencement of tropical cyclogenesis.

Evidently, our most important finding, contributing to both the fundamental science and practical issues of the earlier warning of the population about the hurricane danger, is connected to the fact that the onset of cyclogenesis is now getting the precise interpretation - it coincides with the beginning of new found instability. Moreover, our results show that this instability starts earlier (from a few hours up to a few tens of hours) than the formation of a tropical depression vortex takes place – the event, for which the regional storm tracking services usually declare a tropical cyclone emergency.

References

[1] Moiseev S S, Sagdeev R Z, Tur A V, Khomenko G A and Yanovsky V V 1983 *Sov. Phys. JETP* **58** 1149
[2] Moiseev S S., Sagdeev R Z, Tur A V, Khomenko G A and Shukurov A M 1983 *Sov. Phys. Dokl.* **28** 925
[3] Moiseev S S, Rutkevich P B, Tur A V and Yanovskii V V 1988 *Sov. Phys. JETP* **67** 294
[4] Gray W M 1979 *Meteorology Over the Tropical Oceans*, ed D B Shaw (Reading, UK: R. Meteorol. Soc.) p 155
[5] Moffatt H-K 1969 *J. Fluid Mech.* **35** 117
[6] Moffatt H-K 1978 *Magnetic Field Generation in Electrically Conducting Fluids* (Cambridge-London-New York-Melbourne: Cambridge University Press)
[7] Frisch U 1995 Turbulence: *the Legacy of A N Kolmogorov* (Cambridge UK: Cambridge University Press)
[8] Moffatt H-K 2014 Proc. Natl. Acad. Sci. USA **111** 3663
[9] Lilly D K 1986 *J. Atmos. Sci.* **43** 126
[10] Montgomery M T, Nicholls M E, Cram T A and Saunders A B 2006 *J. Atmos. Sci.* **63** 355
[11] Steenbeck M, Krause F and Rädler K-H 1966 *Zeitschrift für Naturforschung* **21** A 369
[12] Krause F and Rädler K-H 1980 *Mean-Field Magnetohydrodynamics and Dynamo Theory* (Berlin: Academie-Verlag)
[13] Fortov V E, Ivanov M F, Ivlev A V, Gnedin Yu N and Klumov B A 1996 *Phys. Usp* **39** 363
[14] Ivanov M F, Galburt V A and Fortov V E 1996 *JETP Lett.* **63** 813
[15] Levina G V, Moiseev S S and Rutkevich P B 2000 *Nonlinear Instability, Chaos and Turbulence* vol 2, ed L Debnath and D N Riahi, Adv. Fluid Mech. Series, (Southampton: WIT Press) p 111
[16] Levina G V 2018 *Open J. Fluid Dyn.* **8** 86
[17] Rutkevich P B 1993 *JETP* **77** 933
[18] Levina G V 2006 *Dokl. Earth Sci.* **411** A 1417
[19] Levina G V and Burylov I A 2006 *Nonlin. Processes Geophys.* **13** 205
[20] Kurgansky M V 1998 *Izv. Atmos. Ocean. Phys.* **34** 156
[21] Levina G V and Montgomery M T 2010 *Dokl. Earth Sci.* **434** 1285
[22] Levina G V and Montgomery M T 2011 *J. Phys.: Conf. Ser.* **318** 072012
[23] Levina G V and Montgomery M T 2014 *J. Phys.: Conf. Ser.* **544** 012013
[24] Levina G V and Montgomery M T 2014 *Dokl. Earth Sci.* **458** 1143
[25] Levina G V and Montgomery M T 2015 *Procedia IUTAM* **17** 59
[26] Hendricks E A , Montgomery M T and Davis C A 2004 *J. Atmos. Sci.* **61** 1209
[27] Reasor P D, Montgomery M T and Bosart L F 2005 *J. Atmos. Sci.* **62** 3151
[28] Houze R A Jr 2010 *Mon Wea Rev* **138** 293
[29] Montgomery M T, Davies C, Dunkerton T, Wang Z, Velden C *et al.* 2012 *Bull. Amer. Meteor. Soc.* **93** 153
[30] Kilroy G, Smith R K and Montgomery M T 2017 *Q J R Meteorol Soc.* **143** 450
[31] Riehl H and Malkus J S 1958 *Geophysica* **6** 503
[32] Houze R A Jr, Lee W C and Bell M M 2009 *Mon Wea Rev* **137** 2778
[33] Molinari J and Vollaro D 2008 *Mon Wea Rev* **136** 4355
[34] Molinari J and Vollaro D 2010 *J. Atmos. Sci.* **67** 274

A new MPI/OpenMP code for numerical modeling of relativistic hydrodynamics by means adaptive nested meshes

Igor Kulikov[1], **Igor Chernykh**[1], **Anna Sapetina**[1], **Vladimir Prigarin**[1]

[1] Institute of Computational Mathematics and Mathematical Geophysics SB RAS, 6-th Lavrentyev avenue, Novosibirsk, Russian Federation

E-mail: kulikov@ssd.sscc.ru

Abstract. Many astrophysical phenomena are associated with gas motion at relativistic velocities. The source of such currents are active galactic nuclei, microquasars, pulsars, gamma bursts, black holes, neutron stars, and gravitational waves. To study such phenomena, it is necessary to perform simulation within the scope of special relativistic hydrodynamics. One of the difficulties of modeling relativistic flows is the different scale of processes, which requires the use of both parallel computing and adaptive meshes.

1. Introduction

Many astrophysical phenomena are associated with gas motion at relativistic velocities. The source of such currents are active galactic nuclei, microquasars, pulsars, gamma bursts, black holes, neutron stars and gravitational waves. One of the difficulties of modeling relativistic flows is the different scale of processes, which requires the use of both parallel computing and adaptive meshes. Today, many numbers of program codes were developed for simulation of relativistic astrophysical flows. Of particular interest is framework The Einstein Toolkit .

Einstein Toolkit [1] is a framework designed to solve the equations of special and general relativistic hydrodynamics with and without a magnetic field and other factors. The framework consists of several blocks. These are a parallel computing infrastructure for meta-calculations called Cactus [2], a parallel transformation to HDF5 format, a web-interface, and visualization tools. There are two methods of organizing calculations on adaptive nested grids: One is an approach called PUGH, which is based on a local refinement of a regular grid with a scalability of up to 130,000 processors. The other is an approach of multi-level grid nesting called Carpet [3, 4]. Both approaches meet modern requirements for accuracy and stability [5]). To implement the framework, a concept called "simulation factory" [6] with a set of abstractions to organize supercomputer simulations is used. These are remote access, configuration, assembly, shipment, and simulation management. Such abstractions may conceal some low-level constructions, allowing one to focus on solving applied problems. To translate the tensor systems of partial differential equations written in the computer algebra package Mathematica into a code for the C or Fortran languages, a package called Kranc is used [7]. The numerical methods employ a reconstruction of the primitive variables by total variation diminishing (TVD) of the solution,

Content from this work may be used under the terms of the Creative Commons Attribution 3.0 licence. Any further distribution of this work must maintain attribution to the author(s) and the title of the work, journal citation and DOI.
Published under licence by IOP Publishing Ltd

IOP Conf. Series: Journal of Physics: Conf. Series **1336** (2019) 012008 doi:10.1088/1742-6596/1336/1/012008

the piecewise-parabolic method (PPM), and essentially non-oscillatory (ENO) methods. The HLLE, Roe, and Marquina methods are used as solvers for the Riemann problem [8].

GENESIS [9] code was developed for the simulation of relativistic jets. The code has been adapted for an arbitrary equation of state [10], which allows simulating chemical kinetics of gas reactions and stellar equations of state [10]. The code includes a detailed description of a parallel implementation of the calculations, a data distribution scheme, and memory use. A procedure for restoring the primitive (physical) variables is described separately. A problem of the breakdown of a discontinuity with different values of the Lorentz factor has been solved as a test one. The available detailed description of the code allows the readers to create their own implementations on its basis.

This paper describes the author's approach to solving the problems of relativistic hydrodynamics and its application to the jet modeling. In the second section, the equations of special relativistic hydrodynamics are written down and the numerical solution scheme is briefly described. The description of the HydroBox3D code architecture and the study of acceleration on the IBM Power 9 processor are provided in the third section. The fourth section deals with modeling a relativistic jet. The conclusion is given in the fifth section.

2. The Numerical Model

In the paper the light speed is taken $c \equiv 1$. The equations of relativistic hydrodynamics in the form of conservation laws are written in the form:

$$\frac{\partial D}{\partial t} + \frac{\partial (D v_k)}{\partial x_k} = 0, \tag{1}$$

$$\frac{\partial M_j}{\partial t} + \frac{\partial (M_j v_k + p\delta_{jk})}{\partial x_k} = 0, \tag{2}$$

$$\frac{\partial E}{\partial t} + \frac{\partial (E + p) v_k}{\partial x_k} = 0, \tag{3}$$

where

$$D = \Gamma \rho, \quad M_j = \Gamma^2 \rho h v_j, \quad E = \Gamma^2 \rho h - p, \tag{4}$$

where ρ is the density, v_k are the components of the velocity vector, δ_{jk} is the Kronecker symbol, p is the pressure and h is the special enthalpy:

$$h = 1 + \frac{\gamma}{\gamma - 1} \frac{p}{\rho} \tag{5}$$

γ is the adiabatic index. The Lorentz factor Γ is determined by the formula:

$$\Gamma = \frac{1}{\sqrt{1 - v^2}} \tag{6}$$

the speed of sound is determined by the formula:

$$c_s^2 = \gamma \frac{p}{\rho h} \tag{7}$$

It shall be noted that in the numerical method, calculations are carried out in conservative variables D, M_j, E.

With respect to nonlinear relationship between conservative variables and primitive ρ, v, p, a special procedure for their recovery is required. To recovery of primitive variables, we use the Newton method to determine the root of the equation:

$$f(p) = \Gamma^2 \rho h - p - E = 0 \tag{8}$$

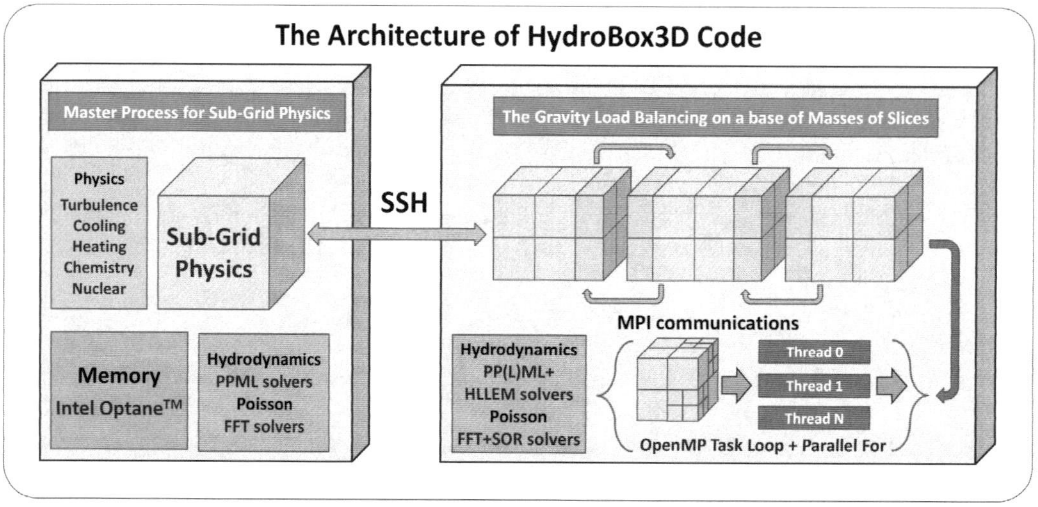

Figure 1. The principal architecture of HydroBox3D code.

according to the classical iterative scheme:

$$p_{m+1} = p_m - \frac{f(p_m)}{f'(p_m)} \tag{9}$$

where

$$f'(p) = \frac{\gamma}{\gamma - 1}\Gamma^2 - \frac{M^2\Gamma^3}{(E+p)^3}\left(D + 2\frac{\gamma}{\gamma - 1}p\Gamma\right) - 1 \tag{10}$$

3. The Architecture of HydroBox3D Code

The principal architecure of HydroBox3D code (see Figure 1). To simulate the hydrodynamic evolution of astrophysical objects, the technology of nested meshes is used, which is described in detail in [11]. The implementation is adapted for the use on distributed memory architectures. To synchronize the overlap regions, the MPI technology is used. An important component of the code is the load balancing procedure for distributed nested meshes. For this, a one-dimensional domain decomposition and an estimate of the number of nested meshes cells in each slice of the root grid are used. This quantity is uniformly distributed among the processes. In each MPI-process, calculations are performed for each nested mesh using a two-level mechanism for distributing calculations among threads based on the "task loop" and "parallel for" pragmas of the OpenMP library. To solve the hydrodynamics equations, the piecewise-parabolic method on a local stencil [12, 13] or its modification based on the piecewise linear function is used. A modification of the HLLEM / HLLI method [14, 15] has also been implemented, whose study we intend to do separately. To solve the Poisson equation, a method based on a combination of the fast Fourier transform and successive over-relaxation is used. When the trigger snaps into action, the subgrid modeling starts with a new process initiation using SSH. The subgrid modeling uses a regular grid and a large set of subgrid processes.

For computational experiments we use 2 × 12-core Typical 2.7 to 3.8 GHz (max) IBM Power 9 Processor, 32 × 32 GB DDR4 Memory, and one 300GB 15K RPM SAS SFF-3 Disk Drive (OS Linux). In all configurations, the root mesh size of 128^3 was chosen as the base mesh. Three configurations of nested grids were considered:

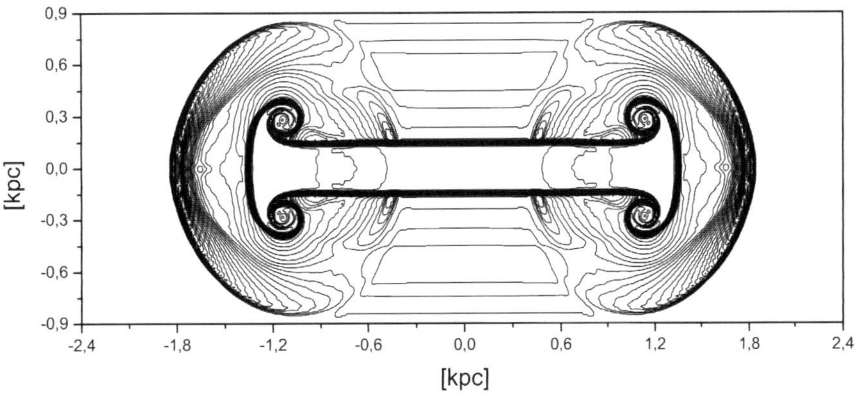

Figure 2. Density isolines and flow structure in equatorial plane of galactic jet.

(i) All nested grids are size of 4^3 (uniform grid with an effective resolution of 512^3).
(ii) 75 % of nested grids are size of 2^3 and 25 % are size of 8^3 (effective resolution is 1024^3).
(iii) 75 % nested grids are size of 2^3, 15 % are size of 8^3 and 10 % are size of $32^{\ 3}$ (effective resolution is 4096^3).

When using a regular grid with an effective resolution of 512^3, a 60-fold acceleration was obtained using 196 streams of IBM Power 9. A 42-fold acceleration was obtained using 64 streams of IBM Power 9 using a nested grid with an effective resolution of 1024^3. When using a nested grid with an effective resolution of 4096^3, a 34-fold acceleration was obtained using 48 streams of IBM Power 9. As can be seen from the results, the hyper-threading system works quite effectively with uniform loading of cores. In case of non-uniform loading the system has local optimum of number of used cores that is several times less than the total number of streams.

4. The Numerical Simulation of Relativistic Jet

Let us simulate a galactic jet of density $\rho_J = 10^{-3}$ cm^{-3} and radius $R_J = 200$ parsec. The jet moves at Lorenz factor $\Gamma = 5$, relativistic Mach number $\mathcal{M} = 8$, the temperature of the galactic atmosphere $T_A = 10^7$ K, and the density $\rho_A = 10^{-2}$ cm^{-2}. The adiabatic index γ is taken to be equal to 5/3. Figures (2) show the results of a simulation of galactic jet evolution. The results of the simulation show that a shock wave is moving ahead with a propagation speed that corresponds to the speed of light. Behind the shock front there is a shell region. It separates the shock front and the hotspot where a maximum temperature is reached. The inner part of the cocoon flow is bounded by a contact surface, and the cocoon, in turn, contains a jet. On the outside of the cocoon, closer to the base, there propagate backflows which, in turn, interact with the jet flow.

5. Conclusion

In the paper, a new code for modeling special relativistic hydrodynamic flows on supercomputer architectures with distributed memory using the technology of nested grids was described. A 60-fold acceleration was obtained using 196 streams of IBM Power 9 using a regular grid with an effective resolution of 512^3. 42-fold acceleration was obtained using 64 streams of IBM Power 9 Using a nested grid with an effective resolution of 1024^3. A 34-fold acceleration was obtained using 48 streams using a nested grid with an effective resolution of 4096^3 IBM Power 9. In terms of computational experiments, the results of mathematical modeling of relativistic hydrodynamics of galactic jets are presented.

Acknowledgments

This work was supported by Russian Science Foundation (project no. 18-11-00044).

References

[1] Loffler F, et al 2012 *Classical and Quantum Gravity* **29** (11) 115001
[2] Goodale T, et al 2003 *Lecture Notes in Computer Science* **2565** 197-227
[3] Schnetter E, Hawley S, and Hawke I 2004 *Classical and Quantum Gravity* **21** (6) 1465-1488
[4] Schnetter E, Diener P, Dorband E, and Tiglio M 2006 *Classical and Quantum Gravity* **23** (16) S553-S578
[5] Lehner L, Liebling S, and Reula O 2006 *Classical and Quantum Gravity* **23** (16) S421-S446
[6] Thomas M, and Schnetter E 2010 *11th IEEE/ACM International Conference on Grid Computing* 369-378
[7] Husa S, Hinder I, and Lechner C 2006 *Computer Physics Communications* **174** (12) 983-1004
[8] Donat R, Font J A, Ibanez J M, and Marquina A 1998 *Journal of Computational Physics* **146** 58-81
[9] Aloy M, Ibanez J M, Marti J, and Muller E 1999 *The Astrophysical Journal Supplement Series* **122** 122-151
[10] Timmes F X, and Arnett D 1999 *The Astrophysical Journal Supplement Series* **125** 277-294
[11] Kulikov I 2018 *Journal of Physics: Conference Series* **1103** 012011
[12] Kulikov I, and Vorobyov E 2016 *The Journal of Computational Physics* **317** 318-346
[13] Kulikov I, Chernykh I, and Tutukov A 2019 *The Astrophysical Journal Supplement Series* **243** 4
[14] Balsara D, Li J, and Montecino G 2018 *Journal of Computational Physics* **375** 1238-1269
[15] Deng X, Boivin P, and Xiao F 2019 *Physics of Fluids* **31** 046102

IOP Conf. Series: Journal of Physics: Conf. Series **1336** (2019) 012009 doi:10.1088/1742-6596/1336/1/012009

Formation of filaments and dense cores during molecular clouds collision

B P Rybakin[1,2], **S G Moiseenko**[3]
[1]Lomonosov Moscow State University, Moscow, Russian Federation
[2]Scientific Research Institute for System Studies, Russian Academy of Sciences, Moscow, Russian Federation
[3]Space Research Institute, Russian Academy of Sciences, Moscow, Russian Federation

E-mail: rybakin1@mail.ru, moiseenko@iki.rssi.ru

Abstract. The paper presents the results of mathematical modeling of the collision process of two molecular clouds(MC). The density distribution over the radius of the MC is modeled by various laws, the density in the center varies. The forces of gravity, thermal conductivity and radiative losses are not taken into account. The processes of deformation, fragmentation and the formation of superdense regions that occurs during the collision of molecular clouds are analyzed. Visualization of the calculation results made it possible to find the features of the emerging flow the formation of vortices, filaments, and superdense nuclei in molecular clouds.

1. Introduction

A large number of works have been devoted to the study of the behavior of molecular dust clouds and their role in the formation of stars. In pioneering works [1, 2], the processes of interaction of flows with various inhomogeneities were analytically studied. Issues related to the influence of external flows on the properties of inhomogeneous regions were studied.

As computing power increased, work appeared in which the inclusion of gravity and the influence of magnetic fields were added. In the papers [3] - [16], the problem of the interaction of the flow of matter and shock waves with a single MC was investigated. These studies were carried out in one-dimensional and two-dimensional formulations.

In the papers [17]-[21] two and three-dimensional formulations of the problem were considered. In these and in a number of other studies, it became possible to identify complex interactions that arise during the interaction of strong shock waves with molecular dust clouds. The appearance of the Rayleigh-Taylor instability, the entrainment of matter (ablation) during the formation of a turbulent flow and radiative cooling of the cloud substance are studied.

In [18] a single-phase gas model that accurately reflects the physical picture of the flow was used. In the calculations, it was assumed that the unperturbed interstellar matter(ISM) consists of relatively warm matter ($\sim 10^4$ K) and small nonuniformly distributed cold clouds ($\sim 10^2$ K), which have a high density [13], [17], [22]. Initially, the clouds are in dynamic equilibrium with the background gas [18].

This paper discusses the processes of interaction of two molecular clouds, with different gas-dynamic characteristics. The process of the emerging hypersonic gas-dynamic flow is simulated and the interaction of the processes of compression, fragmentation and the formation of turbulent flows in the clouds and the environment are analyzed.

Content from this work may be used under the terms of the Creative Commons Attribution 3.0 licence. Any further distribution of this work must maintain attribution to the author(s) and the title of the work, journal citation and DOI.
Published under licence by IOP Publishing Ltd

To carry out the calculations, a parallel algorithm for solving the problems of non-stationary gas dynamics [21] was developed, which was tested in detail when solving the problem of the interaction of an incident shock wave with a region of lower density.

To speed up the calculations, OpenMP technology was used. A parallel code based on high-resolution difference schemes was developed. The code solves three-dimensional Euler equations for a compressible gas. To improve the quality of the parallel code, Intel V Tune Amplifier XE was used.

2. Formulation of the problem

The interstellar medium consists of a relatively warm matter $\sim 10^4 K$. There is a molecular cloud in the ISM, the density is much higher than the density of the ISM. The temperature in the molecular cloud is lower than the temperature of the ISM $T_{cloud} \sim 10^2 K$. At the initial moment the molecular clouds are in contact with each other. Inside each molecular cloud, a radial density distribution is given. In fig. (1) is a schematic representation of the initial moment of impact. We use a single-phase gas model.

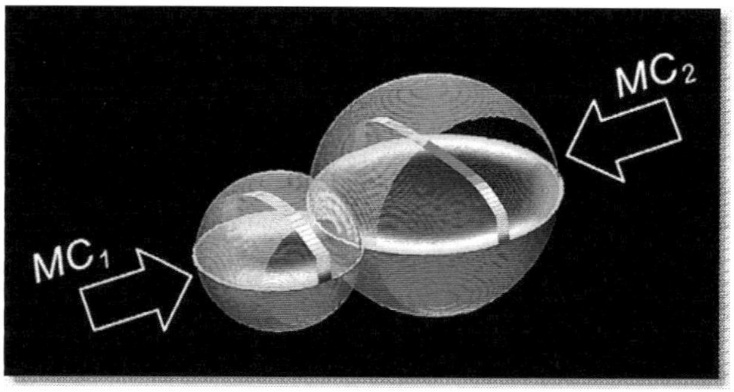

Figure 1. Initial stage of molecular clouds collision

Initially, the cloud is in dynamic equilibrium with the background gas. The thermal conductivity and radiation losses during the interaction can be neglected. The work uses the equation of state of an ideal gas with $\gamma = 5/3$.

ISM density is $\rho_a = 2.15 * 10^{-25}\ g/cm^3$, temperature $T_a = 10^4 K, u_a = 0.0$. Cloud's density is $\rho_c = 1.075 * 10^{-22} g/cm^3$, temperature $T_c = 100K$, Mutual impact speeds ranged from ± 2 km / s to ± 15 km / s. The collision time of two MCs is approximately equal to one million years.

3. Governing equations

3.1. Governing Equations

The gas dynamic process computations in complex problems are not the only factors requiring high performance computers to be used. Additional intensive computations are required, they being used to account for the effects of gravitational and magnetic fields, viscosity, heat conductivity, etc. The time spent for the additional computations by far exceeds the time of gas dynamic phase computations [25], [28].

The problems being solved consider the flows of ideal compressible gas in nonsteady definition. Viscosity and heat conductivity effects on mass transfer are ignored. The system of conservation laws of mass, momentum, and energy in 3-D Cartesian coordinate system with self-gravitation taken into account can be written in the following way:

$$\frac{\partial \mathbf{Q}}{\partial t} + \frac{\partial \mathbf{F_x}}{\partial x} + \frac{\partial \mathbf{F_y}}{\partial y} + \frac{\partial \mathbf{F_z}}{\partial z} = \mathbf{S} \tag{1}$$

where quantity $\mathbf{Q} = (\rho; \rho u; \rho v; \rho w; \rho E)^T$ is the state vector, the components of which are conservative variables – densities of mass, momentum and total energy.

Numerical flux $\mathbf{F_x}$, source term and gravity vectors are written as follows:

$$\begin{aligned} \mathbf{F_x} &= \left(\rho u; \rho u^2 + P; \rho uv; \rho uw; \rho uE + uP\right)^T, \\ \mathbf{S} &= \left(0; \rho g_x; \rho g_y; \rho g_z; \rho \mathbf{g} \cdot \mathbf{v}\right)^T, \\ \mathbf{g} &= \left(g_x; g_y; g_z\right)^T = -\bigtriangledown \Phi. \end{aligned} \tag{2}$$

In the above equations ρ is density, $\mathbf{v}$ is velocity, g is gravity, P is pressure, E is total energy. The vectors $\mathbf{F_y}, \mathbf{F_z}$ can be written similarly. These equations calculate gravitational potential Φ with the solution of Poisson equation taken into account:

$$\bigtriangledown^2 \Phi = 4\pi G\rho. \tag{3}$$

To close the system of equations the equation of state for ideal gas is used in the following form:

$$P = \rho(\gamma - 1)[E - 1/2(u^2 + v^2 + w^2)], \tag{4}$$

where γ is the ratio of heat capacities.

Initially the molecular clouds show a thermodynamic balance with the environment [28]. Heat conduction and radiation losses are ignored in computations [20]. The equation of state for adiabatic gas is used with $\gamma = 5/3$. The density of interstellar medium is taken equal to $2.15 * 10^{-25} g/cm^3$, the temperature T=10000 K, u=0. The density of MC gas is equal to $\rho_a = 1.075 * 10^{-22} g/cm^3$, the temperature T=100 K.

3.2. Non-Dimensional Procedure

Numerical computations are done with the minimum accuracy loss if initial values of design quantities range from ± 1. But the density of interstellar media, the density beyond the shock wave front and in MCs have the quantities of approximately 10^{-25}, the values of temperature and velocity have the quantities of approximately 10^5. The use of such magnitudes in computations results in significant loss of computation accuracy. The research uses nondimensionalization to have the values of all the quantities in the range specified. This requires each function used in the problem definition to be presented as follows: $f = f_0 f'$, where f' is a dimensionless function and f_0 is a dimension factor which can help nondimensionalize the quantities used in computations [26]. The system of equations written in a dimensionless form gives a number of dimensionless fixed factors as Reynolds number, Euler number, etc.

The quantities describing the interstellar medium are used as nondimensionalizing multiplying factors. For example ISM density is $\rho_a = 2.15 * 10^{-25}\ g/cm^3$, temperature $T_a = 10^4 K, u_a = 0.0$. Let us nondimensionalize the system of equations (1-4). The dimension theory provides that there are only four quantities among all dimensional ones which have an independent dimension. If nondimensionalized values are substituted in the system of equations $(1-4)$, and the required transformations are done, we will have a dimensionless system of equations, with the form of the system not changing. After completing computations we can go back to dimensional quantities.

For a good resolution of the processes occurring in the MC, it is necessary that the radius of the cloud should be at least 128 calculated cells. The size of the difference grid was chosen from these considerations; the main calculations were carried out on a grid with dimensions of $1024 \times 1024 \times 1024$.

3.3. Initial and boundary conditions

The calculated area is presented by parallelepiped with sizes $1024 \times 1024 \times 1024$. The density radial distribution formulas for clouds C_1 and C_2 are the following:

$$\rho(r) = \rho_{ism} + \frac{\rho_{cl} - \rho_{ism}}{1 + (r/R_{cl})^{2.7128}}, \tag{5}$$

$$\rho(r) = \rho_{ism}\big(\chi + \frac{\alpha}{\alpha+1}(1-\chi)\big), \quad (6)$$

where $\chi = \frac{\rho_{cl}}{\rho_{ism}}$ – density contrast.

Form factor α in (6) can be calculated by formula:

$$\alpha = exp\{min[20.0, 10 \cdot ((\frac{r}{R_{cl}})^2 - 1)]\}. \quad (7)$$

Several parameters in these formulas controlling the steepness of clouds border were changed to improve density smoothing. Boundary conditions at all boundaries are free outflow conditions.

4. Numerical methods

In this paper we use the schemes of type TVD. This class of schemes is characterised by usage of non-linear monotonicity condition. Such condition permits one to limit appearance of new maximums and minimums which inevitably arise in the high order accuracy schemes. Monotonous against-streaming scheme for conservative laws, used in calculations, is the modification of the first order monotonous against-streaming scheme. This scheme fits very well to the three-dimensional Euler equations solution.

The steps by time are calculated with the help of Courant-Friedrichs-Lewy stability condition. Using partitioning by spatial variables allows effective application of parallel processing technology OpenMP.

Testing of the proposed algorithm and program had been carried out. The testing showed good coincidence of analytical and experimental data with numerical modeling results. The main testing results are brought in [25].

5. Results

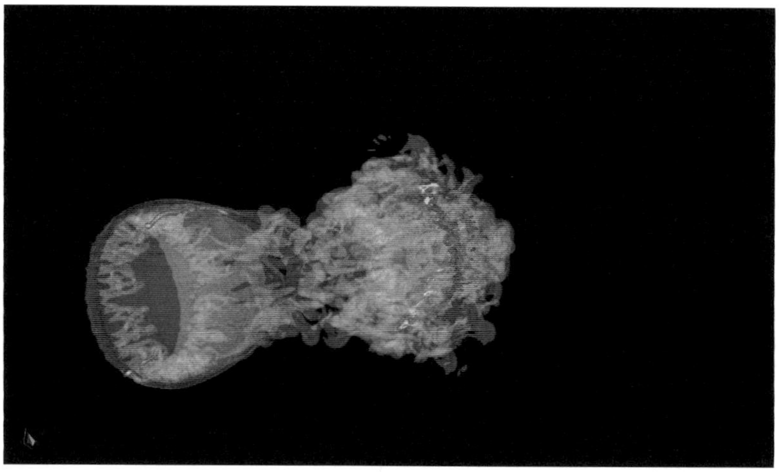

Figure 2. Formation of the vortex and ring-like structures at the developed stage of MC collision process

The formation of superdense clumps resulting from the collision of molecular clouds is an interesting mechanism that explains the appearance of new stars. These dense clumps are gravitationally unstable according to Jeans and are the precursors of new stars and star clusters. Thermal pressure in the MC is not enough to prevent gravitational collapse.

The calculation results showed that the mutual collision of the MC leads to the formation of a superdense region on the surface of the collision, in which ever denser gas clumps are constantly formed. Small irregular pulsating structures appear on the shells of both MCs due to instabilities of the Rayleigh-Taylor type. Strong shock waves appear at the boundary between the clouds, forming a thin dense disk of compressed gas (Fig. 2 is shown in red). Inside this disk, the pulsation structure in thin surface shells is enhanced due to the instability of the thin shell, while the thin shell is limited on both sides by continued compression and accelerated by the continuous movement of the clouds. The density of these clumps exceeds the initial density by several orders of magnitude, which leads to the formation of gravitationally unstable regions. In figures 2, 3, and 4, green indicates the dimensionless density 5, red color indicates 4000.

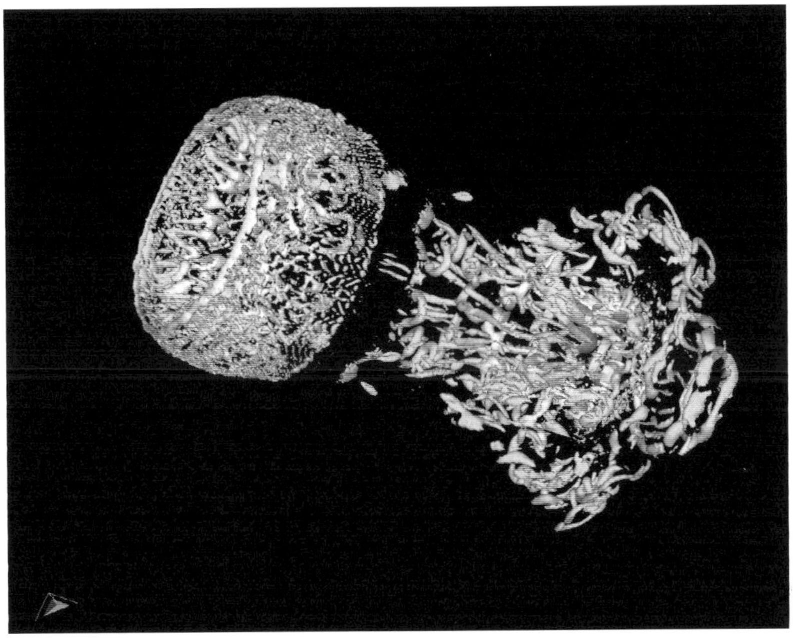

Figure 3. Formation of the vortex and ring-like structures at the developed stage of MC collision process

Vortices form ring structures on the surface of the cloud. Over time, they turn into vortex tracks and initiate vortex pins.

In order to find out the structural features of the vortex flow, the calculated fields of the Q-criterion [27] are presented. The second invariant of the velocity gradient tensor is used to identify areas of unevenly scaled vortex concentrations. Maps of the color palette of local speed, shown on the surfaces of the Q-test, illustrate the intermittency of the gas flow. In Fig. 3 and Fig. 4 figures of the Q-criterion (Q=1000) for moments of dimensionless time t = 225 and t = 250 are shown. Vortices have a lower distribution density inside the mixing region; on the boundaries and surfaces of elongated filamentous buds, the vortex distribution densities are much higher and show a local velocity slope in different MS regions.

In situations with insufficient smoothness of pressure/energy fields in areas with high gradients, the spatial intermittency of the outer layers of the clouds and their lenticular lumpy (core) deformation lead to coherent and wave perturbations inside and out. The process is accompanied by Kelvin-Helmholtz (KN) instability and a violation of the gas density above the perturbed surface layers of the clouds. Coherent increasing oscillations caused by extremely

IOP Conf. Series: Journal of Physics: Conf. Series **1336** (2019) 012009 doi:10.1088/1742-6596/1336/1/012009

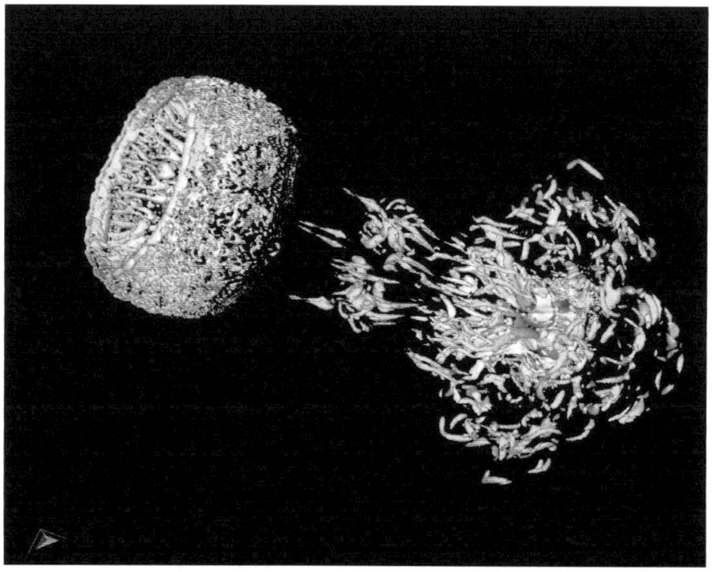

Figure 4. Formation of the vortex and ring-like structures at the developed stage of MC collision process

strong contractions of the core of lenticular clouds become clearly observable temporary pulsations as one cloud penetrates another. Fig. 5. It is very likely that this process is generated through the exchange of energy in high-gradient outer layers and the deformation of a lumpy membrane wave. The nonlinear instability of thin shells (NTSI) in the gas phase can play a decisive role in starting this process.

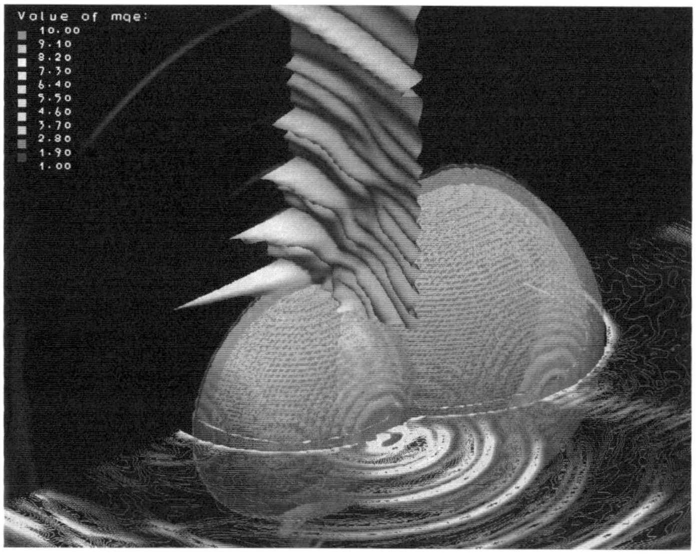

Figure 5. The formation of coherent vibrations arising from the collision of molecular clouds

During the calculations, it was found that temporary pulsations appear during the penetration of one cloud into another. A structure is formed that leads to the interaction between the clouds and the interstellar medium by means of coherent pulsation (Fig.5). This process is generated through the exchange of energy in highly gradient outer layers and deformation of the membrane structure. The nonlinear instability of thin shells in the gas phase can play a key role in starting this process. The Kelvin-Helmholtz instability leads to the deformation of MCs, the appearance and formation of protonuclei - gravitationally bound structures.

Acknowledgments

This work has been supported by the Program of Fundamental Research of the Presidium of the Russian Academy of Sciences No. 26

References

[1] T W Hartquist, J E Dyson 1988 *Ap&SS* **144** 615
[2] T W Hartquist, J E Dyson, R J R Williams 1997 *ApJ* **482** 182
[3] Klein R., McKee C., and Colella P.: On the hydrodynamics interaction of shock waves with interstellar clouds. Nonradiative shocks in small clouds. ApJ 420, 213-236 (1994).
[4] Pittard J.M., Falle S.A.E.G., Hartquist T.W., Dyson J.E.: The turbulent destruction of clouds. MNRAS, 394, 1351-1378 (2009).
[5] Tasker, E.J., Brunino, R., Mitchell, N. L., Michielsen, D., Hopton, S., Pearce, F. R., Bryan, G. L., Theuns, T.: A test suite for quantitative comparison of hydrodynamics codes in astrophysics. Mon. Not. R. Astron. Soc. (2008).
[6] Matsumoto, T.: Self-gravitational Magnetohydrodynamics with Adaptive Mesh Refinement for Protostellar Collapse. Publ. Astron.Soc. Japan, 59 (2007).
[7] Rybakin B., Goryachev V.: The supersonic shock wave interaction with low-density gas bubble. Acta Astronautica 94 (2), 749-753 (2014).
[8] I. M. Kulikov, Igor Chernykh. A New Parallel Intel Xeon Phi Hydrodynamics Code for Massively Parallel Supercomputers. Article in Lobachevskii Journal of Mathematics, November 2018.
[9] Andre, Ph., Menshchikov, A., Bontemps, S. et.al.: From lamentary clouds to prestellar cores to the stellar IMF: initial highlights from the Hersche lGould Belt Survey. Astron. Astrophys. L102, 518-523. doi:10.1051/0004-6361/201014666) (2010)
[10] Ferriere, K.M. (2001). The interstellar environment of our galaxy. Reviews of Modern Physics, 73, 10311066.
[11] P R Woodward, P Colella, 1984 *J. Comp. Physics* **54** 115
[12] M J Berger, J Oliger, 1984 *J. Comp. Physics* **53** 484
[13] Klein R, McKee C, and Colella P (1994). On the hydrodynamics interaction of shock waves with interstellar clouds, 1. Nonradiative shocks in small clouds. ApJ, vol. 420, pp 213-236.
[14] J. Kelly Truelove, Richard I. Klein, Christopher F. Mckee, John H. Holliman, Louish H. Howell, Jeffrey A. Greenough, and D. Tod Woods. 1997. Self-Gravitational Hydrodynamics With Three-Dimensional Adaptive Mesh Refinement: Methodology And Applications To Molecular Cloud Collapse And Fragmentation.
[15] Anderson, M.C., Jones, T.W., Rudnick, L., Tregillis, I.L., Kang, H. 1994, ApJ, 421, L31
[16] Gregori, G., Miniati, F., Ryu, D., Jones, T.W. 1999, ApJ, 527, L113
[17] Melioli C., de Gouveia Dal Pino E., Raga A. Multidimensional hydro dynamical simulations of radiative cooling SNRs-clouds interactions: an application to starburst environments // Astronomy & Astrophysics, 2005. Vol. 443. pp. 495-508.
[18] Frank A., Poludnenko A., Gardiner T., Lebedev V., Drake R (2003), Stellar outflows with new tools: advanced simulations and laboratory experiments. RevMexAA (Serie de Conferencias), 15, pp. 85-91.
[19] Stone J.M., Norman M.L. The three-dimensional interaction of a supernova remnant with an interstellar cloud // APJL, 1992. Vol. 390. pp. 1719.
[20] J. Kelly Truelove, Richard I. Klein, Christopher F. Mckee, John H. Holliman, Louish H. Howell, Jeffrey A. Greenough, and D. Tod Woods. Self-Gravitational Hydrodynamics with Three-Dimensional Adaptive Mesh Refinement: Methodology and Applications to Molecular Cloud Collapse and Fragmentation, 1997.
[21] Rybakin B. Modeling of III-D Problems of Gas Dynamics on Multiprocessing Computers and GPU // Computers and Fluids, 2013. Vol. 80. pp. 403407.
[22] McKee C.F. Supernova remnant shocks in an inhomogeneous interstellar medium // Supernova Remnants and the Interstellar Medium, 1988. IAU Coll. 101. pp. 205-222.
[23] Rybakin B.P., Betelin V.B., Smirnov N.N., Moiseenko S.G., Stamov L.I. 3D numerical simulation of molecular clouds collision process. Journal of Physics: Conference Series, Institute of Physics (United Kingdom), 2018, v. 1103, 1, pp. 012007 DOI
[24] E.F. Toro Riemann Solvers and Numerical Methods for Fluid Dynamics. Springer-Verlag, 2009, 724 p.

[25] B.P. Rybakin, L.I. Stamov, E.V. Egorova. (2014) Accelerated solution of problems of combustion gas dynamics on GPUs. Computers & Fluids 90 164171.

[26] Samarskiy, A.A., Popov, Yu.P.: Raznostnyye metody resheniya zadach gazovoy dinamiki. - M.: Nauka, (1973). (in russian)

[27] Kritsuk A. G., Norman M. L., Padoan P., Wagner R. Turbulence and Nonlinear Processes in Astrophysical Plasmas, American Inst. of Physical Conference Series, Shaikh D., Zank G.P. (Eds.) 2007; ApJ, 665, 416.

[28] Rybakin B., Goryachev V. Parallel Algorithms for Astrophysics Problems. Lobachevskii Journal of Mathematics, Kazanskii Gosudarstvennyi Universitet/Kazan State University (Russian Federation), 39(4), . 562-570 DOI

IOP Conf. Series: Journal of Physics: Conf. Series **1336** (2019) 012010 doi:10.1088/1742-6596/1336/1/012010

Chemistry of ion injection in supernova remnant shocks: hybrid simulations in the light of He/C/O data from AMS-02

Adrian Hanusch[1], **Tatiana V. Liseykina**[2,4], **Mikhail A. Malkov**[3]

[1] University of Rostock, Institute for Physics, Albert-Einstein Str. 24, 18059, Rostock, Germany

[2] Institute of Computational Mathematics and Mathematical Geophysics SB RAS, Lavrent'jev ave. 6, 630090, Novosibirsk, Russia

[3] CASS and Department of Physics, University of California, San Diego, La Jolla, California 92093, USA

E-mail: tatiana.liseykina@gmail.com

Abstract. The high precision spectrometry of galactic cosmic rays (CR), e.g. PAMELA experiment [1], accurately determined an approximately 0.1 difference between the rigidity spectral indices of protons and He ions. Similar deviations have been indicated earlier by other experiments [2] confirmed by the recent high-fidelity AMS-02 measurements [3]. These findings widen the window into a complicated selection process whereby collisionless shocks, such as supernova remnant (SNR) blast waves, extract different chemical elements from an interstellar matter mix. The similarity of He/p, C/p, and O/p rigidity spectra demonstrated by AMS-02 has provided new evidence that injection is a mass-to-charge (A/Z) dependent process. We investigate the particle injection into the diffusive shock acceleration using one- and two-dimensional self-consistent hybrid simulations. Our 1D simulations prove that an SNR shock can modify the chemical composition of accelerated CRs by preferentially extracting them from a homogeneous background plasma without additional, largely untestable assumptions. Our results confirm the earlier theoretical predictions of how the efficiency of injection depends on the shock Mach number M_A. They show that selection rate of different ion species increases with A/Z, saturates, and peaks as a function of M_A. The 2D simulations also evidence a deviation from the linear selection rate vs A/Z trend, pointing towards a saturation for higher A/Z. The integrated SNR rigidity spectrum for proton-to-helium flux ratio, obtained by the convolution of the time-dependent injection rates of protons and He ions with a decreasing shock strength over the active lives of SNRs, compares well with the AMS-02 and PAMELA data. The numerical results [4] correctly predict the decrease in p/He flux ratio with increasing rigidity at exactly the rate measured in the experiments for $\mathcal{R} > 10$ GV. Only at lower rigidities, $\mathcal{R} < 10$ GV, the difference between the data and our predictions becomes noticeable. Whether this deviation comes from the propagation through the interstellar medium or solar modulation, remains unclear. Except for this uncertainty, the suggested mechanism for A/Z-dependence of the injection fully explains the measured p/He ratio.

[4] Present address: University of Rostock, Institute for Physics, Albert-Einstein Str. 24, 18059, Rostock, Germany

Content from this work may be used under the terms of the Creative Commons Attribution 3.0 licence. Any further distribution of this work must maintain attribution to the author(s) and the title of the work, journal citation and DOI.
Published under licence by IOP Publishing Ltd

1. Introduction

The difference between rigidity spectral indices of protons and He ions[5], first measured by the balloon-born experiment ATIC-2 [2], is now thoroughly vetted by the high-fidelity, more than 115-billion cosmic ray AMS-02 database [3], Fig.1 (left frame). These findings are perceived to undermine the leading hypothesis of galactic cosmic ray (CR) origin, whereby they are accelerated in supernova remnant shocks. The specific mechanism behind the acceleration, known as the first order Fermi acceleration (or diffusive shock acceleration, DSA), is electromagnetic in Nature. It has been studied for more than four decades and its fundamental aspect regarding the particle composition is the exclusive rigidity dependence of their spectra. In fact, following [5], consider the equations of motion of a charge particle in arbitrary electromagnetic fields $\vec{E}(\vec{r},t), \vec{B}(\vec{r},t)$

$$\frac{d\vec{\mathcal{R}}}{dct} = \vec{E}(\vec{r},t) + \frac{\vec{\mathcal{R}} \times \vec{B}(\vec{r},t)}{\sqrt{\mathcal{R}_0^2 + \mathcal{R}^2}}, \quad \frac{d\vec{r}}{dct} = \frac{\vec{\mathcal{R}}}{\sqrt{\mathcal{R}_0^2 + \mathcal{R}^2}}, \tag{1}$$

with $\vec{\mathcal{R}}$ and $\vec{r}$ being particle rigidity and position correspondingly and $\mathcal{R}_0 = (A/Ze)m_pc^2$. It follows from Eq.(1) that all particles with $\mathcal{R} \gg \mathcal{R}_0$ independent on the particles species have the same orbits in the phase space $(\vec{r}, \vec{\mathcal{R}})$. In other words, protons and He ions with the same rigidity should be accelerated alike, given the shock compression. For this reason, most of the explanations for the "paradox" appeal to SNR environmental factors, such as inhomogeneous p/He mixes in the shock upstream medium, variable ionization states of He, or even a multi-SNR origin of the observed spectra. The question now is whether such special conditions are vital for the explanation or an SNR shock can modify the chemical composition of accelerated CRs by preferentially extracting them from a homogeneous background plasma without additional and untestable assumptions. It has been argued [5] that such an intrinsic, mass-to-charge ratio (A/Z) based, elemental selectivity should indeed operate at the initial (so-called injection) phase of the Fermi acceleration. The coincidence in particle spectral slopes of three different elements (He, C, and O) discovered recently by the AMS-02 [6] rules out incidental selection mechanisms and points to the A/Z-based one.

In this paper we investigate the particle injection into the DSA using self-consistent hybrid simulations. We provide sufficiently detailed Mach number scans of the p/He injection ratio, that is needed to test the injection bias. Reducing the spatial dimensionality allowed us to perform the modeling with adequate particle statistics, which is vital, when exploring the high-energy tails of the distribution functions, and high grid resolution. The possibility of self-consistent treatment of He ions without need to dramatically reduce the He abundance to keep the noise level low, made it possible to study the more realistic plasma composition. It should be noted, that when the species abundance is relatively high (10% of He ions in a number density is indeed a high abundance) the test-particle approximation, well justified for ions with very low abundance, is probably not the best choice, provided the alternative treatment is possible. Indeed, our simulations with He dynamics included self-consistently show the difference in particle spectra not only for the He ions, but also for protons.

2. Hybrid simulations

The fully consistent description of the nonlinear shock-wave structure belongs to one of the most challenging problems in numerical physics. This is because the most important and interesting phenomena are multiscale in nature and a hydrodynamic treatment is generally not applicable. While the "ab initio" full electromagnetic particle-in-cell (PIC) codes provide

[5] Particle rigidity is defined as $\vec{\mathcal{R}} = \vec{p}c/eZ$, with $c, \vec{p}, eZ$ being the speed of light, particle momentum and the particle charge correspondingly.

IOP Conf. Series: Journal of Physics: Conf. Series **1336** (2019) 012010 doi:10.1088/1742-6596/1336/1/012010

the most reliable numerical description of plasmas, they are computationally very expensive. Following the evolution of a collisionless shock over many ion cyclotron times and with realistic electron-ion mass ratio is even not feasible by means of PIC simulations. If the focus is on the time and spatial scales determined by the ions and the electron scales do not need to be resolved, hybrid modeling (see [7] and references therein) has been proven to be a valuable tool.

Within the hybrid approach, the evolution of the ion distribution function $f(\vec{x},\vec{v})$ is governed by the kinetic Vlasov equation:

$$\frac{\partial}{\partial t} f + \vec{v}_i \nabla f + \frac{q_i}{m_i}\left(\vec{E} + \frac{1}{c}\vec{v}_i \times \vec{B} - \eta \vec{J}\right)\frac{\partial}{\partial \vec{v}} f = 0, \tag{2}$$

where $\vec{E}, \vec{B}$, and $\vec{J}$ are the electric and magnetic fields, and the current density, $q_i = Z e$ and $m_i = A m_p$ denote the ion charge and mass (e and m_p are the proton charge and mass, respectively). The charge neutralizing electronic plasma component in turn obeys the MHD equations. We adopt a massless electron approximation, in which the equation of motion of the electron fluid is reduced to

$$n_e m \frac{d\vec{v}_e}{dt} = 0 = -e\, n_e \left(\vec{E} + \frac{1}{c}\vec{v}_e \times \vec{B}\right) - \nabla p_e + e\, n_e\, \eta\, \vec{J}. \tag{3}$$

Here $-e$, m_e, n_e, and $\vec{v}_e$ are the electron charge, mass, number density and bulk velocity. In the following we assume that the pressure is isotropic and can be calculated using an adiabatic equation of state, $p_e \propto n_e^{\gamma}$, with adiabatic index $\gamma = 5/3$. We allow for the resistive coupling between electron and ions by introducing a phenomenological anomalous resistivity η. It gives rise to electron Ohmic heating and smoothes the fields on the resistive scale-length. A detailed study of the influence of η was performed in [8], where it was found in particular, that the structure of the shock does not change as long as η is sufficiently small, so that the density and magnetic field remain well coupled. The fluid equation (3) can be solved for the electric field under the assumption of quasi-neutrality $n_e = n_i = n$, with the electron bulk velocity from $\vec{J} = \vec{J}_i - \vec{J}_e = en(\vec{v}_i - \vec{v}_e)$:

$$\vec{E} = \frac{1}{e\,n}\left(\frac{(\vec{J} - \vec{J}_i) \times \vec{B}}{c} - \nabla p_e\right) + \eta\, \vec{J}. \tag{4}$$

For the solution of the equation (2) for the evolution of the distribution function of the ion plasma-component the PIC method is used. In this method the distribution function f is sampled by a large number $N^{\max}$ of macro-particles, whose motion is governed by the following non-relativistic equations[6]:

$$m_i \frac{d\vec{v}_j}{dt} = q_i\left(\vec{E} + \frac{1}{c}\vec{v}_j \times \vec{B} - \eta\, \vec{J}\right), \qquad \frac{d\vec{x}_j}{dt} = \vec{v}_j, \qquad j = 1, 2, 3 \ldots N^{\max} \tag{5}$$

The plasma density $n = n_i$ and ion current density $\vec{J}_i$ are reconstructed from ion positions and velocities

$$n = n_i(\vec{x}, t) = \sum_{j=1}^{N^{\max}} R(\vec{x}, \vec{x}_j) \tag{6}$$

$$\vec{J}_i(\vec{x}, t) = \sum_{j=1}^{N^{\max}} q_i\, \vec{v}_j(t)\, R(\vec{x}, \vec{x}_j). \tag{7}$$

[6] As we are interested in the injection into the DSA $|\vec{v}| \ll c$ holds during the simulation.

IOP Conf. Series: Journal of Physics: Conf. Series **1336** (2019) 012010 doi:10.1088/1742-6596/1336/1/012010

Here $R(\vec{x}, \vec{x}')$ is a normalized weighting function $\int_V R(\vec{x}, \vec{x}')\, d\vec{x} = 1$.

The equations for the electric and magnetic fields are solved using finite differences on the Eulerian grid in space. We use a first order weighting to obtain the fields at the particle position as well as when depositing ion density and current to the grid. For the numerical calculation equations (5) are discretized and centered in time

$$\vec{x}_j(t) \to \vec{x}_j^{\,m} = \vec{x}_j(m \cdot \Delta t); \quad \vec{v}_j(t) \to \vec{v}_j^{\,m+1/2} = \vec{v}_j(m \cdot \Delta t + \Delta t/2), \tag{8}$$

here Δt is the time-step. The particle positions are updated straightforwardly

$$\vec{x}_j^{\,m+1} - \vec{x}_j^{\,m} = \vec{v}_j^{\,m+1/2} \cdot \Delta t, \tag{9}$$

the Boris algorithm [9] is applied to update their velocities. For the solution of the Maxwell equations a magnetostatic (Darwin) model is employed,

$$\vec{J} = \vec{J}_i - \vec{J}_e = \frac{c}{4\pi} \nabla \times \vec{B} \tag{10}$$

in which the displacement current is neglected. This inhibits the propagation of high frequency waves. The magnetic field evolves according to Faraday's law,

$$\frac{\partial \vec{B}}{\partial t} = -c \nabla \times \vec{E}. \tag{11}$$

The equations for the electric field (4) as well as for the propagation of the magnetic field (11) are discretized using second order finite difference stencils. The electric field is advanced from the time level (m) to $(m+1)$ in following steps: first $\vec{E}$ is evaluated at $(m + 1/2)$ using the discretized Eq. (4)

$$\vec{E}^{m+1/2} = \vec{F}\left(\vec{B}^{\,m+1/2}, n^{m+1/2}, \vec{J}_i^{\,m+1/2}\right), \tag{12}$$

with $\vec{B}^{m+1/2}$ calculated from the discretized Eq. (11): $\vec{B}^{\,m+1/2} = \vec{B}^{\,m} - c\frac{\Delta t}{2}(\nabla \times \vec{E}^m)$. Then $\vec{E}^{m+1/2}$ is used to calculate $\vec{B}^{m+1}$. Finally the electric field value $\vec{E}^{m+1}$ at time level $(m + 1)$ is to be obtained from $\vec{E}^{m+1} = F(\vec{B}^{\,m+1}, n_i^{m+1}, \vec{J}_i^{\,m+1})$. This last step is not straightforward since is requires the knowledge about the ion current at time level $(m + 1)$.

Several methods have been proposed in the literature (see *e.g.* [10] for a review) to overcome the apparent difficulty of solving $\vec{E}^{m+1} = F(\vec{B}^{\,m+1}, n_i^{m+1}, \vec{J}_i^{\,m+1})$ with an unknown $\vec{J}_i^{\,m+1}$. In our work we used two of them: the predictor-corrector method and the Bashford-Adams extrapolation of the ion current. In the predictor-corrector method the "predicted" electric field

$$\vec{E'}^{\,m+1} = 2\,\vec{E}^{m+1/2} - \vec{E}^m \tag{13}$$

is used to propagate the ions and obtain the source terms $n_i'^{\,m+3/2}$ and $\vec{J}_i'^{\,m+3/2}$ needed for the calculation of $\vec{E'}^{\,n+3/2}$. The corrected field at time level $(m + 1)$ is obtained then from

$$\vec{E}^{m+1} = \frac{1}{2}\left(\vec{E'}^{\,m+3/2} - \vec{E}^{m+1/2}\right). \tag{14}$$

This algorithm is simple and has good energy conserving properties but it suffers from higher computational costs, since the particles have to be propagated twice per time step. Hence, we also implemented a Bashford-Adams extrapolation of the ion current in our code,

$$\vec{J}_i^{\,n+1} = 2\,\vec{J}_i^{\,n+1/2} - \frac{3}{2}\,\vec{J}_i^{\,n-1/2} + \frac{1}{2}\,\vec{J}_i^{\,n-3/2}. \tag{15}$$

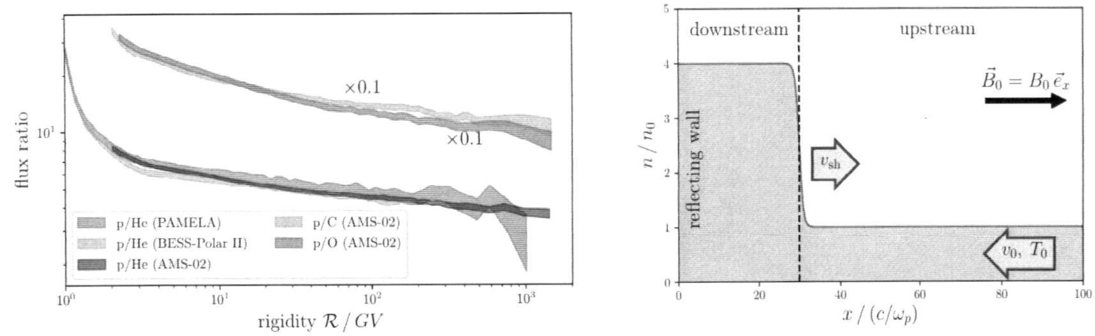

Figure 1. *(left) The similarity of p/He, p/C, and p/O rigidity spectra demonstrated by recent AMS-02 measurements has provided new evidence that injection is a mass-to-charge dependent process. Data are taken from [1, 3, 6, 11]. (right) Simulation set-up: a shock is created by sending a super sonic plasma flow from the right against a reflecting wall. The shock is traveling to the right, parallel (1D) or quasi-parallel (2D) to the background magnetic field.*

It is used by default when performing two-dimensional simulations, the magnetic field in this case is updated with a fourth order Runge-Kutta scheme.

A simulation is initialized by sending a super-alfvénic (and super-sonic) flow of plasma with mean velocity $\vec{v_0} = (-v_0, 0, 0)$ against a reflecting wall, placed at $x = 0$, Fig.1. A shock wave forms due to the interaction of the counter-propagating streams and is then traveling in x-direction in our simulation set-up. In order to increase the particle statistics the spatial dimension is reduced to one $\vec{E} = \vec{E}(x)$, $\vec{B} = \vec{B}(x)$ or two $\vec{E} = \vec{E}(x, y)$, $\vec{B} = \vec{B}(x, y)$ but all components of the velocity and fields are included. In the two-dimensional case periodic boundary conditions are applied in the y-direction. We measure time in units of inverse proton gyrofrequency, $\omega_c^{-1} = (e\,B_0/m_p\,c)^{-1}$, where B_0 is the magnitude of the background magnetic field. Lengths are normalized to the ion inertial length, c/ω_p, with $\omega_p = \sqrt{4\pi\,n_0\,e^2/m_p}$ being the proton plasma frequency, where n_0 is the upstream density. Finally, the velocity is given in units of the Alfvén velocity, $v_A = B_0/\sqrt{4\pi\,n_0\,m_p}$. In all simulations presented below the background magnetic field is set to be (quasi-)parallel to the shock normal. The electron fluid is assumed to be initially in thermal equilibrium with the ions with $\beta_e = \beta_p = 1$.

3. Simulation results

To investigate the mass-to-charge dependence of the ion injection into the DSA we have performed a series of one-dimensional simulations with large simulation boxes with lengths up to $L_x = 48 \times 10^3\,c/\omega_p$ and the spatial resolution of $\Delta x = 0.25\,c/\omega_p$. The time step was $\Delta t = 0.01\,(v_0/v_A)\,\omega_c^{-1}$. Besides protons also helium, carbon and oxygen ions were included in the simulations with charge states $Z = 1$ and $Z = 2$. We used 100 particles per each ion-species per numerical cell[7]. All ion species were treated *self-consistently* with relative abundances $\sim 10\%$ for helium and $\sim 0.04\%$ for carbon and oxygen. Note, that the self-consistent treatment is especially important for helium, which contributes with approximately 10%, and hence influences significantly the dynamics of the waves excited in the upstream medium. The simulations were performed on $360 \div 480$ cores of the HLRN Cluster Module (Berlin, Germany). The largest 1D simulation took approximately 40 hours.

We follow the shock evolution over up to $2000\,\omega_c^{-1}$. During this time window ions are

[7] The total number of particles by the end of the simulation was $\sim 5 \times 10^7$.

IOP Conf. Series: Journal of Physics: Conf. Series **1336** (2019) 012010 doi:10.1088/1742-6596/1336/1/012010

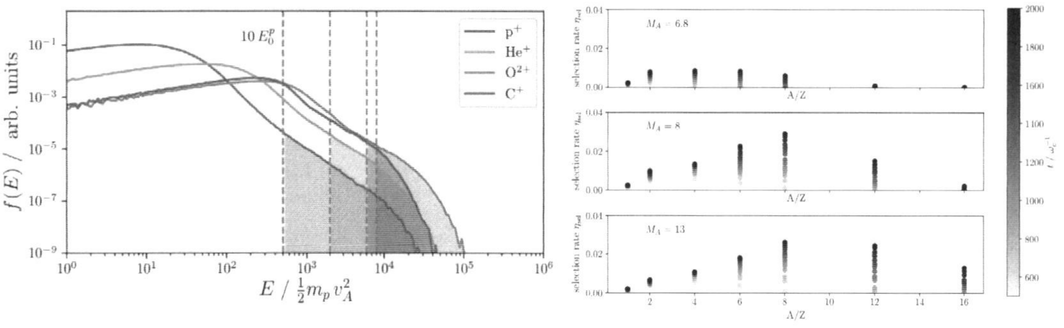

Figure 2. *(left) Downstream energy spectra of different ion species present in 1D simulations. The transition from the thermal Maxwellian distribution $f_{th} \propto \sqrt{E}\exp(-E/T)$ to the power-law $f_{pow} \propto E^{-q}\exp(-E/E_{cut})$ is clearly visible for all ion species. (right) Fraction of accelerated ions in the tail (shaded areas in the left frame) of the distribution function for different shock Mach numbers. This quantity turns to be a non-monotonic function of the mass-to-charge ratio with a position of the maximum being a function of M_A.*

accelerated and in the downstream energy spectra a power-law tail becomes clearly visible, Fig.2 (left frame). After the power-law tail has been certainly formed we compute the number injection efficiency (or selection rate), i.e. fraction of particles in the tail of the distribution function $f^\alpha(E)$

$$\eta^\alpha_{sel} = \int_{10E_0^\alpha}^{\infty} f^\alpha(E)\,\mathrm{d}E \Big/ \int_0^{\infty} f^\alpha(E)\,\mathrm{d}E \qquad (16)$$

for each ion species α. The resulting dependence of the number efficiency on A/Z is depicted in Fig. 2 (right frame) for simulations with far-upstream flow velocities $v_0 = 5,\, 7,\, 10\, v_A$, which correspond to the (alfvénic) shock Mach numbers $M_A = v_{\mathrm{shock}}/v_A = 6.8$, 8, and $M_A = 13$. Different colors indicate the times at which the number efficiency is obtained. In all cases the fraction of accelerated particles increases as function of A/Z, saturates and then decreases, recovering a physically correct $A/Z \to \infty$ asymptotic behavior expected for neutral particles. The position of the maximum depends in an increasing manner on the shock Mach number M_A [4]. Thus the observed $\eta_{sel}(M_A)$ behavior of the selection rate confirms the earlier theoretical predictions [5]. By performing a series of simulations for the two ion species plasma (protons and He ions) we obtained a sufficiently detailed scan of the the injection efficiency for both species over a range of Mach numbers. By combining in [4] the injection efficiency from the simulation with DSA predicted power-law and accounting for the evolution of a SNR during the Sedov-Taylor stage, we calculated the proton-to-helium ratio, extending in energy beyond the capability of any simulation. The resulting rigidity spectra are presented in Fig. 3 in the whole range in rigidity. The detailed

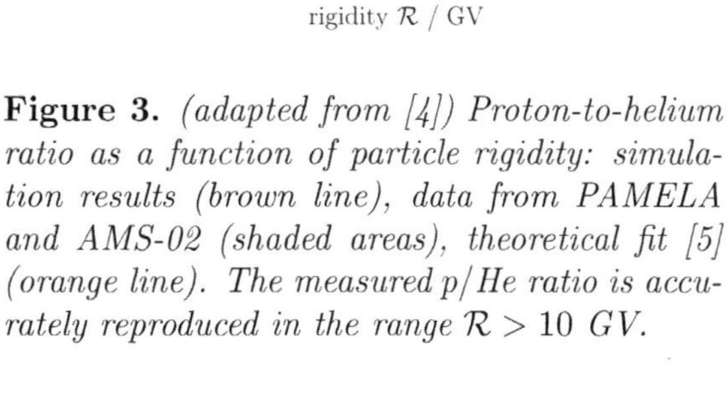

Figure 3. *(adapted from [4]) Proton-to-helium ratio as a function of particle rigidity: simulation results (brown line), data from PAMELA and AMS-02 (shaded areas), theoretical fit [5] (orange line). The measured p/He ratio is accurately reproduced in the range $\mathcal{R} > 10$ GV.*

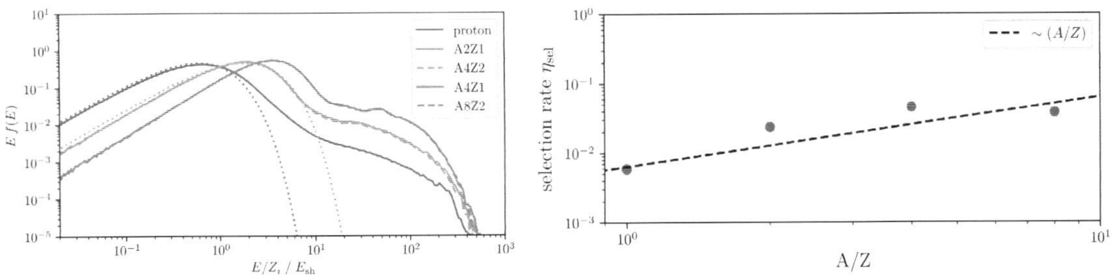

Figure 4. *(left) Downstream energy spectra at* $t = 1000\,\omega_c^{-1}$ *for selected ion species present in the 2D simulation. (right) Mass-to-charge dependence of the selection rate.*

description of rigidity spectra reconstruction procedure can be found in [4]. Most importantly, we found that p/He decreases with rigidity for $\mathcal{R} > 10$ GV with almost the same rate $\Delta q \approx 0.1$ as observed by the high accuracy measurements[8]. A deviation from the measurements is visible at lower rigidities, $\mathcal{R} \ll \mathcal{R}_0 = (A/Ze)m_p c^2$, where the equations of motion are different for p and He. The most likely cause of this deviation is particle interaction with the turbulent solar wind in the Heliosphere, but the interaction with the interstellar medium turbulence may also contribute. By contrast, the deviation from the AMS-02 data in the high-rigidity range where the equations of motion for p and He become identical, is insignificant, but whether it comes from a simplified integration over the SNR or it is a mixing effect from different SNRs or spallation in the interstellar medium, remains unclear.

Now, to confirm the 1D prediction [4] about the A/Z behavior of η_{sel} we have performed 2D Cartesian simulations with $v_0 = 10\,v_A$, including different ion species with A/Z up to 8. The background magnetic field is initialized to form a $\theta_{\rm Bn} = 20^\circ$ angle with the shock normal, *i.e.*, the shock is quasi-parallel. Note, that in the 1D set-up $\theta_{\rm Bn} = 0^\circ$ was used. To follow the shock for a long time, we use a very elongated box with dimensions $L_x \times L_y = 20000 \times 20\,(c/\omega_p)^2$ and a grid spacing of $\Delta x = \Delta y = 0.5\,c/\omega_p$. We have checked that the results do no change significantly with the transverse box size for $L_y \geq 20\,c/\omega_p$. The simulations were performed on 480 cores of the HLRN Cluster (Berlin, Germany) and took approximately 42 hours.

In the top panel of Fig. 4 the downstream energy spectra obtained at $t = 1000\,\omega_c^{-1}$ are shown. The energy is measured in terms of $E_{\rm sh} = m_p v_0^2/2$. The Maxwellian part at low energies (the dotted line in Fig. 4 depicts a Maxwellian distribution) as well as the power-law tails at high energies are clearly visible. Following [4] we compute the fraction of energetic particles in the tail of the distribution function. The results for ion species with charge state $Z = 1$ are plotted in the lower panel of Fig. 4. They confirm the prediction of the 1D modelling: for small mass-to-charge values the selection rate increases with A/Z, but deviates from the linear trend for $A/Z > 8$, pointing toward a saturation for higher mass-to-charge ratios. The decrease of the selection rate at even higher A/Z could not be observed in our 2D simulations, because only ions with mass-to-charge ratios up to 8 were included.

4. Summary

We have investigated the ion injection into the DSA by means of 1D and 2D self-consistent hybrid simulations. Our results confirm the earlier theoretical predictions as well as the results of numerical modeling [12], that the injection depends on the mass-to-charge ratio as well as on the shock Mach number. Our 1D simulations show that a selection rate increases with A/Z, saturates, and peaks as a function of shock Mach number. The question of the exact position of

[8] The measured [1, 3] spectra are shown in Fig. 3 by shaded areas.

maximum of the selection rate as function of A/Z is debatable and there is indeed no consensus yet. Physically, its position should depend on the current maximum energy (Larmor radius of protons) since resonant waves produced by them may scatter particles with larger A/Z. But the particle scattering rate in general decays with the growing wave length, so this effect should not be overestimated. We find that for our simulation parameters ($M_A \sim 10$) the selection rate turns over in the region of $A/Z = 8 - 12$. This depends on the shock Mach number and to some extend also on the simulation time, as the power-law tails of the high A/Z ions form at later times. The results of 2D hybrid modeling confirm the A/Z selection rate trend. The rigidity dependent p/He ratio, obtained by a convolution of the time-dependent injection rates of protons and He ions with a shock strength which decreases with time, accurately reproduce the AMS-02 data. In particular, it correctly predict the decrease in proton-to-helium ratio at exactly the rate, measured in the experiments in the rigidity range > 10 GV. Our interpretation of the elemental "anomaly" is intrinsic to collisionless shock acceleration mechanism and does not require additional largely untestable assumptions, such as the contributions from several different SNRs or their inhomogeneous environments.

Acknowledgments
The development of a highly adaptable hybrid code was partially supported by the Russian Science Foundation within research grant 19-71-20026. T.L. acknowledges the Siberian Supercomputer Center ICMMG SB RAS (Novosibirsk) for providing the computing time on NKS-1P cluster. A.H. acknowledges the support by DFG project 278305671TL and North-German Supercomputing Alliance (HLRN) for providing the computational resources for the simulations (the project mvp00015).

References

[1] Adriani O *et al.* 2011 *Science* **332** 69–72 ISSN 0036-8075 URL `http://science.sciencemag.org/content/332/6025/69`
[2] Panov A D *et al.* 2009 *Bulletin of the Russian Academy of Sciences: Physics* **73** 564–567 ISSN 1934-9432 URL `https://doi.org/10.3103/S1062873809050098`
[3] Aguilar M *et al.* (AMS Collaboration) 2015 *Phys. Rev. Lett.* **115**(21) 211101 URL `https://link.aps.org/doi/10.1103/PhysRevLett.115.211101`
[4] Hanusch A, Liseykina T V and Malkov M 2019 *The Astrophysical Journal* **872** 108 URL `https://doi.org/10.3847`
[5] Malkov M A, Diamond P H and Sagdeev R Z 2012 *Phys. Rev. Lett.* **108**(8) 081104 URL `https://link.aps.org/doi/10.1103/PhysRevLett.108.081104`
[6] Aguilar M *et al.* (AMS Collaboration) 2017 *Phys. Rev. Lett.* **119**(25) 251101 URL `https://link.aps.org/doi/10.1103/PhysRevLett.119.251101`
[7] Lipatov A S 2002 *The Hybrid Multiscale Simulation Technology. An Introduction with Application to Astrophysical and Laboratory Plasmas* (Springer, Berlin, Heidelberg) URL `https://doi.org/10.1007/978-3-662-05012-5`
[8] Leroy M M, Winske D, Goodrich C C, Wu C S and Papadopoulos K 1982 *Journal of Geophysical Research: Space Physics* **87** 5081–5094 URL `https://agupubs.onlinelibrary.wiley.com/doi/abs/10.1029/JA087iA07p05081`
[9] Boris J P and Shanny R A 1970 *Proceedings: Fourth Conference on Numerical Simulation of Plasmas, November 2, 3, 1970* (Naval Research Laboratory)
[10] Winske D, Yin L, Omidi N, Karimabadi H and Quest K 2003 *Hybrid Simulation Codes: Past, Present and Future—A Tutorial* (Berlin, Heidelberg: Springer Berlin Heidelberg) pp 136–165 ISBN 978-3-540-36530-3
[11] Abe K *et al.* 2016 *The Astrophysical Journal* **822** 65 URL `http://stacks.iop.org/0004-637X/822/i=2/a=65`
[12] Caprioli D, Yi D T and Spitkovsky A 2017 *Phys. Rev. Lett.* **119**(17) 171101 URL `https://link.aps.org/doi/10.1103/PhysRevLett.119.171101`

Estimation of the change in the gas flow rate from the surface of a tungsten plate under pulsed heat load

G G Lazareva[1,2,4,5], **A S Arakcheev**[3,5], **V A Popov**[3,4] **and A G Maksimova**[2,3,4]
[1]Peoples' Friendship University of Russia (RUDN University) 6, Miklukho-Maklaya str., Moscow, 117198, Russian Federation
[2]Institute of Computational Mathematics and Mathematical Geophysics SB RAS 6, Ac. Lavrentieva Pr., Novosibirsk, 630090, Russian Federation
[3]Budker Institute of Nuclear Physics SB RAS 11, Ac. Lavrentieva Pr., Novosibirsk, 630090, Russian Federation
[4] Novosibirsk State University 1, Pirogova str., Novosibirsk, 630090, Russian Federation
[5]Novosibirsk State Technical University 20, Prospekt K. Marksa, Novosibirsk, 630073, Russian Federation

E-mail: lazareva@ssd.sscc.ru, asarakcheev@gmail.com, v.a.popov94@gmail.com, maksimova@oapmg.sscc.ru

Abstract. The work is devoted to the numerical implementation of the tungsten evaporation process model. The tungsten evaporation model is based on solving the two-phase Stefan problem for temperature in the sample area and gas dynamics equations over the sample. The calculation results for parameters corresponding to those used on the BETA facility at BINP SB RAS show that it is possible to use the boundary homogeneous conditions of the Neumann for the velocity when setting the gas density on the plate surface.

1. Introduction

On the experimental stand Beam of Electrons for materials Test Applications (BETA) created in the INP SB RAS [1], the results of heating the tungsten plate by the action of a high-speed electron beam on it were obtained. Currently, due to the development of thermonuclear fusion technologies, new practical problems of melting of refractory metals are emerging. To plan and analyze full-scale experiments it is necessary to have mathematical models that accurately take into account all the details and features of the process. Construction of a complete model of electron beam heating enables better assessment of the relevance of the simulation and determination of mechanisms that will occur in the case of plasma flow heating in modern plasma and future thermonuclear facilities. The work is devoted to the numerical implementation of the tungsten evaporation process model. The tungsten evaporation model is based on solving the two-phase Stefan problem for temperature in the sample area and gas dynamics equations over the sample. To calculate the velocity and mass flow of a substance evaporating from the sample surface, the system of equations of gas dynamics is solved numerically. The boundary conditions describing the heating and evaporation of the material have a great influence on the solution of the problem. They are discontinuous and nonlinear in time and space. The aim of the study is to simulate the erosion and droplets emission perpendicular to the

Content from this work may be used under the terms of the Creative Commons Attribution 3.0 licence. Any further distribution of this work must maintain attribution to the author(s) and the title of the work, journal citation and DOI.
Published under licence by IOP Publishing Ltd

surface of the sample surface as a result of evaporation and penetration of heat flux into the material taking into account microcracks [2].

Modeling the process of metal evaporation is a complex and not sufficiently studied problem. The problem of setting the boundary conditions for the process of evaporation of a solid substance has been posed for a long time [3]. However, they are still not precisely defined. Mathematical modeling of tungsten evaporation is usually associated with the study of industrial processing of materials [4-6]. The specifics of the problem under consideration does not allow the use of this data. Known models of heat transfer of the divertor wall for the ITER project do not take into account the characteristics of the material and are limited to determining the thermal field in the sample [7-8].

2. Problem definition

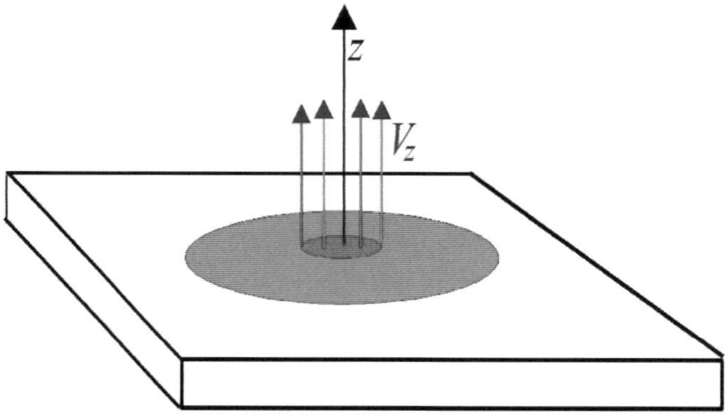

Figure 1. Scheme of the computational domain. Tungsten plate with melt region (orange), evaporation region (red), gas flow (blue arrows) along the axis Z.

In the mathematical model of melting in a tungsten plate exposed to a pulsed electron beam [9], the calculation of the surface gas flow is added. Consider the following gas dynamics equations in a cylindrical coordinate system to calculate the evaporation in vacuum (Fig. 1):

$$\begin{cases} \dfrac{\partial \rho}{\partial t} + \dfrac{1}{r}\dfrac{\partial (r\rho v_r)}{\partial r} + \dfrac{\partial (\rho v_z)}{\partial z} = 0, \\ \dfrac{\partial (\rho v_r)}{\partial t} + \dfrac{1}{r}\dfrac{\partial (v_r r(\rho v_r))}{\partial r} + \dfrac{\partial (v_r(\rho v_z))}{\partial z} = -\dfrac{\partial P}{\partial r}, \\ \dfrac{\partial (\rho v_z)}{\partial t} + \dfrac{1}{r}\dfrac{\partial (v_z r(\rho v_r))}{\partial r} + \dfrac{\partial (v_z(\rho v_z))}{\partial z} = -\dfrac{\partial P}{\partial z}, \\ \dfrac{\partial T}{\partial t} + v_r\dfrac{\partial T}{\partial r} + v_z\dfrac{\partial T}{\partial z} - \dfrac{2T}{3}\left(\dfrac{1}{r}\dfrac{\partial (r v_r)}{\partial r} + \dfrac{\partial v_z}{\partial z}\right) = 0, \\ P = \dfrac{R}{M}\rho T, \\ T\big|_\gamma = T^*(r,t), \quad \rho\big|_\gamma = \dfrac{M}{2RT^*(r,t)} exp\left(a1 - \dfrac{a2}{T^*(r,t)}\right), \quad \dfrac{\partial v_z}{\partial z}\bigg|_\gamma = 0, \\ (n, \nabla f) = 0 \quad \text{at other boundaries}, \\ T\big|_{t=0} = T(r,z,0), \quad (v_r, v_z)\big|_{t=0} = 0, \quad \rho\big|_{t=0} = 0. \end{cases} \tag{1}$$

where ρ is the gas density, $\bar{v}=(v_r,v_z)$ is the gas velocity, P is the gas pressure, $T(r,z,t)$ is the temperature, $R=8{,}31\cdot10^6\,W\cdot\mu s/mol\cdot K$ is the gas constant, M is the molar mass, $T^*(r,t)$ is the temperature on the surface γ obtained from the calculation of Stefan problem [9]. The numerical values of the parameters are given in Table 1.

Table 1. Dimensional parameters of nondimensionalization.

Parameter	Typical value	Units
r_0	10^{-3}	m
t_0	10^{-4}	s
T_0	10^3	K
ρ_0	10	kg/m^3
$a2$	83.9713	
$a1$	26.191	

3. Simulation results

The sample for shot 1228 has a 25mm × 25mm cross section and a thickness of 4mm. Since the sample is heated to a depth of several hundred microns in such a short time, the simulation region was a transverse section of the sample, a region of 12mm × 4mm. As for the time, the numerical simulation continued until the moment of the last measurement. Comparison of the calculated and experimental temperature on the surface of the plate is given in papers [9, 11, 12]. Calculation results for the radial temperature distribution over the sample surface are consistent with experimental data and analytical estimates.

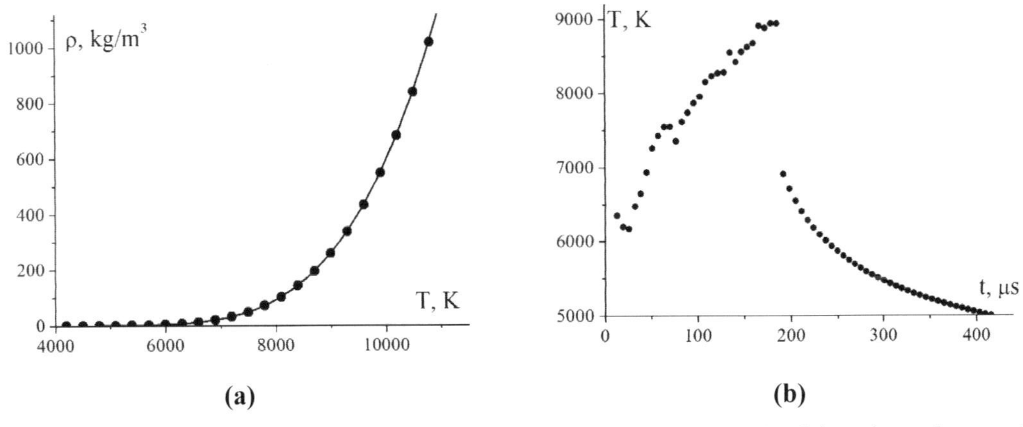

Figure 2. Graph of temperature dependence of gas density (a) and graph of time dependences of surface temperature (b) obtained from calculation for shot 1228.

The evaporation of a substance is set by the boundary condition for ρ (Fig. 2a), which is an estimate of the surface gas density:

$$\rho|_\gamma=\frac{1}{2}\left(\frac{R}{M}\right)^{-1}\frac{1}{T^*(r,t)}\exp\left(a1-\frac{a2}{T^*(r,t)}\right). \qquad (2)$$

To calculate the boundary conditions, we used the temperature values in the most heated central part of the plate surface. First is linear growth and second is the surface temperature obtained from numerical simulation [9] for shot 1228. Dependencies $T(t) = T^*(r,t)\big|_{r=0}$ (Fig. 2b) are used in numerical simulation. The calculation points show the temperature increase in the first 190 μs from the beginning of the experiment. The pulse has such a high power that the temperature immediately rises to 6000K. The temperature rises to 9000K. The temperature graph is not smooth due to the type of total beam power (Fig. 3b). As soon as the pulse stops, the temperature drops sharply to 7000K and continues to smoothly decrease to 5000K within 200 μs.

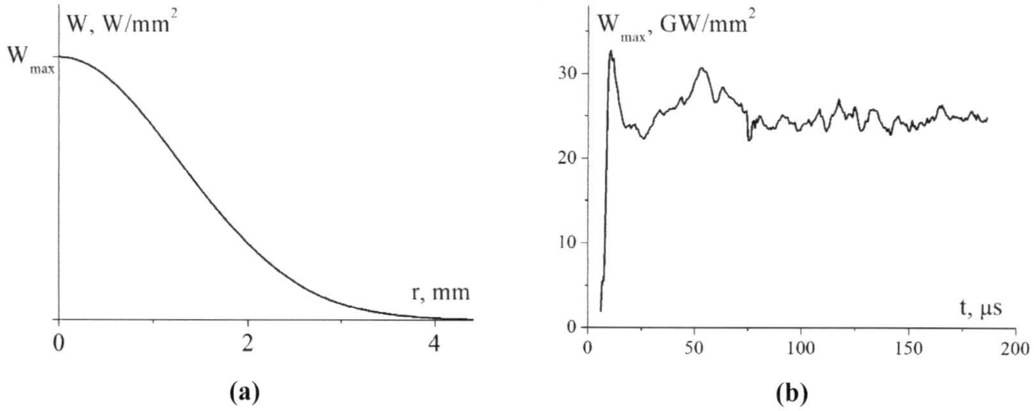

Figure 3. Radial distribution of heat flux power (a) and time dependence of total beam power for shot 1228 (b).

To calculate the temperature for shot 1228 used the melting point T_m = 3695K, the heat of fusion $L_m = 51.1 \cdot 10^5 \frac{\text{W}\mu\text{s}}{\text{mm}^3}$. The process of evaporation at the boundary is taken into account via a given resultant energy flux $W_*(t,r) = W(t,r) - N(T\big|_\gamma)$, where $W(t,r)$ is the power of the heat flux, $N(T\big|_\gamma) = L_e \cdot \frac{1}{S}\frac{dm}{dt}$ is the power loss. The power of the heat flux is $W(t,r) = W_{\max}(t) \cdot \exp(-A \cdot r^2)$, $A = 0.03088523$ (Fig. 3a). At each time step in the numerical simulation, the variable $W_{\max}(t)$ is taken from the experimental data file (Fig. 3b). The distribution of the heating power density over the surface was measured using X-ray visualization [4].

The numerical implementation for equations (1) is based on the Rusanov finite difference scheme [13]. This method well approximates the flow of gas into a vacuum. The Dirichlet boundary condition for the temperature and the gas density on the heated surface and Neumann boundary conditions in all other cases are used. Of particular interest is the formulation of boundary conditions for gas velocity on the heated surface. There is an estimate for the gas velocity:

$$v_z\big|_\gamma = \sqrt{\frac{5}{3}\frac{R}{M}}\sqrt{T^*(r,t)}. \tag{3}$$

This estimate is exactly carried out with Neumann homogeneous boundary condition for gas density. In this case, the density of the gas on the heated surface increases as the root of the temperature, as does the velocity of the gas. This is contrary to the existing theoretical estimates.

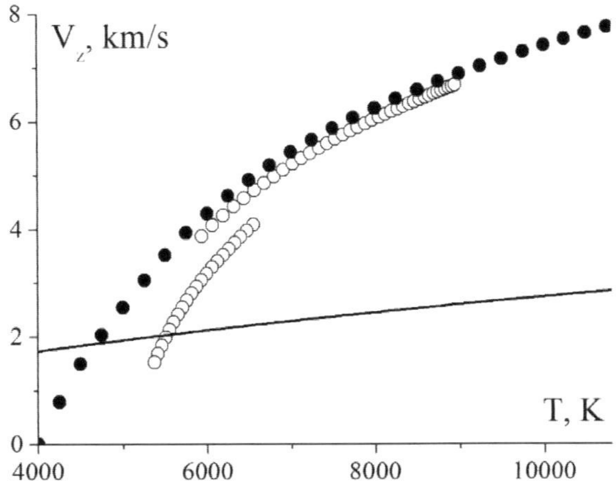

Figure 4. Temperature dependence of the surface gas velocity for linear temperature growth (black dots) and for calculation temperature for shot 1228 (white dots). Estimate for the gas velocity (line).

At temperatures up to 4000 K, the gas density is zero. At high temperatures, the calculated gas velocity is 2-3 times higher than the analytical estimate. A linear temperature increase $T^*(r,t) = kt$ gives the same temperature dependence of the velocity of the outgoing gas at various values of the coefficient k (Fig. 4, black dots). The value of the coefficient k affects only the heating rate.

The calculation for the model linear temperature increase showed that at surface temperatures of 11000 K the speed can reach $v_z|_\gamma = 7.5\,km/s$. It is impossible to heat the surface more strongly, since the evaporation process will prevail at such high temperatures. To calculate the 1228 shot, the surface temperature dynamics (Fig. 2b) was approximated by two polynomials for heating and cooling. As a calculation result, two curves were obtained (Fig. 4, white dots). The first dependence is obtained by heating the sample to 9000K. This curve almost coincides with the result of the calculation of the model problem. The second curve was obtained in the process of cooling to 5000K. It is expected that during the cooling process, the gas velocity on the surface is lower than during heating. The temperature of the evaporating gas above the surface is lower than the surface temperature. It can be assumed that we must calculate the boundary condition (2) for the gas density from a lower temperature. Given this, it is possible to obtain results that satisfy both estimates (2) and (3).

4. Conclusion

The calculation results for parameters corresponding to those used on the BETA facility at BINP SB RAS show that it is possible to use the boundary homogeneous conditions of the Neumann for the velocity when setting the gas density on the plate surface. The calculation for the parameters taken from the experimental data shows close dependencies to the calculations of the model problem with a linear increase in temperature.

The constructed mathematical model is planned to be integrated with the calculation of the current distribution in the gas. The construction of a complete model of electron beam heating will enable better assessment of the relevance of the simulation and finding out which mechanisms will occur in the case of plasma flow heating in modern plasma and future thermonuclear installations.

5. Acknowledgements

Experimental data have been obtained under support of the Russian Science Foundation (project No 17-79-20203). The publication has been prepared with the support of the "RUDN University Program 5-100" (Lazareva G.G., mathematical model development).

6. References

[1] Vyacheslavov L, Arakcheev A, Burdakov A, Kandaurov I, Kasatov A, Kurkuchekov V, Mekler K, Popov V, Shoshin A, Skovorodin D, Trunev Y, Vasilyev A 2016 Novel electron beam based test facility for observation of dynamics of tungsten erosion under intense ELM-like heat loads *American Institute of Physics* CP1771, paper 060004.

[2] Arakcheev A S, Arakcheev S A, Kandaurov I V, Kasatov A A, Kurkuchekov V V, Lazareva G G, Maksimova A G, Mashukov V I, Popov V A, Trunev Yu A, Vasilyev A A, Vyacheslavov L N 2019 On the mechanism of surface-parallel cracks formation under pulsed heat loads *Nuclear Materials and Energy* **20** 100677

[3] Kucherov R.Ya., Rickenglaz R.E. On hydrodynamic boundary conditions during evaporation and condensation 1959 *ZhETP*, 37, Issue 1 (7), p. 125-126. [in Russian]

[4] Tsydypov B D 2007 The effect of plasma-generating gas pressure on the parameters of activated thermionic cathode *High Temperature*, 45, Issue 5, p. 595-600. [in Russian]

[5] Mordyuk B C, Bochkanov P V, Ivanov Yu I 1984 Modeling of tungsten evaporation in strong electric fields *Physics and Chemistry of Materials Processing,* 3, p.41. [in Russian]

[6] Aleksandrov L N, Kogan A N, Bochkova R V, Troostina N P 1985 Monte Carlo study of the effect of substrate surface imperfections on the kinetics of the growth and evaporation of semiconductor films 1985 *Crystallography,* 30, 2, p. 236-242. [in Russian]

[7] Tanchuk V, Grigoriev S, Divavin V, Lipko A, Makhankov A 2001 Thermal analysis of the tile impacted by concentrated heat loads caused by the loss of an upstream *tile Fusion Engineering and Design*, 56-57, p. 225-231.

[8] Grigoriev S, Engelko V, Tanchuk V 2002 Influence of Pulsed Electron Beam Parameters on the Thickness of Modified Layer *Plasma Devices and Operations*, 10, 2, p. 117-127.

[9] Arakcheev A S, Apushkinskaya D E, Kandaurov I V, Kasatov A A, Kurkuchekov V V, Lazareva G G, Maksimova A G, Popov V A, Snytnikov A V, Trunev Yu A, Vasilyev A A, Vyacheslavov L N 2018 Two-dimensional numerical simulation of tungsten melting under pulsed electron beam *Fusion Engineering and Design* **132**, pp. 13-17

[10] Taluts S G 2001 Experimental study of the thermophysical properties of transition metals and alloys based on iron at high temperatures *abstract, PhD Thesis*, code 01.04.14, p. 38 [in Russian]

[11] Lazareva G G, Arakcheev A S, Burdakov A V, Kandaurov I V, Kasatov A A, Kurkuchekov V V, Maksimova A G, Popov V A, Shoshin A A, Snytnikov A V, Trunev Yu A, Vasilyev A A and Vyacheslavov L N 2018 Numerical model of high-power transient heating of tungsten with considering of various erosion effects *IOP Conf. Series: Journal of Physics: Conf. Series* **1103**, 012001

[12] Lazareva G G, Arakcheev A S, Kandaurov I V, Kasatov A A, Kurkuchekov V V, Maksimova A G, Popov V A, Snytnikov A V, Trunev Yu A, Vasilyev A A, Vyacheslavov L N 2018 Computational experiment for solution of the Stefan problem with nonlinear coefficients *AIP Conference Proceedings* **2025**, 080005

[13] Rusanov V V 1968 Third Order Accuracy Difference Schemes for End-to-End Counting Discontinuous Solutions *Dokl. Academy of Sciences of the USSR*, 180: 6, p. 1303-1305. [in Russian]

The grid codes generation for solving problems of the cosmic plasma hydrodynamics on supercomputers

Anna Sapetina[1], **Ivan Ulyanichev**[2] **and Boris Glinskiy**[1]

[1] Institute of Computational Mathematics and Mathematical Geophysics SB RAS, 6 Akad. Lavrentieva pr., Novosibirsk, Russia

[2] Novosibirsk State Technical University, 20 K. Marksa pr., Novosibirsk, Russia

E-mail: afsapetina@gmail.com

Abstract. In this paper an attempt is made to create a system for generating the parallel codes to simulate some astrophysical phenomena on modern computing architectures. A code construction scheme for the numerical solution of various problems of the cosmic plasma hydrodynamics is proposed using a knowledge base with an ontological representation of a given mathematical simulation area. The ontology of numerical methods and parallel algorithms and the ontology of parallel architectures and technologies together with the expert (inference) rules allow selecting an efficient numerical method, a parallel algorithm and a computational architecture for solving a necessary problem from the outlined domain. The corresponding program units that implement various stages of solving the problem on a chosen computational architecture are substituted into scheme of solving the problem constructed in this way. Program units can be based on both the author's implementation and on calling various parallel libraries.

1. Introduction

It is known that more than 90% of the visible matter in the universe consists of plasma, so the plasma-astrophysics discipline has numerous fields of application. For example, plasma is a component of stars, nebulae, interplanetary, interstellar and intergalactic media. In the formation and dynamics of astrophysical objects, the magnetic field can fundamental affect the plasma dynamics.

The study of many astrophysical problems (for example, the formation of jets and magnetized disks around the newborn black holes, etc [1]) requires the investigation of the hydrodynamics of magnetized liquids.

An crucial part in studying of astrophysical processes is played by mathematical modeling, for which a large number of parallel codes have been developed. Among them, we can distinguish the codes based on the SPH method [2]-[4], the grid codes [5]-[7], codes using the adaptive [8]-[10] and the moving [11]-[13] meshes. Most of them are focused on the classical supercomputer architectures usage, but there are also representatives focused on graphics accelerators [14]-[16] and Intel Xeon Phi accelerators [17].

Most of the existing codes for simulating the cosmic plasma hydrodynamics are aimed at solving certain types of problems, but there are also universal codes [18]-[21] with a modular structure that support the variability of the problems being solved and the methods used. For example, the MPI-AMRVAC code has a certain versatility for solving the problems of stellar

Content from this work may be used under the terms of the Creative Commons Attribution 3.0 licence. Any further distribution of this work must maintain attribution to the author(s) and the title of the work, journal citation and DOI.
Published under licence by IOP Publishing Ltd

evolution [22], but, nevertheless, it has a certain restrictions on solving the problems of galactic evolution or the problems of supernova explosions.

There are also attempts to create systems for generating astrophysical codes. For example, a code generation system using the EXCALC package has been created at the University of Costa Rica. This system transforms MHD equations to the Cartesian coordinates, followed by discretizing the equations by a method selected [23]. Similar works are carried out in Russia. Thus, the Institute of Astronomy of RAS, under the leadership of D. V. Bisikalo, Corresponding member of RAS, has created a code generation system for the evolution of the close binary stars [NSCF, 2013]. Still, all these systems are not completely universal.

This paper proposes a concept for creating a system of generating the grid parallel codes for the numerical solution of various hydrodynamic problems of the cosmic plasma based on a generalized representation of the magnetic hydrodynamics equations using the Godunov method. Using the ontological knowledge of the area in question and the modular structure of the code make it easier to prepare the necessary blocks for its generation.

2. A generalized description of the problem statement and solution method

The basic equations of the gravitational magnetic hydrodynamics for the vector of conservative variables $\mathbf{U}$ have the following form:

$$\frac{\partial \mathbf{U}}{\partial t} + \nabla \cdot \mathbf{F}(\mathbf{U}) = Q, \tag{1}$$

$$\Delta \mathbf{\Phi} = P(\mathbf{U}), \tag{2}$$

where $\mathbf{F}(\mathbf{U})$ is the flow vector of conservative variables, Q is the right-hand side vector (a source), $\mathbf{\Phi}$ is the gravitational potential, $P(\mathbf{U})$ is the gravity source.

The operator splitting approach allows us to represent this system as a combination of a hyperbolic system $\frac{\partial \mathbf{U}}{\partial t} + \nabla \cdot \mathbf{F}(\mathbf{U}) = 0$, an elliptic system $\Delta \mathbf{\Phi} = P(\mathbf{U})$ and a right-hand side $\frac{\partial \mathbf{U}}{\partial t} = Q$. Equation (1) can also include a parabolic term describing, for example, the heat conduction effect, but at this stage of our study this was not considered.

For the numerical solution of system (1-2), various discretization methods can be used. In this paper, we consider a finite difference approach, which is more suitable for modeling media without significant curvilinear inclusions, in contrast to the finite volume method or finite element method. To discretize the computational domain, we introduce a uniform grid with a step h, at which the conservative variables are determined at the cell centers using fractional indices, and the conservative variables flows – at the centers of the interfaces between the cells. Then the Godunov scheme for equation (1) in the general form can be written down as follows:

$$\frac{U^{n+1}_{i+\frac{1}{2},j+\frac{1}{2},k+\frac{1}{2}} - U^{n}_{i+\frac{1}{2},j+\frac{1}{2},k+\frac{1}{2}}}{\tau} + \frac{F^{x}_{i+1,j+\frac{1}{2},k+\frac{1}{2}} - F^{x}_{i,j+\frac{1}{2},k+\frac{1}{2}}}{h} + \frac{F^{y}_{i+\frac{1}{2},j+1,k+\frac{1}{2}} - F^{y}_{i+\frac{1}{2},j,k+\frac{1}{2}}}{h} +$$

$$+ \frac{F^{z}_{i+\frac{1}{2},j+\frac{1}{2},k+1} - F^{z}_{i+\frac{1}{2},j+\frac{1}{2},k}}{h} = Q^{n}_{i+\frac{1}{2},j+\frac{1}{2},k+\frac{1}{2}},$$

where τ is the time step, calculated from the CFL condition, F^x, F^y, F^z are the conservative variables flows in terms of x,y,z coordinates.

To implement such a computational scheme, it is necessary to find the flow values at the interfaces between the cells, that is, to solve the Riemann problem. For this, various solvers can be applied, including high-order ones with different ways of representing a solution, for example, such as: MUSCL schemes, piecewise-linear and parabolic approximations, WENO schemes, HLL method, etc.

To numerically solve the Poisson equation (2), one can use any known grid method for the elliptic systems, for example, using the fast Fourier transform, various iterative methods, etc.

Such above-mentioned generalized description of the equations system and the numerical method for solving problems of magnetohydrodynamics allows us to compose a unified upper algorithm level suitable for solving a wide class of astrophysical problems and to propose a code generation scheme based on it.

3. A code generation scheme

Let us describe in greater detail the proposed structure of the generated code and the necessary modules from which it is assembled.

The general code organization scheme for the numerical solution of equations of the forms of (1-2) is shown in Figure 1. The first step is to initialize the necessary grid arrays for a selected data structure, which is determined by the specificity of solving equations and the effect on the parallelization efficiency. The second step is to upload the information about a medium under consideration. Further, in the loop, while the time condition is fulfilled, the following actions are sequentially performed: recalculating the current time step to fulfill the CFL condition, fulfilling boundary conditions for primitive variables, reconstructing primitive variables based on the selected local template, fulfilling boundary conditions for reconstructed variables, starting the non-blocking exchange of boundary values between the adjacent nodes (parallelization by decomposition of the computational domain), solving the Riemann problem, computing the solution to equation (1) according to the Godunov scheme, solving the Poisson equation (2), verification of the exchanges completion, saving the solution at the current time step. Figure 1 also lists the necessary modules for assembling the final code, which, together with the universal function main, based on the described scheme, compound the finished version of the code (MPI-program). Let us note that as specific modules one can use both the library solutions and author's implementations. Moreover, specific solving equations determine only a set of conservative and primitive variables; the general solution scheme remains universal. The choice of specific methods to solving the arising sub problem and the corresponding modules is up to the researchers. Therewith, modules that implement solutions to the Riemann problem and calculations according to the Godunov scheme assume a parallel implementation, for example, using OpenMP technology and vector extensions for existing Intel processors and co-processors or using the CUDA core-function on the threads grid for calculations on graphic accelerators. The main structure can be expanded in the case of using the operator splitting approach or adding the calculation of chemical reactions, adaptive grids. However in any case the main framework and all the above modules will remain the same.

4. Intelligent support ontology for solving compute-intensive astrophysics problems

The development of efficiently parallel codes for modern supercomputer systems requires the knowledge about relevant computational methods, parallel architectures and technologies. To simplify the problem of choosing optimal numerical methods and suitable target architectures, it is proposed to present the accumulated knowledge about solving compute-intensive problems of astrophysics in an obvious ontological form. The ontology of mathematical methods and parallel algorithms and the ontology of parallel architectures and technologies, tied together with the rules of inference, allow one to build an optimal scheme for solving the problem. On this basis, the necessary modules for generating a code are determined according to the above described scheme. Such an approach can essentially simplify the process of developing scientific software and make it accessible to a researcher who is poorly skilled in some blocks of knowledge represented in ontology.

Figure 2 shows the top level of the compiled ontology, some basic concepts of which are supplemented by a finite set of objects. The solution to a problem begins with the determination of a Physical Phenomenon under study described by the corresponding Equation system.

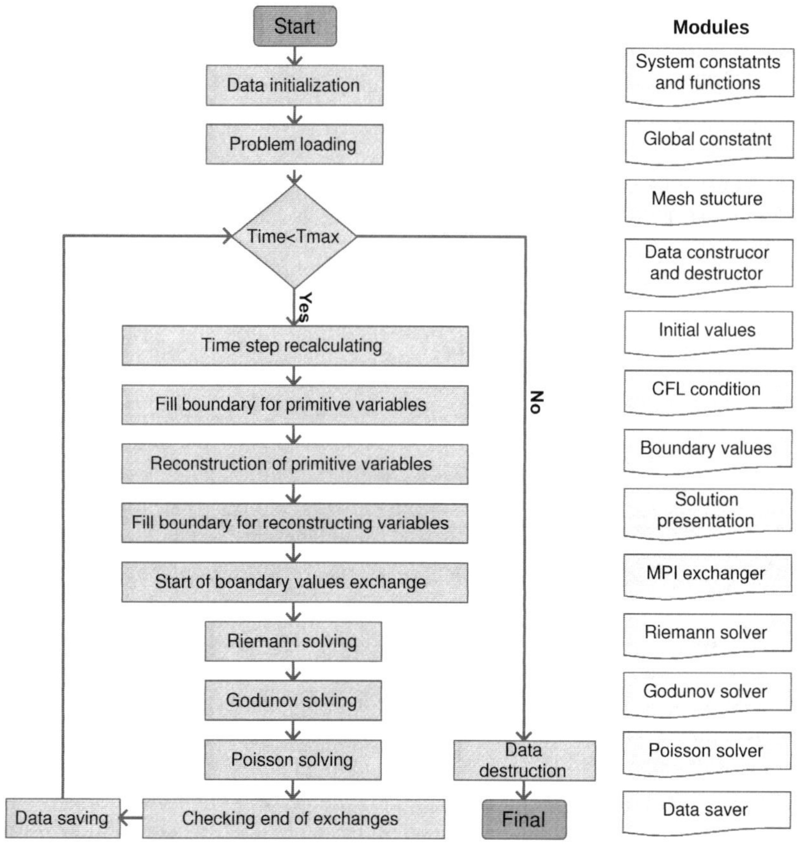

Figure 1. The general scheme of code organization and the necessary modules for the numerical solution of MHD-equations.

The Equation System is solved by a Computational Method constructed using the previously mentioned methods. The final Computational Method, in turn, is implemented by a Parallel Algorithm, which is encoded by program Code optimized for a selected Computing Device using assigned Parallel Programming Technologies. In the future, the ontology will be expanded by expert (inference) rules containing the information not explicitly presented in the ontology about the properties of mapping computational methods to computational architectures. The rules will allow one to come to a comparative assessment of the computation time, the size of the memory used, the accuracy of the solution, scalability for a selected bundle: the computational method – the target architecture.

5. Conclusion

This paper presents the code generation concept for solving the problems of cosmic plasma simulation on parallel computing systems using an ontological representation. The generated code structure and the main modules are described, as well as the top-level ontology of the intelligent support for solving compute-intensive problems of astrophysics.

The proposed generation system maintains multipurposeness with respect to the conservation laws of magnetic hydrodynamics and numerical methods in question. The generated code allows one to simulate the dynamics of cosmic plasma for a wide range of problems.

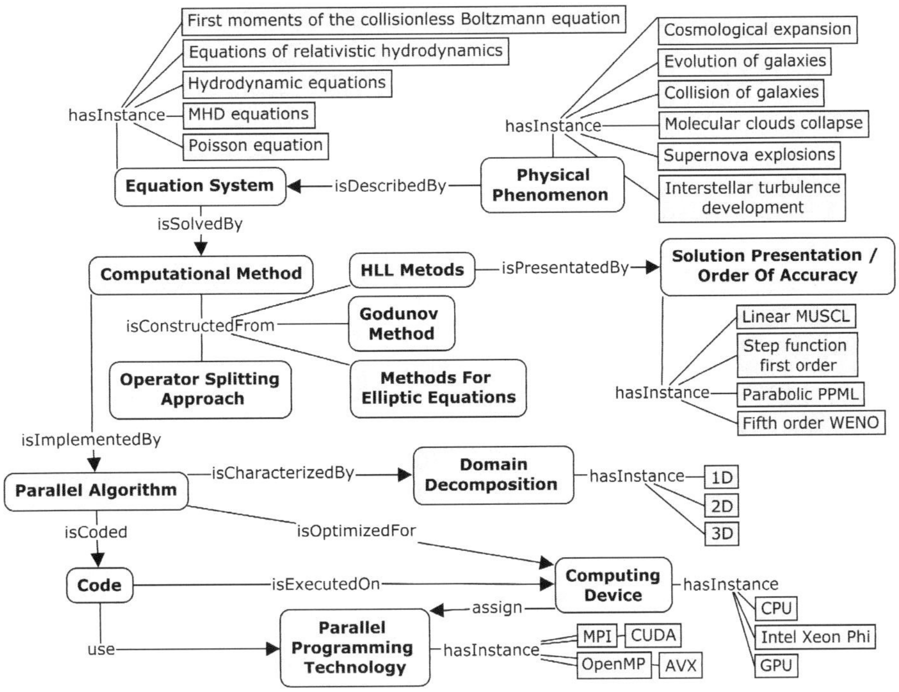

Figure 2. The intelligent support ontology for solving compute-intensive astrophysics problems.

Acknowledgments
This research was conducted within the framework of the budget project No. 0315-2019-0009, for ICMMG SB RAS, and supported in part by the Russian Foundation for Basic Research [grant No. 19-07-00085].

References
[1] Tajima T and Shibata K 2002 *Plasma Astrophysics* (Boulder:Westview Press)
[2] Volker S 2005 *Mon. Not. R. Astron. Soc.* **364** 1105–34
[3] Wadsley J W, Stadel J and Quinn T 2004 *New Astronomy* **9** 137–58
[4] Steinmetz M 1996 *Mon. Not. R. Astron. Soc.* **278** 1005–17
[5] The Pencil code, http://pencil-code.nordita.org/references.php
[6] Stone J M and Norman M L 1992 *ApJS* **80** 753–90
[7] Stone J M, Gardiner T A, Teuben P, Hawley J F and Simon J B 2008 *ApJS* **178** 137–177
[8] Mignone A, Bodo G, Massaglia S, Matsakos T, Tesileanu O, Zanni C and Ferrari A 2007 *ApJS* **170** 228–42
[9] Bryan G L, Norman M L, OShea B W, at al 2014 *ApJS* **211** 19
[10] Teyssier R 2002 *A&A* **385** 337–64
[11] Hopkins P F 2015 *Mon. Not. R. Astron. Soc.* **450** 53–110
[12] Springel V 2010 *Mon. Not. R. Astron. Soc.* **401** 791–851
[13] Murphy J W and Burrows A.2008 *ApJS* **179** 209–41
[14] Hsi-Yu Schive, Yu-Chih Tsai and Tzihong Chiueh 2010 *ApJS* **186** 457–84
[15] Kulikov I 2014 *ApJS* **214** 12
[16] Schneider E E and Robertson B E 2015 *ApJS* **217** 24
[17] Kulikov I M, Chernykh I G, Snytnikov A V, Glinskiy B M and Tutukov A V 2015 *CPC* **186** 71–80
[18] The Einstein Toolkit, https://einsteintoolkit.org/index.html
[19] The MPI-AMRVAC code, http://amrvac.org/
[20] The Athena++ radiation GRMHD code, https://princetonuniversity.github.io/athena/
[21] The PLUTO Code, http://plutocode.ph.unito.it/
[22] Goedbloed J, Keppens R and Poedts S 2003 *Space Science Reviews* **107** 63
[23] Frutos-Alfaro F and Carboni-Mendez R 2016 *Rev. Mat* **23** 41–61

IOP Conf. Series: Journal of Physics: Conf. Series **1336** (2019) 012013 doi:10.1088/1742-6596/1336/1/012013

Computer simulation of diamagnetic regime in open magnetic trap

G I Dudnikova[1], I G Chernykh[1], I S Chernoshtanov[1,2], V A Vshivkov[1]
[1]Institute of Computational Mathematics and Mathematical Geophysics SB RAS, Novosibirsk, Russia
[2]Budker Institute of Nuclear Physics SB RAS, Novosibirsk, Russia
E-mail: gdudnikova@gmail.com

Abstract.
Preliminary results of numerical simulation of transition to regime of diamagnetic confinement in an axially-symmetric mirror trap are presented. Author's two-dimensional modification of hybrid Particle-in-Cell model is used in the simulations. Formation of region with hot dense plasma and very small magnetic field is demonstrated. Evolution of plasma and magnetic field, as well as structure of the formed region are discussed and compared with simple analytical estimations.

1. Introduction

Present work is aimed at numerical simulation of newly proposed regime of plasma confinement in mirror trap, so-called diamagnetic confinement [1]. This method relates to confinement in axially-symmetric mirror trap of plasma with extremely high pressure which is equal to pressure of magnetic field of the trap. The magnetic field is fully displaced from the region occupied by the plasma because of plasma diamagnetism so that cluster of hot dense plasma with very weak magnetic field forms inside the trap (so-called diamagnetic "bubble"). The mirror ratio inside the "bubble" formally is extremely small so good particle and energy confinement is expected. The structure of the magnetic field in the diamagnetic trap has much in common with the structure of the field in elongated Field Reversed Configurations (FRCs) with very small field reversal [2]. So many of methods of MHD stabilization (in particular stabilization by conducting shell) verified in the experiments on FRCs can be used in the diamagnetic trap.

The attractive features of the diamagnetic trap (simple structure of magnetic system, stationarity, principal possibility of developing of compact reactor with high density of thermonuclear power) stimulate experimental investigation of the regime of the diamagnetic confinement. The diamagnetic-confinement-like regimes were demonstrated recently at the machine with neutral beam injection (NBI) C-2W [3]. The Compact Axisymmetric Toroid (CAT) device targeted on studying of formation of FRC and diamagnetic "bubble" in mirror trap with powerful NBI is being constructed in Budker Institute of Nuclear Physics [4]. Models describing equilibrium, evolution and stability of plasma with extremely high pressure are necessary for planning and adequate interpretation of results of experiments targeted to demonstration of diamagnetic confinement.

Content from this work may be used under the terms of the Creative Commons Attribution 3.0 licence. Any further distribution of this work must maintain attribution to the author(s) and the title of the work, journal citation and DOI.
Published under licence by IOP Publishing Ltd

At present one- and two-dimensional models describing equilibrium in the diamagnetic trap in the MHD approximation exist [1], [5]. The models predict the particle and the energy confinement time $\sqrt{\tau_{\|}\tau_{\perp}}$, here $\tau_{\|}$ is the particle lifetime in the gas-dynamic regime and $\tau_{\perp}$ is the time of magnetic field diffusion across the margin of "bubble" [1]. The models also allow calculating the equilibrium magnetic field, as well as the distribution of the plasma density and temperature [5]. At the same time, some assumptions underlying the MHD models cannot be adequate to this task, in particular, the assumption of zero Larmor radius of ions. Therefore, hybrid MHD- kinetic models based on the Vlasov equations for ions and the MHD approximation for magnetized electrons should be developed. Even two-dimensional model without asymmetry on azimuthal angle can help to clarify such questions as prediction of the particles lifetime and the structure of the magnetic field generated by the plasma. The first is important for estimation of reactor perspectives of the diamagnetic trap, the second plays significant role in the investigation of stabilization of MHD instabilities by conducting shell. Also, the simulations can clarify some properties of the trajectories of the ions, for example, the rate of regular and stochastic orbits and the ions spin-up caused by preference losing of ions with positive sign of azimuthal momentum. These aspects of ions dynamic in FRCs are briefly discussed in [6]. Currently, there is a fairly large number of publications devoted to the problems of numerical modelling of nonlinear processes in high-pressure plasma. However, most of the articles are aimed either at the stability of the FRC, or at their formation by pulsed methods or a rotating magnetic field.

Articles devoted to computer simulation of the configuration of high-beta plasma in the mirror traps with a powerful injection are virtually absent. It should be noted, in the paper [7] simulation of FRC formation in a mirror machine with powerful off-axis NBI is described.

The Particle-in-Cell method (PIC) was used for simulation of evolution of ion component but the electron currents were not taken into account. Formation of FRC in a mirror trap with off-axis NBI was simulated numerically with hybrid PIC code in paper [8]. Also, there are articles describing simulation of NBI in FRC, see for example [9]. Therefore, the development of a new numerical code for modeling the regime of diamagnetic plasma confinement in open magnetic systems is necessary.

Preliminary and test results of simulations of the regime of the diamagnetic confinement are presented in this article. Author's modification of hybrid PIC model with kinetic description for the ions and hydrodynamic description for the electrons is used in the simulations. The article is organized as follows. The formulation of the problem is discussed in the second section. Some analytical estimations of the injection power necessary for the transition regime of the diamagnetic confinement and the time required to reach steady-state are given in the third section. The numerical code is described in the fourth section. The results of the preliminary simulations are discussed in the fifth section.

2. Formulation of the problem

To be able to compare the results of our and MHD simulations, we choose formulation of the problem close to that in MHD models [5]. Let's consider a mirror trap $0 < r < R_0, -L/2 < z < L/2$ with axially symmetrical vacuum magnetic field $\boldsymbol{B} = \{B_r, 0, B_z\}$. Ions are continuously injecting into the trap from the point $r = 0, z = 0$. Under the assumption of quasineutrality of the plasma, the electron density inside the trap is equal to the sum of the ion densities of the background plasma and the injected ions. Here we do not consider specific methods of the matter and energy injection into the trap, for example, injection of a neutral beam, granules, etc. The probability of occurrence of ion-electron pair in the trap does not depend on the longitudinal coordinate (coordinate along the axis of the trap). The velocity distribution function of the electrons is Maxwellian with temperature T_e, the injected ions can have arbitrary velocity distribution function. It should be noted, that in the MHD models the velocity distribution

IOP Conf. Series: Journal of Physics: Conf. Series **1336** (2019) 012013 doi:10.1088/1742-6596/1336/1/012013

functions of both ions and electrons are Maxwellian with the same temperatures. The main goal of the simulation is investigation of accumulation of the plasma and formation of the stationary "bubble"-like structure of the magnetic field.

There are several typical times: ion cyclotron gyro-period $2\pi/\Omega$, here $\Omega_0 = eB_0/m_i c$ is ion cyclotron frequency (here B_0 is value of vacuum magnetic field in the center of the trap and m_i is the mass of ions), period of bounce-oscillations of ions $2\pi/\Omega_b$ (which is of the order of the ratio of distance between mirrors to typical ion velocity) and times of ion-ion and ion-electron collisions, τ_{ii} and τ_d. With typical experimental parameters, there is a hierarchy of times $\Omega_0^{-1} \ll \Omega_b^{-1} \ll \tau_{ii} < \tau_d$, for example, in the Gas Dynamic Trap $\Omega_0^{-1} \sim 10^{-8} sec$, $\Omega_b \sim 10^{-6} sec$ and $\tau_d \sim 10^{-3} sec$ [10]. This hierarchy leads to some obvious difficulties in the simulation of formation of the "bubble"-like structure of the magnetic field. In the present model the ions are modelled by the PIC method, while the electrons are modelled as a liquid with density that coincides with the density of the ions. Typical times in this problem are the times of electron drag and L/v_A, here L is the distance between mirrors and v_A is the Alfvén velocity. The times are much greater than typical electron times, which are of the order of inverse electron cyclotron Ω_e^{-1}, plasma ω_{pe}^{-1} and bounce $L/(T_e/m_e)^{1/2}$ frequencies. Thus using the hybrid model allows the simulation to be essentially accelerated. We assume $m_e \to 0$ in the fluid equations. This neglecting implies $\omega_{pe}^{-1} \to 0$ and $\Omega_e^{-1} \to 0$, so excitation of electrostatic and electron-cyclotron waves is impossible in the simulation. These processes are insufficient for formation of the diamagnetic "bubble". This approach implies neglecting of all electron drifts, except of the electric drift. Strictly speaking, such approximation is correct if the electron temperature is much less than the mean energy of the ions.

3. Analytical estimations in the case of low plasma pressure

Let us consider the evolution of ion population in a mirror trap with injected plasma (ion and electron pairs form inside the trap). For simplicity, we limit ourselves to the case when the injection power is small enough and the plasma pressure is small in comparison with the pressure of magnetic field.

The evolution of population of ions satisfies the kinetic equation

$$\frac{\partial f_i}{\partial t} + \boldsymbol{v}\frac{\partial f_i}{\partial \boldsymbol{r}} + \frac{e}{c}(\boldsymbol{v} \times \boldsymbol{B})\frac{\partial f_i}{\partial \boldsymbol{v}} = S(\boldsymbol{r}, \boldsymbol{v}) + St_{ii} + \tau_d^{-1}\frac{\partial}{\partial \boldsymbol{v}}(\boldsymbol{v} f_i). \qquad (1)$$

Here $\boldsymbol{B}$ is magnetic field of the trap, $S(\boldsymbol{r}, \boldsymbol{v})$ is source of ions, term St_{ii} describes ion-ion collisions and τ_d is the time of electron drag. The boundary condition in the mirrors is $\partial_{v_\parallel} f_i = 0$ for trapped particles (here $v_\parallel$ is the component of velocity along magnetic field) and $f_i = 0$ at $\pm v_\parallel > 0$ for passing particles (here signs "+" and "−" correspond the left and right mirrors).

We neglect the density of passing ions further. Let's us neglect the ion-ion collisions now. After integration over speeds and space with the weights of 1 and $v_\perp^2 \equiv \boldsymbol{v}^2 - v_\parallel^2$, the kinetic equation reduces to

$$\frac{dN}{dt} = J, \quad \frac{dW}{dt} + 3\tau_d^{-1} W = m_i w_\perp^2 J, \qquad (2)$$

here N and W are the total number of trapped particles in the trap and their transversal kinetic energy, J is the number of ions appearing in the trap per unit time, and $m_i w_\perp^2/2$ is the mean transversal kinetic energy of appearing trapped ions. So plasma density grows linearly and plasma energy reaches a limit value $\tau_d(m_i w_\perp^2)J/3$ in the time of the order of electron drag time if angle scattering of ions is neglected. So the system reaches a steady-state in a time of order $\tau_d > \tau_{ii}$. Also these calculations allow parameters of the diamagnetic "bubble" to be estimated through injection parameters: $m_i w_\perp^2 J \; (B_0^2/8\pi) L a^2/\tau_d$, here B_0 is the vacuum magnet field in the center, L is the distance between mirrors and a is the radius of the "bubble".

Accumulation of ions with low energy is a non-physical effect which is provoked by neglecting of ion-ion collisions. Actually, the angle scattering restricts the density of the trapped particles by value of the order of J/τ_{ii}. Thus neglecting of the scattering does not prevent the formation of the "bubble"-like stationary structure, but it leads to some non-physical effects.

4. Basic equations

The original system of equations includes the Vlasov equations for the ion component of the plasma and the incoming ions, the equations of magnetic hydrodynamics for the electrons and Maxwell equations [11]:

$$\begin{gathered}
\frac{\partial f_i}{\partial t} + \boldsymbol{v}\frac{\partial f_i}{\partial \boldsymbol{r}} + \frac{\boldsymbol{F}_i}{m_i}\frac{\partial f_i}{\partial \boldsymbol{v}} = 0, \quad \boldsymbol{F}_i = q_i\left(\boldsymbol{E} + \frac{1}{c}\boldsymbol{v}\times\boldsymbol{B}\right) + \boldsymbol{R}_i, \\
e\left(\boldsymbol{E} + \frac{1}{c}\boldsymbol{v}_e\times\boldsymbol{B}\right) + \frac{\nabla p_e}{n_e} - \boldsymbol{R}_e = 0, \\
n_e\left(\frac{\partial T_e}{\partial t} + (\boldsymbol{v}_e\cdot\nabla)T_e\right) + (\gamma-1)p_e\nabla\cdot\boldsymbol{v}_e = (\gamma-1)(Q_e - \nabla\cdot\boldsymbol{q}_e), \\
\nabla\times\boldsymbol{B} = \frac{4\pi}{c}\boldsymbol{j}, \quad \nabla\times\boldsymbol{E} = -\frac{1}{c}\frac{\partial\boldsymbol{B}}{\partial t}, \quad \nabla\cdot\boldsymbol{B} = 0.
\end{gathered} \tag{3}$$

Here, f_i is the ion distribution function, $\boldsymbol{B}$ and $\boldsymbol{E}$ are the magnetic and the electric fields, $\boldsymbol{v}_e$ is the electron velocity, $\boldsymbol{v}_i = \int f_i(\boldsymbol{r},\boldsymbol{v},t)\boldsymbol{v}d\boldsymbol{v}/n_i$ is the average velocity of ions, p_e and T_e are the pressure and temperature of the electrons. The collision force $\boldsymbol{R}_i = -\boldsymbol{R}_e$ takes into account the exchange of pulses between the electron and ion components of the plasma and the beam, $\boldsymbol{R}_e = -m_e(\boldsymbol{v}_e - \boldsymbol{v}_i)/\tau_d$, where τ_d is the characteristic ion-electron collision time.

It is assumed here that the thermal conductivity coefficient k and the collision frequency $\nu = 1/\tau_d$ are independent of the plasma and magnetic field parameters ($\nu = const$ and $k = const$).

The heat generated in the electrons is $Q_e = j^2/\sigma$, the electron heat flux is $q_e = -k\nabla T_e$.

In the calculations, the adiabatic index $\gamma = 5/3$ is used. To solve the Vlasov kinetic equation, the author's modification of PIC is used [12]. To solve the MHD and Maxwell equations, finite-difference schemes of the first order of accuracy are used. To solve the heat balance equation for the electrons the up-wind scheme is used. In hybrid models, the condition of plasma quasi-neutrality is assumed to be $n_e = n_i$ (the scales under consideration are larger than the Debye radius), and the displacement currents are also neglected, only low-frequency processes are considered. The electric field is determined from the equation of motion of the electrons under the assumption $m_e = 0$.

A detailed description of the used numerical methods and algorithms for the problem solution is presented in the report [13].

5. Numerical results

Results of verification of the numerical code are described in this section. All quantities are normalized on the ion cyclotron frequency $\Omega_0 = eB_0/(m_i c)$, the Alfvén velocity $v_A = B_0/\sqrt{4\pi m_i n_0}$ and the ion dispersion size $c/\sqrt{4\pi n_0 e^2/m_i}$, here B_0 is the magnitude of vacuum magnetic field in the center and n_0 is the initial concentration of the background ions. We choose following normalizing parameters: $B_0 = 2kG$ and $n_0 = 10^{12}cm^{-3}$, so that $\Omega_0 \approx 1.9\cdot 10^7 s^{-1}$, $v_A \approx 4.4\cdot 10^8 cm/s$ and the ion dispersion size is $22.8cm$. The electron drag time is chosen non-physically large to decrease time necessary for reaching stationary state. The ion-ion collisions are neglected in this simulation.

IOP Conf. Series: Journal of Physics: Conf. Series **1336** (2019) 012013 doi:10.1088/1742-6596/1336/1/012013

The vacuum magnetic field is created by the two coaxial coils with equal radii and current. The mirror ratio is $r_{mirror} = 2$. The distribution function of the injected ions is proportional to $\exp(-(v-v_0)^2/v_T^2)$, here v is the magnitude of velocity of the ions, v_0 and v_T are constants. The mean transversal energy per one injected ion $m_i w_\perp^2/2$ is equal to $m_i v_T^2/2$ in the case $v_0 \ll v_T$ and $m_i v_0^2/3$ in the case $v_0 \gg v_T$.

An example of evolution of magnetic field during injection of ions is shown in figure 1. Following parameters of injection are used: $v_0 = 0.05 v_A$ and $v_T = 0.01 v_A$. Continuous injection yields the growth of the number of the beam ions with the constant rate as equation (2) predicts. The flow of ions injected into the trap is $J \approx 2.4 \cdot 10^{18} \Omega_0$. On the other hand, the ions continuously lose their energy because of the electron drag and the interaction with the vortex electric field (see figure 3) so the total kinetic energy of the ions reaches a constant value. The ratio of the accumulated kinetic energy of the beam ions $\tau_d J m_i w_\perp^2$ and energy of the extruded magnetic field $(B_0^2/8\pi)a^2 L$ are of the order of the unity which coincides with analytical estimations made in third section. The accumulation of the beam ions and the increasing of their kinetic energy lead to extruding of the magnetic field because of the plasma diamagnetism and formation of the stationary "bubble"-like structure of the magnetic field (see figure 1).

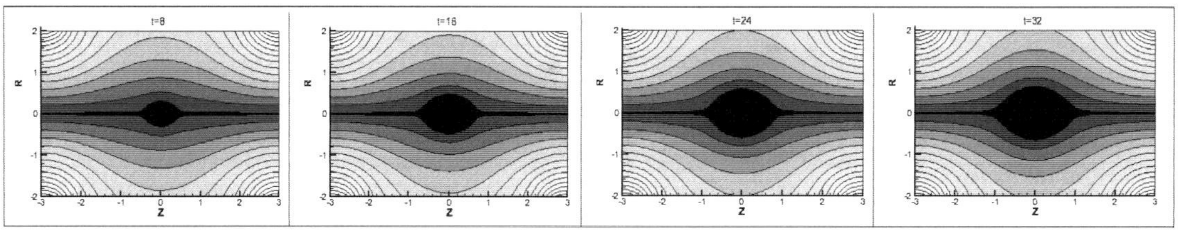

Figure 1. An example of magnetic field lines at different moments of time. Parameters: $v_0 = 0.05 v_A$ and $v_T = 0.01 v_A$.

Formed configuration of the magnetic field consists of inner space with approximately zero magnitude of the field and outer region with weakly modified field (see figure 2). These areas are separated by a transition region where the magnetic field changes sharply. The width of the transition region is of the order of the average Larmor radius of the beam ions. The radius of the transition region, defined as the radius at which the magnitude of the magnetic field is equal to one percent of B_0, monotonically grows and reach a stationary value in a time of the order of τ_d (see figure 2).

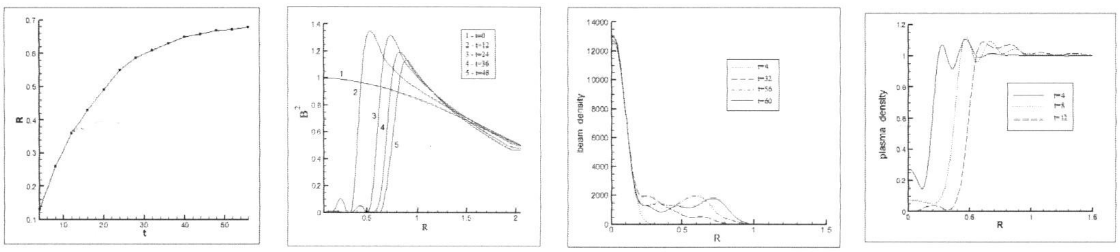

Figure 2. An examples of dependence of radius of transition region on time (left) and dependence of energy of magnetic field (second), density of beam (third) and background ions (right) on radial coordinate at $z = 0$ at different moment of time. Parameters are the same as in the figure 1.

The changing of the magnetic field leads to the generation of the azimuthal component of the electric field. This vortex field decelerates the beam ions so that resulting changing in the kinetic energy of the ions is equal to the changing in the magnetic field energy. The magnitude of the vortex electric field decreases with time and is equal to zero when a stationary state is reached.

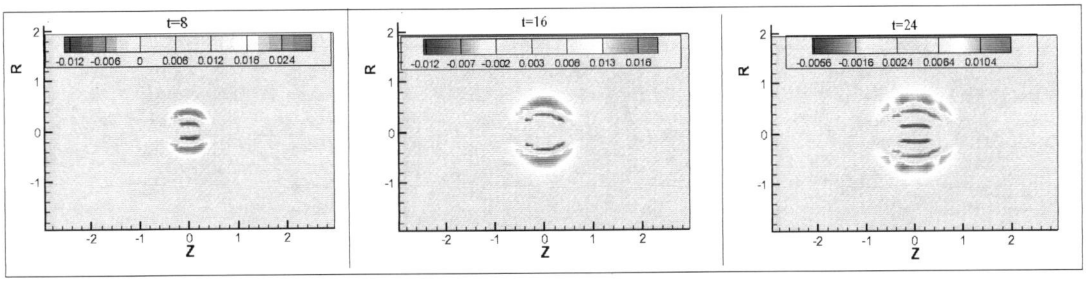

Figure 3. An example of distribution of azimuthal component of electric field at different moment of time. Parameters are the same as in the figure 1.

The azimuthal momentum of the beam ions is close to zero because it was born on the axis. So the density of the beam ions inside the "bubble" decreases approximately inversely with the distance to the axis (see figure 2). The beam ions are reflected from the transition region, so their density is very small outside of the "bubble". The background plasma is extruded from the region occupied by the beam ions (see figure 2). There are two extrusion mechanisms. Firstly, an increase in the beam ion density leads to the formation of an electron density gradient due to quasi-neutrality. The gradient of electron pressure leads to arising potential electric field (see third equation in (3)), which ejects cold background ions along magnetic field lines. This effect is analogical to ejecting of cold ions from an open trap by ambipolar potential. On the other hand, an increase in the pressure of the beam ions leads to a change in the magnetic field and a generation of the azimuthal component of the electric field localized near the edge of the "bubble" (see figure3). This field removes the background ions and electrons in the radial direction due to the electric drift.

6. Conclusion

Preliminary results of simulations of transition to the regime of diamagnetic plasma confinement in open magnetic trap are presented. Formation of quasi-stationary structure of magnetic field with region of extremely small magnetic field is demonstrated. Typical times of processes and structure of magnetic field are consistent with qualitative conception given in [1] and analytical estimations. The numerical model is applicable for simulation with parameters corresponding to real experimental conditions.

Acknowledgments

This work was supported by Russian Science Foundation (Project No. 19-71-20026).

References

[1] Beklemishev A D 2016 *Phys. Plasmas* **23** 082506, doi: 10.1063/1.4960129
[2] Steinhauer L C 2011 *Phys. Plasmas* **18** 070510, doi: 10.1063/1.3613680
[3] Gota H, et al. 2019 *Nucl. Fusion* **59** 112009, doi: 10.1088/1741-4326/ab0be9
[4] Bagryansky P A, et al. 2016 *AIP Conf. Proc.* **1771** 030015, doi: 10.1063/1.4964171
[5] Beklemishev A D and Khristo M S 2019 *Plas. Fus. Res.* **14** 2403007, doi: 10.1585/pfr.14.2403007

[6] Belova E V, Davidson R C, Ji H and Yamada M 2006 *Phys. Plasmas* **13** 056115, doi: 10.1063/1.2179426
[7] Byers J A 1977 *Phys. Rev. Lett.* **39** 23, doi: 10.1103/PhysRevLett.39.1476
[8] Tsidulko Yu A and Chernoshtanov I S 2016 *AIP Conf. Proc.* **1771** 040005, doi: 10.1063/1.4964190
[9] Lifschitz A F, Farengo R and Arist N R 2002 *Nucl. Fusion* **42** 7, doi: 10.1088/0029-5515/42/7/309
[10] Ivanov A A and Prikhodko V V 2017 *Physics-Uspekhi* **60** 5, doi: 10.3367/UFNe.2016.09.037967
[11] Berezin Yu A, Dudnikova G I 1985 *Numerical plasma models and reconnection phenomena* (Moskow: Nauka) p 125 (in Russian)
[12] Berezin Yu A, Vshivkov V A 1980 *Particle method in the dynamics of rarefied plasma* (Novosibirsk: Nauka) p 94 (in Russian)
[13] Vshivkov V A, Boronina M A, Genrikh E A, Dudnikova G I, Vshivkova L V 2019 *J. Phys.: Conf. Series* (these proceedings)

IOP Conf. Series: Journal of Physics: Conf. Series **1336** (2019) 012014 doi:10.1088/1742-6596/1336/1/012014

Hydrodynamical Simulation of Astrophysical Flows: High-Performance GPU Implementation

Elena Akimova[a,b], Vladimir Misilov[a,b], Igor Kulikov[c], Igor Chernykh[c]

[a]N. N. Krasovskii Institute of Mathematics and Mechanics UrB RAS, Ekaterinburg, Russia
[b]Ural Federal University, Ekaterinburg, Russia
[c]Institute of Computational Mathematics and Mathematical Geophysics SB RAS, Novosibirsk, Russia

E-mail: v.e.misilov@urfu.ru

Abstract. We present a new hydrodynamical code GPUPEGAS 2.0 for 3D simulation of astrophysical flows using the GPUs. This code is an extension of GPUPEGAS code developed in 2014 for simulation of interacting galaxies. GPUPEGAS 2.0 is based on the Authors' numerical method of high order of accuracy for smooth solutions with small dissipation of the solution in discontinuities. The high order of accuracy and small dissipation are achieved by using the piecewise-linear representation of the physical variables in each dimension. The Rusanov flux allows one to simply vectorize the solution of the Riemann problem. The code was implemented for the cluster supercomputers NKS-30T (Siberian Supercomputer Center, SB RAS) and Uran (Institute of Mathematics and Mechanics, UrB RAS) using the hybrid MPI+CUDA technology. To avoid the compute capability-specific implementations of reduction routines, the Thrust library was used. The optimal parameters for kernel function were found for the three-dimensional computation grid. The Sedov point blast problem was used as a main test one. The numerical experiment was performed to simulate the hydrodynamics of the type II supernova explosion for the grid size of 256^3. A set of experiments was performed to study performance and scalability of the developed code. The performance of 25 GFLOPS was achieved using a single Tesla M2090 GPU. The speedup of 3 times was achieved using a node with 4 GPUs. By using 16 GPUs, 70% scalability was achieved.

1. Introduction

Studying the hydrodynamics of the supernova explosions is an important astrophysical problem. Naturally, because of the proportion of scales between the stars, the supernova remnant, and the core, the simulation of such explosion requires the most powerful supercomputers. Most of the modern supercomputers use the hybrid architecture, namely, the graphics accelerators (GPUs) and multicore coprocessors. Development of codes for such architectures is a difficult problem. Previously, the Authors had implemented the GPUPEGAS code [1] for supercomputers with graphics processors and the AstroPhi code for Intel Xeon Phi in the offload mode and in native mode using the low-level vectorization [2].

In this work, we present a new code GPUPEGAS 2.0 for 3D simulation of astrophysical flows using the GPUs. The code is based on the efficient piecewise-parabolic method on the local stencil [3, 4], which was successfully applied to a number of astrophysical problems [5, 6]. The main advantages of this method is the simplicity and cost-efficiency while retaining the features of the high order of accuracy and small dissipativity.

Content from this work may be used under the terms of the Creative Commons Attribution 3.0 licence. Any further distribution of this work must maintain attribution to the author(s) and the title of the work, journal citation and DOI.
Published under licence by IOP Publishing Ltd

2. Numerical Model

We will briefly describe the problems and the numerical methods. Let us consider the conservative form of the equations of the gravitational hydrodynamics in the three-dimensional Cartesian coordinates for the law of conservation of mass

$$\frac{\partial \rho}{\partial t} + \nabla \cdot (\rho \vec{u}) = 0, \tag{1}$$

for the law of conservation of momentum

$$\frac{\partial \rho \vec{u}}{\partial t} + \nabla \cdot (\rho \vec{u}\vec{u} + p) = -\rho \nabla \Phi, \tag{2}$$

for the law of conservation of total mechanical energy

$$\frac{\partial}{\partial t}\left[E + \rho\frac{\vec{u}^2}{2}\right] + \nabla \cdot \left(\left[E + p + \rho\frac{\vec{u}^2}{2}\right]\vec{u}\right) = -\left(\rho\nabla\Phi, \vec{u}\right) \tag{3}$$

complemented by the Poisson equation for the gravitational potential

$$\Delta\Phi = 4\pi G\rho, \tag{4}$$

where ρ is the density, $\vec{u}$ is the velocity, p is the pressure, Φ is the gravitational potential, E is the intrinsic energy, G is the gravitational constant.

The initial conditions are drawn from the statement of the problem. In our code, we use the periodic boundary conditions. We will consider two sets of variables: the conservative variables $U = \left(\rho, \rho\vec{u}, E + \rho\frac{\vec{u}^2}{2}\right)^\top$ and physical (or primitive) variables $P = (\rho, \vec{u}, p)^\top$. The conservation laws (1, 2, 3) can be rewritten in the vector form

$$\frac{\partial U}{\partial t} + \nabla \cdot F(U) = Q, \tag{5}$$

where $F(U)$ is the flux vector of the conservative variables, Q is the right part vector describing the work of the gravitational forces. For discretizing the computational 3D domain, let us introduce regular (on each dimension) rectangular grid of $I \times K \times L$ dimension, where h_x, h_y, h_z are the grid steps.

The time step τ is calculated from the Courant condition

$$\frac{\tau \times (c + \|\vec{u}\|)}{h} = CFL < 1, \tag{6}$$

where CFL is the Courant number, $c = \sqrt{\gamma p/\rho}$ is the speed of sound.

The conservative and primitive variables are defined at the centers of cells. As a result, for the vector form of the conservation laws (5), let us write the Godunov type scheme

$$\frac{U^{n+1}_{i+\frac{1}{2},k+\frac{1}{2},l+\frac{1}{2}} - U^{n}_{i+\frac{1}{2},k+\frac{1}{2},l+\frac{1}{2}}}{\tau} + \frac{F^x_{i+1,k+\frac{1}{2},l+\frac{1}{2}} - F^x_{i,k+\frac{1}{2},l+\frac{1}{2}}}{h} + \tag{7}$$

$$\frac{F^y_{i+\frac{1}{2},k+1,l+\frac{1}{2}} - F^y_{i+\frac{1}{2},k,l+\frac{1}{2}}}{h} + \frac{F^z_{i+\frac{1}{2},k+\frac{1}{2},l+1} - F^z_{i+\frac{1}{2},k+\frac{1}{2},l}}{h} = Q^n_{i+\frac{1}{2},k+\frac{1}{2},l+\frac{1}{2}}.$$

For implementing the scheme (7), we need to define the values of fluxes at the interfaces between the cells. To achieve this, we have constructed the piecewise-linear representation of the primitive variables P at each cell of the computational domain for each direction using the minmod limiter.

As a Riemann solver for the right (R) and left (L) cell, we use the Rusanov flux or the Lax–Friedrichs flux

$$\mathcal{R}\left(P^{\mathrm{R}}, P^{\mathrm{L}}\right)=\frac{F\left(P^{\mathrm{L}}\right)+F\left(P^{\mathrm{R}}\right)}{2}+\frac{\max\left(c\left(P^{\mathrm{R}}\right)+\|\vec{u}^{\mathrm{R}}\|, c\left(P^{\mathrm{L}}\right)+\|\vec{u}^{\mathrm{L}}\|\right)}{2}\left(U\left(P^{\mathrm{L}}\right)-U\left(P^{\mathrm{R}}\right)\right). \quad (8)$$

For reconstructing the gravitational potential using the density function, we use the 27 point template for approximating the Poisson equation. This arises from the need for maximum invariance of the solution with respect to rotation. The density function and the potential are written in form of the superposition of eigenfunctions of the Laplace operator. To perform this, the fast Fourier transform is used. After substitution of this superposition to the approximation scheme of the Poisson equation, we get the fairly simple expression for calculation the harmonic magnitudes of the potential

$$\phi_{jmn}=\frac{\frac{2}{3}\pi G h^2 \sigma_{jmn}}{1-\left(1-\frac{2\sin^2\frac{\pi j}{I}}{3}\right)\left(1-\frac{2\sin^2\frac{\pi m}{K}}{3}\right)\left(1-\frac{2\sin^2\frac{\pi n}{L}}{3}\right)}.$$

where I, K, L are the numbers of cells for each coordinate. After this, we need to perform the reverse fast Fourier transform of the potential magnitudes to the functional space of the harmonics.

3. Parallel Implementation

We will describe the architecture of the code and the specifics of implementation using the MPI and CUDA technologies. Developed numerical method for solving the hydrodynamic equations allows one to apply the geometrical domain decomposition using one layer for subdomains overlapping. To solve the Poisson equation on the distributed memory supercomputer on the basis of the fast Fourier transform, the FFTW library was used.

The GPUPEGAS 2.0 code uses the multilevel multidimensional decomposition of domain. The outer one-dimensional decomposition into set of subdomains is performed along one coordinate axes using MPI tools. The decomposition of each subdomain is performed using CUDA tools as shown in Fig 1.

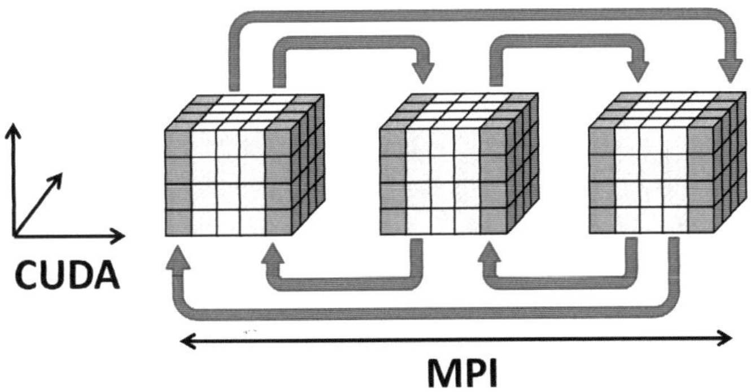

Figure 1. Geometric domain decomposition.

This structure of the computation provides a potentially infinite scalability.

The following algorithm for computing a single time step is used:

(i) Find a minimum time step between all spatial cells (on GPU; the reduction is performed by means of the Thrust library).

IOP Conf. Series: Journal of Physics: Conf. Series **1336** (2019) 012014 doi:10.1088/1742-6596/1336/1/012014

(ii) Solving the Poisson equation in a harmonic space (on CPU; the forward and inverse Fourier transforms are performed by means of the FFTW library).
(iii) Linear reconstruction of the primitive variables (on GPU).
(iv) Transferring reconstructed primitive variables in the overlapped subdomain boundaries between the MPI processes.
(v) Solving the Riemann problem for each interface of each cell by the Riemann solver (on GPU).
(vi) Calculating the values of the primitive variables (on GPU).

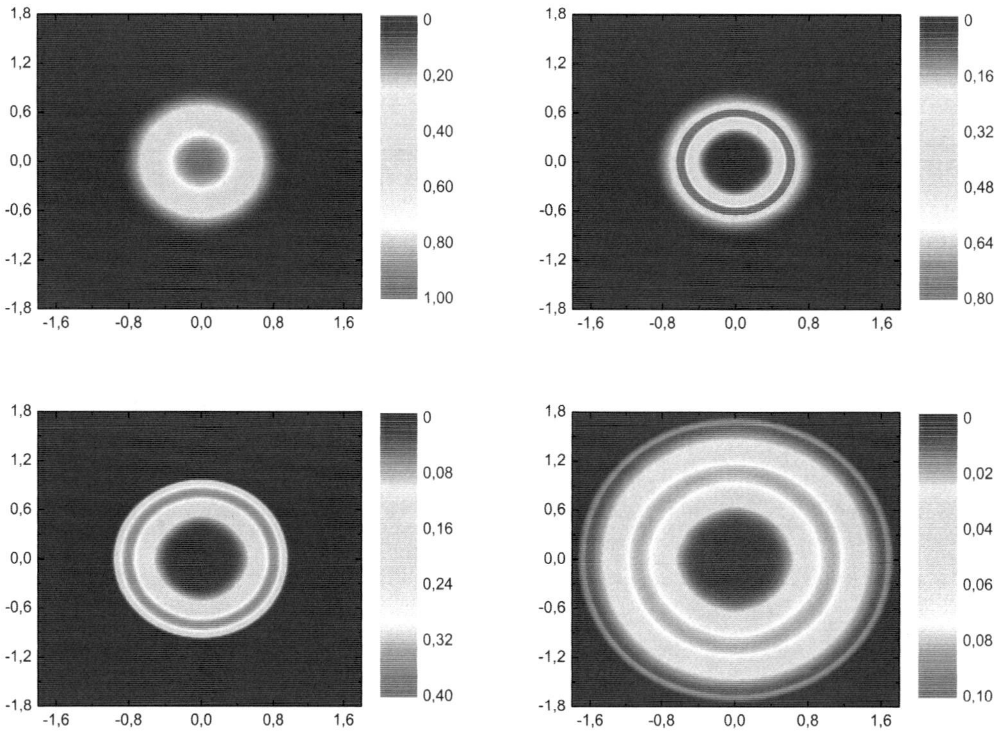

Figure 2. Density of the supernova in the equatorial cut of the star at the dimensionless instants of time 0 (top left), 0.1 (top right), 0.2 (bottom left), and 0.4 (bottom right).

4. Numerical experiments

As a test problem, we use the Sedov blast problem. For simulating the problem of point blast, we will consider the domain $[0;1]^3$, the isentropic exponent is $\gamma = 5/3$, the initial density in the area is $\rho_0 = 1$, and the initial pressure is $p_0 = 10^{-5}$. At the instant $t = 0$, the intrinsic energy $E_0 = 0.6$ is emitted. The blast area is constrained by the radius $r_{\text{central}} = 0.02$. For the numerical experiments, the computational grid size of 256^3 was used. The simulated density profiles at the various instants t are shown in Fig. 2.

The Sedov blast problem is the standard test for checking the capability of the method and its implementation to reconstruct powerful shockwaves with large Mach numbers. The speed of sound in the background is extremely low; thus, the Mach number reaches the value of $M \approx 10^3$.

The Authors' numerical method fairly well reconstructs the location of the shockwave, as well as, the profiles for density and momentum.

The experiments were performed on the Tesla M2090 accelerators incorporated into the NKS-30T cluster of the Siberian Supercomputer Center. For studying the strong scalability, we use the speedup $S_m = T_1/T_m$ and efficiency $E_m = S_m/m$ coefficients, where T_m is calculation time on m devices for the same problem. Table 1 shows the results of solving the problem with the 256^3 mesh on various number of GPUs.

Table 1. Results of strong scaling experiments.

Number of GPUs m	Time per iteration T_m, msec	Speedup S_m	Efficiency E_m
1	1696	—	—
2	924	1.84	0.92
4	581	2.92	0.73

For studying the weak scalability, each of the GPUs stores and processes a portion of the whole mesh. The size of the portion is equal to 256^3. Table 2 shows the efficiency of solving the problem with the fixed portion size per GPU.

Table 2. Results of weak scaling experiments.

Number of GPUs m	Time per iteration T_m, msec	Efficiency T_1/T_m
1	1696	—
2	1787	0.95
4	1862	0.91
8	2055	0.82
16	2420	0.70

5. Conclusion

We presented the novel hydrodynamic GPUPEGAS 2.0 code for the three-dimensional simulation of the astrophysical flows using graphics accelerators. Within one GPU we achieved the performance of 25 GFLOPS. By using single node with 4 GPUs we achieved the 3-fold speedup. By using 16 GPUs, we achieved the scalability of 70%. The numerical experiment was performed for the test Sedov point blast problem.

Acknowledgments

The work of Igor Kulikov and Igor Chernykh was supported by Russian Science Foundation (project no. 18-11-00044).

References

[1] Kulikov I 2014 *Astrophys. J. Suppl. Ser.* **214**(1) 12
[2] Kulikov I M, Chernykh I G, Glinskiy B M, Protasov V A 2018 *Lobachevskii J. Math.* **39**(4) 543–50
[3] Popov M, Ustyugov S 2007 *Comp. Math. & Math. Phys.* **47**(12) 1970–89
[4] Popov M, Ustyugov S 2008 *Comp. Math. & Math. Phys.* **48**(3) 477–99
[5] Kulikov I, Vorobyov E 2016 *J. Comp. Phys.* **317** 318–46
[6] Kulikov I M, Chernykh I G, Tutukov A V 2018 *Lobachevskii J. Math.* *39*(9) 1207–16

IOP Conf. Series: Journal of Physics: Conf. Series **1336** (2019) 012015 doi:10.1088/1742-6596/1336/1/012015

Gas-monodisperse dust mixtures in smoothed particle hydrodynamics: computing of stiff and non-linear drag

O P Stoyanovskaya[1,2], T A Glushko[1,2], F A Okladnikov[1,2], V N Snytnikov[1,2,3] and N V Snytnikov[1,4]

[1] Lavrentyev Institute of Hydrodynamics SB RAS
[2] Novosibirsk State University
[3] Boreskov Institute of Catalysis SB RAS
[4] Institute of Computational Mathematics and Mathematical Geophysics SB RAS

E-mail: o.p.sklyar@gmail.com

Abstract.
Simulation the dynamics of gas-solid mixtures is crucial in many applications: chemical reactor design, evolution of circumstellar discs, etc. Such mixtures where gas is a carrier phase and solid grains are dispersed phase can be simulated as interpenetrating continuum media. The characteristic parameter of this problem is relaxation time between gas and dust velocities. In many applications this parameter varies significantly during the simulation (from small to unity). Moreover, the drag force can be non-linearly dependent on solids to gas relative velocity. These factors place high requirements on the numerical methods for such problems. We propose a novel non-iterative algorithm for Smoothed Particle Hydrodynamics for computing gas-solid mixtures with exchange momentum between phases. Forces (except drag force) are approximated explicitly, while drag force is linearized and velocity relaxation time is approximated explicitly while relative velocity is approximated implicitly. The algorithm was tested on dynamical problems for dusty gas mixtures. It was shown that in the developed algorithm for stiff nonlinear drag we can use temporal and spatial resolution independent of drag parameters.

1. Introduction

Mathematical modeling is widely used for studies of gas-solid mixtures. These studies solve scientific problems and technological tasks, e.g. reactor design for natural gas processing.

Extensive literature [1, 2, 3, 4] is written about mathematical models of two-fluid gas-dispersed phase (gas-dust and gas-droplet) laminar and turbulent flows. In particular, gas-monodisperse dust models were studied in [1, 2, 4], and gas-polydisperse dust models were proposed in [3] (in the local volume, the dispersed fluid is represented by a finite set of fractions, in each solid particles have one characteristic size). A review of current problems and progress of fluid dynamics and thermophysics of two-fluid flows is presented in [4]. In many developed models dynamics of dispersed phase are computed with equations of gasodynamic type separately from gas dynamics. In our paper we also consider this approach of describing gas-solid mixture dynamics where gas is carrier phase and dust is dispersed phase.

Content from this work may be used under the terms of the Creative Commons Attribution 3.0 licence. Any further distribution of this work must maintain attribution to the author(s) and the title of the work, journal citation and DOI.
Published under licence by IOP Publishing Ltd

The inter-phase interaction in gas-solid mixture is described by relaxation terms: velocity (dynamic) and thermal. The presence of these terms leads to the stiff equations if the relaxation time is much less than dynamic time of the problem. We will say that in this case relaxation time is a small parameter.

Finding numerical solution of problems with stiff relaxation terms is an issue. Several authors have shown that in order to reach acceptable accuracy of numerical solution it is necessary to adjust spatial and time steps according to small parameter value. For example simulations in [20] show that in case of smoothed particle hydrodynamics (SPH) where gas and dispersed phase are modelled with different particle sets there are following steps restrictions

$$\tau < t_{\mathrm{stop}}, \tag{1}$$

$$h < c_{\mathrm{s}} t_{\mathrm{stop}}, \tag{2}$$

where t_{stop} - time of particle to gas velocity relaxation (small parameter), τ - time step, h - smoothing length, defining spatial resolution in SPH method, c_{s} - sound speed in the gas. Necessity to satisfy conditions (1)-(2) makes it impossible to run series of computational experiments for many applications.

In some cases, the search for a numerical solution of the original problem can be replaced with a solution of the simplified problem obtained from the original by expansion with respect to a small parameter (for example, [5]). Particularly it means that velocity of dispersed phase for gas-solid mixture dynamics is found not from initial differential equation but from algebraic relation

$$u = v + g_{\mathrm{rel}} t_{\mathrm{stop}}, \tag{3}$$

where u - particles velocity, v - gas velocity, g_{rel} - difference of accelerations that affect gas and dispersed phase. The simplified problem becomes non-stiff, but such a transition to the simplified problem is validated only for the case when the relaxation time is much less than the dynamic time of the carrier phase. Simplified problem does not allow simulating a transient regime with a relaxation time is close to the dynamic time.

At the same time practice of solving problems with stiff relaxation terms has shown that there are methods which do not require spatial and time steps dependant on a small parameter. Number of cases has succeeded in acquiring methods that reached appropriate solution accuracy with substantially bigger τ and h. Such methods for stiff or multi-scale problems are known as asymptotic-preserving schemes [9].

Main ideas for constructing finite-difference asymptotic-preserving schemes are provided in [9, 10, 11]. Finite-difference schemes for gas-monodisperse particles mixture dynamics with intense inter-phase interaction are reported and studied in [6, 7], gas-polydisperse particles - in [8]In Euler finite-difference methods [6, 7, 8] the ability of scheme to keep asymptotic is achieved with the way of time approximation. In Lagrangian methods where gas and dust is simulated with different sets of particles the situation is more complex. To compute the relaxation term (drag force) we have to interpolate gas and dust quantities defined in different spatial points. So in two-fluid Lagrangian approaches keeping the asymptotic is defined not only by way of computing time differentials but also with space interpolation scheme [15]. For two-fluid SPH authors of [12, 13, 14] have suggested different ways of such interpolation. Their quantitative comparison when drag force linearly depends on the relative velocity between the gas and solids is given in [16].

It should be noted that for simulation of dusty gas dynamics with SPH authors of [17]-[20] have developed numerical schemes for an alternative mathematical model. Their model of the dynamics of a gas-dust medium is based on a one-fluid approach in contrast to the approach, where gas and disperse particles are modelled as interpenetrating continua and with their own system of equations for each continuum. One-fluid approach means equations are solved for the properties of the gas-dust medium as a whole, herewith the diffusion equation is solved for the concentration of the solid phase. In the one-fluid approach, each SPH particle is a carrier of the characteristics of both gas and dust. Such an approach allows computing intense inter-phase interaction using finite-difference scheme ideas [9]. Moreover it is free of numerical over-dissipation appearing in the implementation of several methods in the two-fluid approach. On the other side, in the one-fluid approach, the fulfillment of the mass conservation law for the dispersed phase is not guaranteed. In addition, there is a hard upper limit on the size of the simulated bodies as compared with the two-fluid approach.

This paper is devoted to the method for computing intense inter-phase interaction in a two-fluid SPH. The approach we have presented in [14] is developed here for the case when drag force depends non-linearly on relative velocity between gas and solids. In the section 2 the statement of the test problem of the dynamics of the medium of gas-monodisperse dust with intense and moderate interphase interaction is given. Section 3 describes methods of computing inter-phase interaction: Monaghan - Kocharyan (MK) [13] and Implicit Drag in Cell (IDIC) [14]. Section 4 presents the results of computing the test problem using MK and IDIC methods. It was shown that for the intense inter-phase interaction MK method does not preserve assymptotics when IDIC does. It also shows that IDIC method gives appropriate solution accuracy of the non-linear problem at spatial and time steps determined by solving the equations of pure gas dynamics. Conclusions are given in the section 5.

2. Problem statement

We will study the properties of numerical schemes on one-dimensional shock tube problem for two-fluid gas-dust mixture:

$$\frac{\partial \rho_{\mathrm{g}}}{\partial t} + \nabla(\rho_{\mathrm{g}} v) = 0, \quad \rho_{\mathrm{g}} \left[\frac{\partial v}{\partial t} + (v \cdot \nabla) v \right] = -\nabla p - f_{\mathrm{D}}, \tag{4}$$

$$\frac{\partial \rho_{\mathrm{d}}}{\partial t} + \nabla(\rho_{\mathrm{d}} u) = 0, \quad \rho_{\mathrm{d}} \left[\frac{\partial u}{\partial t} + (u \cdot \nabla) u \right] = f_{\mathrm{D}}, \tag{5}$$

$$\rho_{\mathrm{g}} \left(\frac{\partial e}{\partial t} + v \frac{\partial e}{\partial x} \right) = -p \frac{\partial v}{\partial x}, \tag{6}$$

where ρ_{g} and ρ_{d} are volume densities of gas and dust, v and u are velocities of gas and dust, p is the gas pressure, f_{D} is a drag force acting on a unit volume of the medium, e is an internal gas energy associated with pressure as follows:

$$p = \rho_{\mathrm{g}} e (\gamma - 1), \tag{7}$$

here γ is adiabatic exponent.

Drag force is defined as

$$f_{\mathrm{D}} = n_{\mathrm{d}} F_{\mathrm{D}}, \tag{8}$$

where n_{d} is a volume concentration of dust particles, F_{D} is a drag force acting on a separate particle

$$F_{\mathrm{D}} = \frac{1}{2} C_{\mathrm{D}} s \rho_{\mathrm{g}} \|v - u\| (v - u), \tag{9}$$

where s is a cross-section area of the particle (e.g. for a sphere of radius a $s = \pi a^2$), C_{D} is a dimensionless drag coefficient, which is a function of two dimensionless quantities - Mach number

$$\mathrm{Ma} = \frac{\|v - u\|}{c_s} \tag{10}$$

and Knudsen number

$$\mathrm{Kn} = \frac{\lambda}{a}, \tag{11}$$

here λ is mean free path of a gas molecule. Ma and Kn can be tied together in Reynolds number

$$\mathrm{Re} = 4\frac{\mathrm{Ma}}{\mathrm{Kn}}. \tag{12}$$

Following [21] we will use piece-wise approximation of drag coefficient

$$C_{\mathrm{D}} = \begin{cases} \frac{8}{3}\mathrm{Ma}^{-1}, \ \mathrm{Kn}^{-1} < \frac{9}{4} \text{ (Epstein regime; I)}, \\ 24\mathrm{Re}^{-1}, \ \mathrm{Re} < 1 \text{ (Stokes regime; II)}, \\ 24\mathrm{Re}^{-0.6}, \ 1 < \mathrm{Re} < 800 \text{ (transient regime or nonlinear Stockes regime; III)}, \\ 0.44, \ \mathrm{Re} > 800 \text{ (Newton regime; IV)}. \end{cases} \tag{13}$$

We will make system (4)-(6) dimensionless by the values of length l_*, mass m_* and time t_*, then the derived dimensional quantities have the form $\rho_* = \frac{m_*}{l_*^3}$, $v_* = \frac{l_*}{t_*}$, $p_* = \frac{m_*}{t_*^2 l_*}$, $e_* = \frac{l_*^2}{t_*^2}$. The system will be solved in dimensionless variables.

For the resulting system (4)–(5), (6)–(7) at the boundaries of the interval, we define the non-leakage conditions. At the initial moment of time, we set the initial zero dust and gas velocities, the gas pressure jump, the gas and dust density jump:

$$\left[\frac{\rho_{\mathrm{g}}}{\rho_*}, \frac{p}{p_*}, \frac{e}{e_*}, \frac{\rho_{\mathrm{d}}}{\rho_*}\right]_l = [1, 1, 2.5, 1],$$

$$\left[\frac{\rho_{\mathrm{g}}}{\rho_*}, \frac{p}{p_*}, \frac{e}{e_*}, \frac{\rho_{\mathrm{d}}}{\rho_*}\right]_r = [0.125, 0.1, 2, 0.125],$$

$$\gamma_l = \gamma_r = 7/5.$$

We define $\lambda = 5 \cdot 10^{-6} l_* \frac{\rho_*}{\rho_{\mathrm{g}}}$, $\rho_{\mathrm{s}} = 2.3\rho_*$ and we will use two particle sizes: small dust $a = 5 \cdot 10^{-6} l_*$ and big bodies $a = 5 \cdot 10^{-3} l_*$.

3. Methods description

3.1. Monaghan - Kocharyan method

The classical SPH approximation for the equations of motion (4)-(5) for the gas-dust medium is based on Monaghan-Kocharyan (MK) method [13]. We have implemented it in explicit way, where the term describing drag uses the values of velocities from the previous time step:

$$\frac{dv_a^n}{dt} = -\sum_b m_{\mathrm{g}} \left(\frac{p_b}{(\rho_{b,\mathrm{g}}^n)^2} + \frac{p_a}{(\rho_{a,\mathrm{g}}^n)^2} + \Pi_{ab} \right) \nabla_a W_{ab}^n - \sigma m_{\mathrm{d}} \sum_j \frac{K_{aj}}{\rho_{a,\mathrm{g}}^n \rho_{j,\mathrm{d}}^n} \frac{(v_a^n - u_j^n, r_{ja})}{r_{ja}^2 + \eta^2} r_{ja} W_{ja}^n,$$

$$\frac{du_j^n}{dt} = \sigma m_{\mathrm{g}} \sum_a \frac{K_{aj}}{\rho_{a,\mathrm{g}}^n \rho_{j,\mathrm{d}}^n} \frac{(v_a^n - u_j^n, r_{ja})}{r_{ja}^2 + \eta^2} r_{ja} W_{ja}^n, \tag{14}$$

$$K_{aj} = \frac{\rho_{j,\mathrm{d}}^n \rho_{a,\mathrm{g}}^n c_{a,\mathrm{s}}^n}{s_j^n \rho_{j,\mathrm{s}}^n}, \tag{15}$$

where $r_{ja} = r_j - r_a$, η is a clipping constant, $\eta^2 = 0.001h^2$ and σ is a factor, determined by the dimensionality of the problem; for 1D problems $\sigma = 1$. Here Π_{ab} is the tensor of artificial viscosity [22], $W_{ab}^n = W(h, r_{ab})$ is the smoothing kernel.

3.2. Implicit "drag in cell" method.
The second scheme is different from MK by terms describing drag.

For each time moment let us split the whole computational domain into nonoverlapping cells, so that the union of these sets coincides with the whole domain. Let a certain cell contains N gas particles of the same mass m_{g} and L dust particles with the same mass m_{d}, with $N > 0$, $L > 0$.

For convenience, we introduce the relaxation time of the particle velocity to the gas velocity:

$$t_{\mathrm{stop}} = \frac{m_{\mathrm{p}}\|v-u\|}{\|F_{\mathrm{D}}\|} = \frac{8}{3}\frac{a\rho_{\mathrm{s}}}{C_{\mathrm{D}}\rho_{\mathrm{g}}\|v-u\|}, \tag{16}$$

where m_{p} is a particle mass, ρ_{s} is a dust particle material density.

Then let us introduce volume-averaged values of t^*_{stop} and ρ^*_{d}, and set

$$\varepsilon^* = \frac{m_{\mathrm{d}}L}{m_{\mathrm{g}}N}, \quad \rho^*_{\mathrm{g}} = \frac{\rho^*_{\mathrm{d}}}{\varepsilon^*}. \tag{17}$$

We assume, that in the computing of the drag force, affecting the dust from the gas, the gas velocity is the same within the whole cell and equal to v_*, but the dust particles have different velocities (and vice versa). Besides that, we shall compute the drag coefficient and densities, using calculated values from the previous time step, while the relative velocity will be used from the next step. The scheme thus has the following form:

$$\frac{\mathrm{d}v^n_a}{\mathrm{d}t} = -\sum_b m_b\left(\frac{p_b}{(\rho^n_{b,\mathrm{g}})^2} + \frac{p_a}{(\rho^n_{a,\mathrm{g}})^2} + \Pi_{ab}\right)\nabla_a W^n_{ab} - \frac{\varepsilon^*}{t^{*,n}_{\mathrm{stop}}}(v^{n+1}_a - u^{n+1}_*), \tag{18}$$

$$\frac{\mathrm{d}u^n_j}{\mathrm{d}t} = \frac{(v^{n+1}_* - u^{n+1}_j)}{t^{*,n}_{\mathrm{stop}}}, \tag{19}$$

$$v_* = \frac{\sum_{i=1}^N v_i}{N}, \quad u_* = \frac{\sum_{j=1}^L u_j}{L}. \tag{20}$$

If the time derivative in (18)-(19) is approximated by the first order finite difference, then a fast way for computing of u^{n+1}, v^{n+1} can be proposed.

We compute averaged $\rho^*_{\mathrm{d}} = \dfrac{\sum_{j=1}^L \rho_{\mathrm{d,j}}}{L}$ and t^*_{stop} as follows

$$t^*_{\mathrm{stop}} = \frac{8}{3}\frac{a\rho_{\mathrm{s}}}{C^*_{\mathrm{D}}\rho^*_{\mathrm{g}}||v^* - u^*||}, \tag{21}$$

where for C^*_{D} we use formula (13) with volume-averaged $\mathrm{Re}^*, \mathrm{Kn}^*$ and Ma^* . This quantities compute using u^*, v^* and $c^*_s = \dfrac{\sum_{a=1}^N \sqrt{\gamma \dfrac{p_a}{\rho_{\mathrm{g,a}}}}}{N}$.

4. Results

Calculations were performed for the cases of small ($a = 5\cdot 10^{-6}$) and big ($a = 5\cdot 10^{-3}$) dust particles with a high content of dust in the gas ($\varepsilon = \dfrac{\rho_{\mathrm{d}}}{\rho_{\mathrm{d}}} = 1$). For the smoothing lengths $h = 0.02$ and $h = 0.01$, $N_{total} = 2\cdot 990$ particles was taken, for $h = 0.001$ $N_{total} = 2\cdot 9900$ particles. The calculation results are given at time $t = 0.2$. The black line in the figures 1-2 shows the reference solution. We used cubic spline kernel.

Figure 1 shows solution by MK scheme for small dust particles. The upper panels show the values for gas, the lower panels show dust. On the left panels are the densities, on the right - the velocities. In this case, the time step was $\tau = 0.0001$ (from stability condition (1)). It can be noticed that at $h = 0.02$ and $h = 0.01$ distinct diffuse spreading is present which disappears after reducing smoothing length to $h = 0.001$.

Figure 2 shows solution by IDIC scheme for small dust particles. The arrangement of panels in this case is the same as for the previous figure. Time step is defined from Courant condition:

$$\tau < \frac{h \cdot \mathrm{CFL}}{max(c_s, u, v)}, \tag{22}$$

with Courant parameter CFL $= 0.1$, cell size is $h_{cell} = \frac{1}{2}h$. In the current calculation, there is no amplitude attenuation, but at $h = 0.02$ solution nonmonotonicity that disappears at $h = 0.01$ is visible.

Figure 3 shows solution by IDIC scheme for the big dust particles case with $h = 0.01$ and $h = 0.001$. Time step is defined from Courant condition with CFL $= 0.1$, cell size is $h_{cell} = h$. The upper panels show the values for gas (from left to right: density, velocity, pressure), the lower panels show density and velocity of dust, Reynolds numbers (also from left to right). The bottom right panel (Re numbers) shows that in this case nonlinear drag regimes are present. Indeed, f_{D} is nonlinearly dependant on relative velocity as seen from formula (13) at $\mathrm{Re} > 1$.

Thus, unlike MK, the IDIC method preserves the asymptotic behavior of the solution and is suitable for finding a solution for both small dust and large solids.

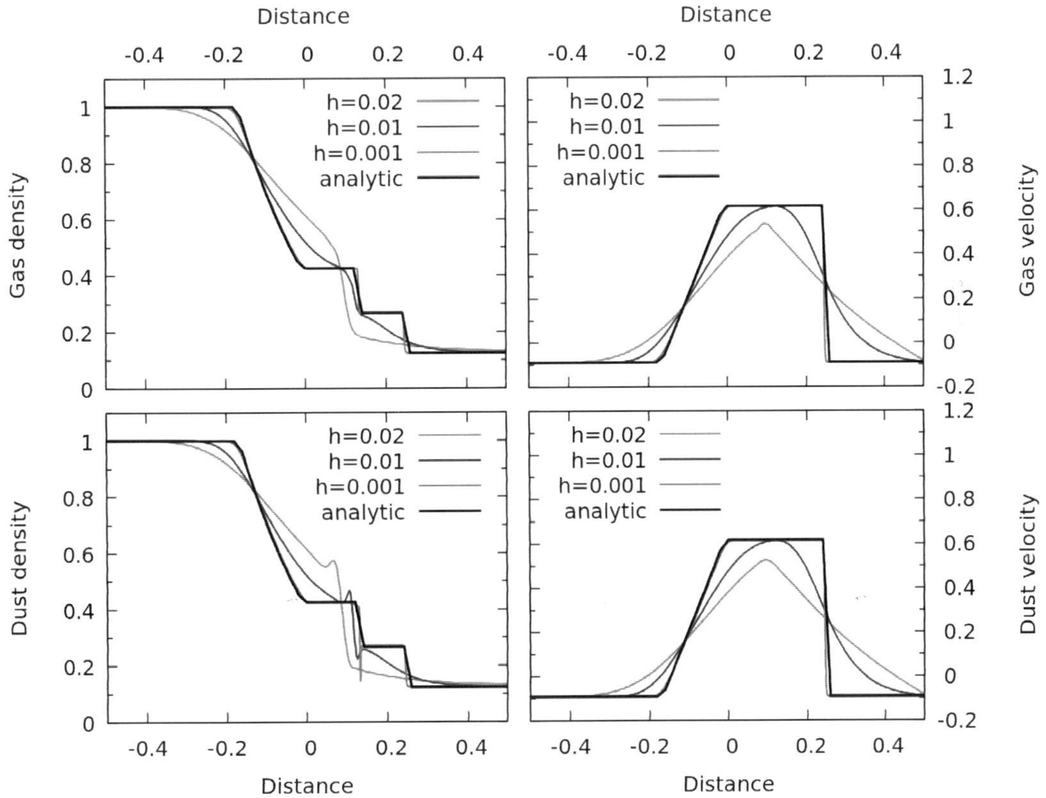

Figure 1. Solution of test problem with MK method with different numerical resolution. Small grains, linear drag.

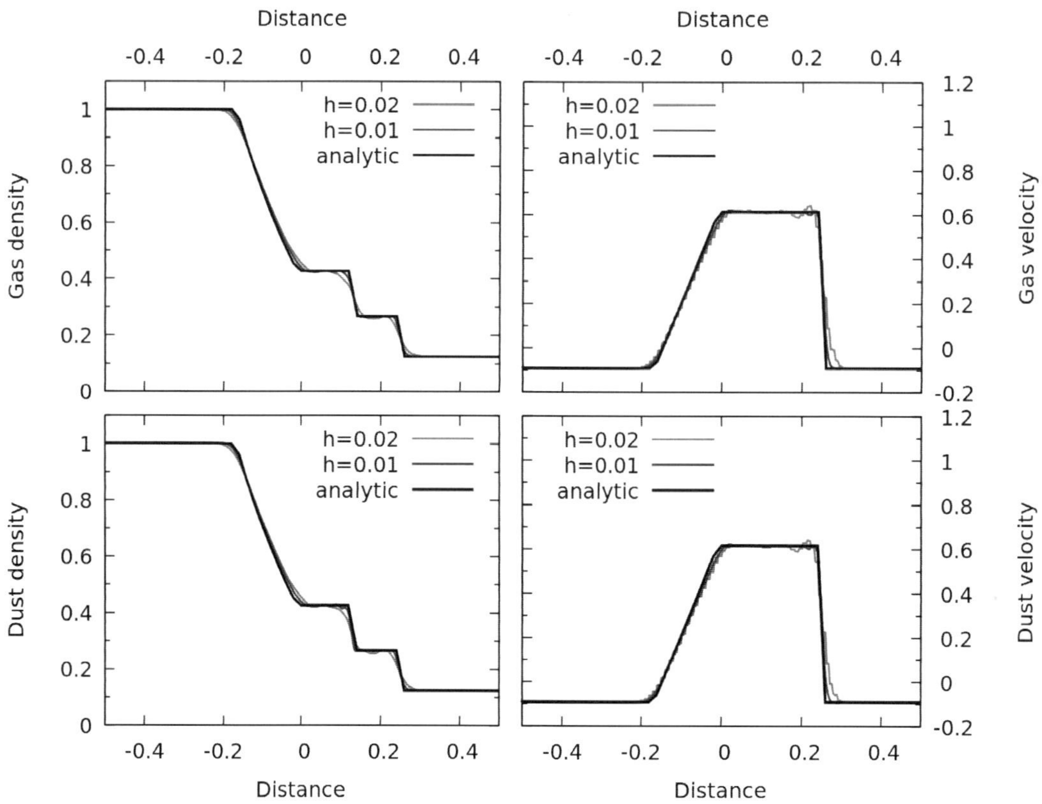

Figure 2. Solution of test problem with assymptotic preserving IDIC method with different numerical resolution. Small grains, linear drag.

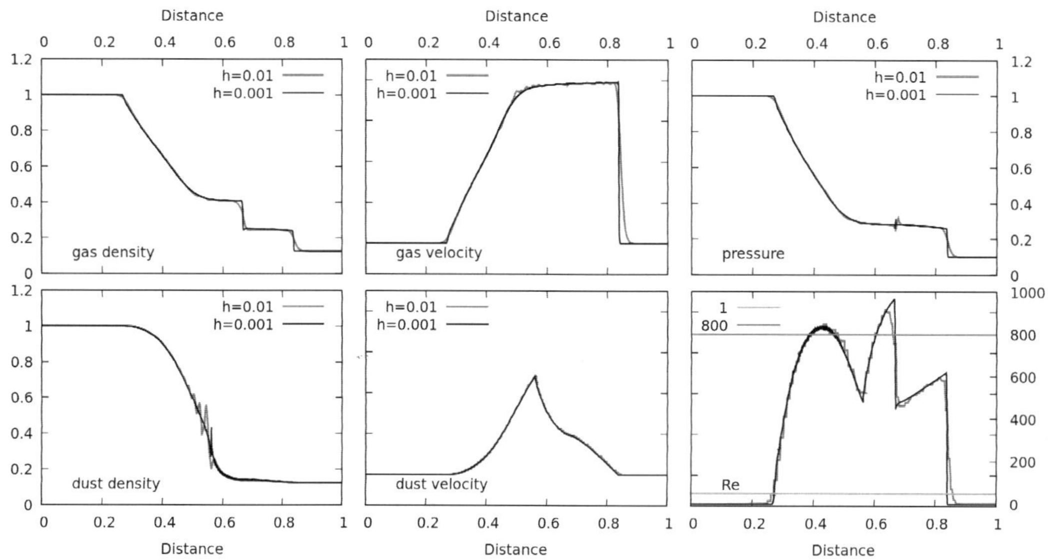

Figure 3. Solution of test problem with IDIC method with different numerical resolution. Large solids, nonlinear drag.

5. Conclusions

To simulate the dynamics of dusty gas mixtures with two-fluid SPH we propose fast and implicit method IDIC. We found that IDIC method for computing the inter-phase interaction is asymptotic preserving. This means that when computing the intense interaction, the spatial and temporal resolution can be chosen independently of the drag parameters. This property of IDIC method distinguishes it from previously proposed methods for computing inter-phase interaction in the two-fluid SPH. Moreover, we found that IDIC method allows to obtain appropriate accuracy for the case when the drag force non-linearly depends on the relative velocity between the gas and the bodies with spatial and time steps determined by the equations of pure gas dynamics. This means that the proposed method can be recommended for problems with intense and moderate inter-phase interaction and those where the size of the solid phase changes by orders of magnitude.

Acknowledgements. This work was supported by the Russian Science Foundation grant 19-71-10026.

References

[1] I. M. Vasenin, V. A. Arkhipov, V.G. Butov and others. Gas dynamics of two-phase flows in the nozzles. Tomsk: TSU. 264 pp. (1986)
[2] M. E. Deutsch, G. A. Filippov Gas dynamics two-phase media. M .: Energoizdat. 472 pp. (1981)
[3] N. E. Sternin, B. N. Maslov, A. A. Schreiber, A. M. Podvysotsky. Two-phase mono-and polydisperse gas flows with particles. M .: Mashinostroenie. 172 pp. (1980)
[4] A. Yu. Varaksin. Fluid dynamics and thermal physics of two-phase flows: problems and achievements (Review). High Temperature, Vol. 51, Issue 3, 421–455 pp. (2013)
[5] G.-Q. Chen, C. D. Livermore and T.-P. Liu. Hyperbolic Conservation Laws with Stiff Relaxation Terms and Entropy. Communications on Pure and Applied Mathematics, Vol. XLVII. 787-830 pp. (1994)
[6] D.V. Sadin. A method for computing heterogeneous wave flows with intense phase interaction. Comp. Maths Math. Phys., Vol. 38, no. 6, 987–993 pp. (1998)
[7] D. V. Sadin. TVD scheme for stiff problems of wave dynamics of heterogeneous media of nonhyperbolic nonconservative type. Computational Mathematics and Mathematical Physics. 016 Vol. 56, no. 12, 2068–2078 pp. (2016)
[8] P. Benítez-Llambay, L. Krapp, M.E. Pessah. Asymptotically stable numerical method for multispecies momentum transfer: gas and multifluid-dust test suite and implementation in FARGO3D, The Astrophysical Journal Supplement Series, Volume 241, Issue 2, article id. 25, 22 pp. (2019)
[9] S. Jin. Asymptotic preserving schemes for multiscale kinetic and hyperbolic equations: a review. Riv. Mat. Univ. Parma, 3, 177–216 (2012)
[10] O.P.Stoyanovskaya, V.N.Snytnikov, E.I.Vorobyov. Analysis of Methods for Computing the Trajectories of Dust Particles in a Gas-Dust Circumstellar Disk. Astronomy Reports, Volume 61, Issue 12, pp.1044-1060 (2017)
[11] O.P.Stoyanovskaya, E.I.Vorobyov, V.N.Snytnikov. Analysis of Numerical Algorithms for Computing Rapid Momentum Transfers between the Gas and Dust in Simulations of Circumstellar Disks. Astronomy Reports, Volume 62, Issue 7, pp.455-468 (2018)
[12] P. Lorén-Aguilar, M. R. Bate. Two-fluid dust and gas mixtures in smoothed particle hydrodynamics: a semi-implicit approach. Monthly Notices of the Royal Astronomical Society. Vol.443 (2014)
[13] J. Monaghan, A. Kocharyan. SPH simulation of multi-phase flow. Computer Physics Communications, Volume 87, Issue 1-2, p. 225-235. (1995)
[14] O. P. Stoyanovskaya, T. A. Glushko, N. V. Snytnikov, V. N. Snytnikov. Two-fluid dusty gas in smoothed particle hydrodynamics: Fast and implicit algorithm for stiff linear drag. Astronomy and Computing, Volume 25, p. 25-37. (2018)
[15] O. P. Stoyanovskaya, V. V. Akimkin, E. I. Vorobyov, T. A. Glushko, Ya. N. Pavlyuchenkov, V. N. Snytnikov, N. V. Snytnikov. Development and application of fast methods for computing momentum transfer between gas and dust in supercomputer simulation of planet formation. Journal of Physics: Conference Series, Volume 1103, Issue 1, article id. 012008 (2018)
[16] O. P. Stoyanovskaya, T. A. Glushko, V. N. Snytnikov, N. V. Snytnikov. Monodisperse gas-solid mixtures with intense interphase interaction in two-fluid Smoothed Particle Hydrodynamics. Proceedings of International Conference on Particle-based Methods - Fundamentals and Applications PARTICLES 2019. arXiv 190801682S. (2019)

[17] M. Hutchison, D.J. Price, G. Laibe. MULTIGRAIN: A smoothed particle hydrodynamics algorithm for multiple small dust grains and gas. Monthly Notices of the Royal Astronomical Society, 476, 2186 (2018)
[18] G. Laibe, D.J. Price. Dust and gas mixtures with multiple grain species — a one-fluid approach, MNRAS 444, 1940 (2014)
[19] G. Laibe, D.J. Price. Dusty gas with one fluid in smoothed particle hydrodynamics, MNRAS 444, 1940 (2014)
[20] G. Laibe, D.J. Price. Dusty gas in SPH: II - Implicit timestepping and astrophysical drag regimes. MNRAS, 420, 2365 (2012)
[21] S. J. Weidenschilling. Aerodynamics of solid bodies in the solar nebula. Monthly Notices of the Royal Astronomical Society, vol. 180, p. 57-70. July (1977)
[22] J.J. Monaghan, Smoothed particle hydrodynamics. Annual Review of Astronomy and Astrophysics. 30, 543–574. (1992)

IOP Conf. Series: Journal of Physics: Conf. Series **1336** (2019) 012016 doi:10.1088/1742-6596/1336/1/012016

Numerical implementation of the Lame equation with complex boundary conditions

S B Sorokin[1,4], A G Maksimova[1,3,4], G G Lazareva[1,2,4,5] and A S Arakcheev[3,5]
[1] Institute of Computational Mathematics and Mathematical Geophysics SB RAS 6, Ac. Lavrentieva Pr., Novosibirsk, 630090, Russian Federation
[2]Peoples' Friendship University of Russia (RUDN University) 6, Miklukho-Maklaya str., Moscow, 117198, Russian Federation
[3]Budker Institute of Nuclear Physics SB RAS 11, Ac. Lavrentieva Pr., Novosibirsk, 630090, Russian Federation
[4] Novosibirsk State University 1, Pirogova str., Novosibirsk, 630090, Russian Federation
[5]Novosibirsk State Technical University 20, Prospekt K. Marksa, Novosibirsk, 630073, Russian Federation

E-mail: sorokin@sscc.ru, maksimova@oapmg.sscc.ru, lazareva@ssd.sscc.ru, asarakcheev@gmail.com

Abstract. A discrete model is constructed for calculating the Lame equation with complex boundary conditions. The model is tested on an analytical solution. A complex boundary condition arises when a microcrack is specified on one of the boundaries. Calculation of microcracks will enable better assessment of the relevance of the simulation and finding out which mechanisms will occur in the case of plasma flow heating in modern plasma and future thermonuclear installations.

1. Introduction

One of the most critical problems in the international experimental thermonuclear reactor (ITER project) is design of first wall and divertor plates. Plates material should conduct heat well, spray a small amount of particles into plasma, accumulate little hydrogen and so on. These materials must not be destroyed mechanically, melt and spray out under action of the expected effects in a tokamak. Problems of durability of materials exposed to the high-power plasma streams relevant for fusion reactors on the basis of a different geometry of the magnetic field. Results on the heating of a tungsten plate by a pulsed electron beam were obtained on the experimental setup BETA (Beam of Electrons for materials Test Applications) at BINP SB RAS [1]. The full-scale experiment goes in parallel with computational ones [2]. The calculation of displacements around microcracks perpendicular to the surface is very relevant. The work is devoted to the numerical implementation of the Lame equation with complex boundary conditions. A complex boundary condition arises when a microcrack is specified on one of the boundaries. The aim of the study is to simulate the erosion of the sample surface as a result of evaporation and penetration of heat flux into the material taking into account microcracks [3].

Content from this work may be used under the terms of the Creative Commons Attribution 3.0 licence. Any further distribution of this work must maintain attribution to the author(s) and the title of the work, journal citation and DOI.
Published under licence by IOP Publishing Ltd

2. Problem definition

Consider a mathematical two-dimensional model of the linear static theory of elasticity [4]. The model includes the equation of equilibrium, Hooke's law and relations "displacement-deformation":

$$\begin{cases} \operatorname{div}\bar{\boldsymbol{\sigma}} = \mathbf{f}, \\ \bar{\sigma}_{ij} = 2\mu\bar{\varepsilon}_{ij} + \lambda\left(\bar{\varepsilon}_{11} + \bar{\varepsilon}_{22}\right)\delta_{ij}, \quad i,j = 1,2, \\ \bar{\varepsilon}_{ij} = \dfrac{1}{2}\left(\dfrac{\partial u_i}{\partial x_j} + \dfrac{\partial u_j}{\partial x_i}\right), \quad i,j = 1,2. \end{cases}$$

where $\bar{\boldsymbol{\sigma}}$ – is stress tensor with components $\bar{\sigma}_{ij}$; $\mathbf{f}$ – is mass force vector; μ, λ – are Lame constants; $\bar{\varepsilon}_{ij}$ – are components of the strain tensor $\bar{\boldsymbol{\varepsilon}}$; δ_{ij} – is Kronecker symbol; u_i – are components of the displacement vector $\mathbf{u}$.

As well as in [5, 6] we represent mathematical model in conjugate-operator form:
equation of equilibrium:

$$R^*\boldsymbol{\sigma} \equiv \begin{bmatrix} \dfrac{\partial}{\partial x_1} & 0 & \dfrac{\partial}{\partial x_2} \\ 0 & \dfrac{\partial}{\partial x_2} & \dfrac{\partial}{\partial x_1} \end{bmatrix} \boldsymbol{\sigma} = \mathbf{f},$$

Hooke's law:

$$\boldsymbol{\sigma} = K\boldsymbol{\varepsilon}, \text{ where } K = \begin{bmatrix} \lambda+\mu & \lambda & 0 \\ \lambda & \lambda+\mu & 0 \\ 0 & 0 & \mu \end{bmatrix},$$

and relations "displacement-deformation":

$$\boldsymbol{\varepsilon} = R\mathbf{u} \equiv \begin{bmatrix} \dfrac{\partial}{\partial x_1} & 0 \\ 0 & \dfrac{\partial}{\partial x_2} \\ \dfrac{\partial}{\partial x_2} & \dfrac{\partial}{\partial x_1} \end{bmatrix} \mathbf{u}.$$

The last three equations introduce vectors

$$\boldsymbol{\sigma} = (\bar{\sigma}_{11}, \bar{\sigma}_{22}, \bar{\sigma}_{12})^T, \quad \boldsymbol{\varepsilon} = (\bar{\varepsilon}_{11}, \bar{\varepsilon}_{22}, 2\bar{\varepsilon}_{12})^T,$$

composed of components of tensors $\bar{\boldsymbol{\sigma}}$ and $\bar{\boldsymbol{\varepsilon}}$.

To solve the problem, we will use the statement "in displacements". We obtain the necessary equation for determining the displacements by substituting the expression for $\boldsymbol{\varepsilon}$ from the "displacement-deformation" relation in Hooke's law, then we substitute the result into the equation of equilibrium. As a result, we get:

$$R^*KR\mathbf{u} = \mathbf{f}.$$

The main difficulty in discretization this problem is the construction of an approximation of the boundary conditions specified in terms of stresses. The need to preserve the basic properties of the operator of a differential problem at a discrete level requires cumbersome and far from obvious arguments. The discrete analogue construction scheme described in the next section is quite simple to implement and obviously leads to self-conjugate positive definite approximations of the static problem of the theory of elasticity in the formulation of "displacement". In this case, the boundary conditions specified in terms of stresses are approximated automatically.

IOP Conf. Series: Journal of Physics: Conf. Series **1336** (2019) 012016 doi:10.1088/1742-6596/1336/1/012016

3. The difference scheme

The mathematical model of the problem of the theory of elasticity belongs to the class of models having a conjugate-operator structure [5, 6]. To maintain this structure at the difference level, the following procedure is used to construct a discrete analogue [6, 7]:

- The operator R is selected as the support.
- An approximation of R_h is constructed.
- The approximation of R^* conjugate to the selected operator is constructed as the operator conjugate to R_h.

Consider a rectangular domain $\Omega = [a,b]\times[c,d]$. We introduce two rectangular grids (Fig. 1):

$$\bar{\omega} = \mathbf{x}_{i,j} = (x_i, y_j): \qquad \begin{array}{ll} x_i = a + ih_x, & i = 0,\ldots N_x + 1, \\ y_j = c + jh_y, & j = 0,\ldots N_y + 1, \end{array}$$

$$\omega_{\frac{1}{2}} = \mathbf{x}_{i+\frac{1}{2},j+\frac{1}{2}} = \left(x_{i+\frac{1}{2}}, y_{j+\frac{1}{2}} \right): \qquad \begin{array}{ll} x_{i+\frac{1}{2}} = a + \left(i + \frac{1}{2} \right) h_x, & i = 0,\ldots N_x, \\ y_{j+\frac{1}{2}} = c + \left(j + \frac{1}{2} \right) h_y, & j = 0,\ldots N_y, \end{array}$$

where $h_x = \dfrac{b-a}{N_x + 1}$, $h_y = \dfrac{d-c}{N_y + 1}$ – are space steps. The values of the displacement vector $\mathbf{u}$ are determined at the nodes of the first grid. In the nodes of the second grid, the values of the strain tensor $\boldsymbol{\varepsilon}$ and stress tensor $\boldsymbol{\sigma}$ are determined.

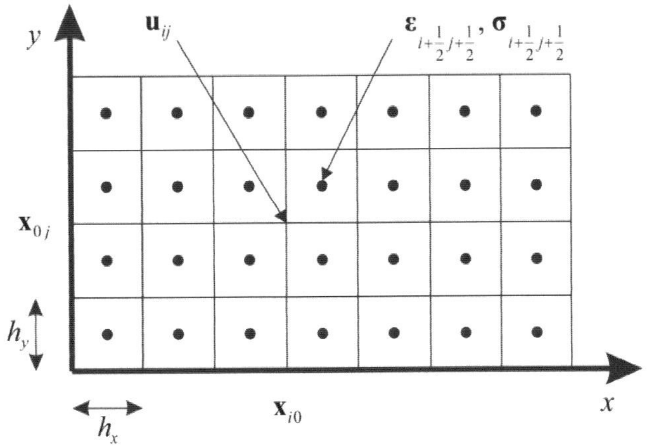

Figure 1. The grid scheme.

We introduce:

H_h – the Hilbert space of grid vector-functions $\mathbf{u}^h = \left(u_1^h, \ u_2^h \right)^T$ defined on the grid $\bar{\omega}$: $\mathbf{u}^h(x_{i,j}) = \left(u_1^h(\mathbf{x}_{i,j}), \ u_2^h(\mathbf{x}_{i,j}) \right)^T \equiv \mathbf{u}_{i,j}^h = \left(u_{1,i,j}^h, \ u_{2,i,j}^h \right)^T$, with the scalar product:

$$(\mathbf{u}^h,\mathbf{v}^h)_{H_h} = \sum_{x_{i,j}\in\bar{\omega}} [u_1^h(\mathbf{x}_{i,j})v_1^h(\mathbf{x}_{i,j}) + u_2^h(\mathbf{x}_{i,j})v_2^h(\mathbf{x}_{i,j})]\hat{h}_{1,i}\hat{h}_{2,j},$$

$$\forall\, \mathbf{u}^h \in H_h,\ \mathbf{v}^h \in H_h.$$

$$\hat{h}_{x,m} = \begin{cases} 0{,}5\,h_x, & m=0,\\ h_x, & 1\le m\le N_x,\\ 0{,}5\,h_x, & m=N_x+1,\end{cases} \qquad \hat{h}_{y,m} = \begin{cases} 0{,}5\,h_y, & m=0,\\ h_y, & 1\le m\le N_y,\\ 0{,}5\,h_y, & m=N_y+1.\end{cases}$$

H_h^* – is the Hilbert space of grid vector-functions $\boldsymbol{\sigma}^h = (\sigma_{11}^h,\sigma_{22}^h,\sigma_{12}^h)^T$ defined on the grid $\omega_{\frac{1}{2}}$:

$$\boldsymbol{\sigma}^h(\mathbf{x}_{i+\frac12,j+\frac12}) = (\sigma_{11}^h(\mathbf{x}_{i+\frac12,j+\frac12}),\sigma_{22}^h(\mathbf{x}_{i+\frac12,j+\frac12}),\sigma_{12}^h(\mathbf{x}_{i+\frac12,j+\frac12}))^T \equiv \boldsymbol{\sigma}^h_{i+\frac12,j+\frac12} = (\sigma^h_{11,i+\frac12,j+\frac12},\sigma^h_{22,i+\frac12,j+\frac12},\sigma^h_{12,i+\frac12,j+\frac12})^T$$

with the scalar product:

$$(\mathbf{w}^h,\boldsymbol{\sigma}^h)_{H_h^*} = \sum_{x_{i-\frac12,j-\frac12}\in\omega_{\frac12}} [w_{11}^h(\mathbf{x}_{i-\frac12,j-\frac12})\sigma_{11}^h(\mathbf{x}_{i-\frac12,j-\frac12}) + w_{22}^h(\mathbf{x}_{i-\frac12,j-\frac12})\sigma_{22}^h(\mathbf{x}_{i-\frac12,j-\frac12}) + w_{12}^h(\mathbf{x}_{i-\frac12,j-\frac12})\sigma_{12}^h(\mathbf{x}_{i-\frac12,j-\frac12})]h_x h_y,$$

$$\forall\, \mathbf{w}^h \in H_h^*,\quad \boldsymbol{\sigma}^h \in H_h^*.$$

To approximate R, we will use the following operator $R_h : H_h \to H_h^*$ [8]:

$$(R_h\mathbf{u}^h)_{i+\frac12,j+\frac12} = \frac12\begin{bmatrix} \dfrac{u^h_{1i+1,j}-u^h_{1i,j}}{h_x} + \dfrac{u^h_{1i+1,j+1}-u^h_{1i,j+1}}{h_x} \\ \dfrac{u^h_{2i,j+1}-u^h_{2i,j}}{h_{h_y}} + \dfrac{u^h_{2i+1,j+1}-u^h_{2i+1,j}}{h_y} \\ \dfrac{u^h_{1i,j+1}-u^h_{1i,j}}{h_y} + \dfrac{u^h_{1i+1,j+1}-u^h_{1i+1,j}}{h_y} + \dfrac{u^h_{2i+1,j}-u^h_{2i,j}}{h_x} + \dfrac{u^h_{2i+1,j+1}-u^h_{2i,j+1}}{h_x}\end{bmatrix}.$$

The operator R_h^* was built as conjugate to R_h, namely $R_h^* = (R_h)^* : H_h^* \to H_h$

$$(\mathbf{w}^h, R_h\mathbf{u}^h)_{H_h^*} = (R_h^*\mathbf{w}^h,\mathbf{u}^h)_{H_h}.$$

Using the summation formulas in parts, we obtain

$$R_h^*\mathbf{w}^h(\mathbf{x}_{i,j}) = -\frac12\begin{bmatrix} \dfrac{w_{11,i+\frac12,j-\frac12}-w_{11,i-\frac12,j-\frac12}}{h_x} + \dfrac{w_{11,i+\frac12,j+\frac12}-w_{11,i-\frac12,j+\frac12}}{h_x} + \dfrac{w_{12,i-\frac12,j+\frac12}-w_{12,i-\frac12,j-\frac12}}{h_y} + \dfrac{w_{12,i+\frac12,j+\frac12}-w_{12,i+\frac12,j-\frac12}}{h_y} \\ \dfrac{w_{12,i+\frac12,j-\frac12}-w_{12,i-\frac12,j-\frac12}}{h_x} + \dfrac{w_{12,i+\frac12,j+\frac12}-w_{12,i-\frac12,j+\frac12}}{h_x} + \dfrac{w_{22,i-\frac12,j+\frac12}-w_{22,i-\frac12,j-\frac12}}{h_y} + \dfrac{w_{22,i+\frac12,j+\frac12}-w_{22,i+\frac12,j-\frac12}}{h_y}\end{bmatrix}$$

The resulting operator R_h^* can only be used to calculate values at internal nodes of the grid ($i=1,\dots N_x$, $j=1,\dots N_y$). In this case, the scheme is applicable only if a displacement vector $\mathbf{u}$ is specified at the boundary, which imposes significant restrictions on the statement of the problem.

For problems where the stress tensor $\boldsymbol{\sigma}$ is specified at some boundaries, the operator needs to be expanded. The following is an example for calculating at the boundary of the domain for $i=0$:

$$R_h^*\mathbf{w}^h\left(\mathbf{x}_{0j}\right) \equiv \begin{bmatrix} -\dfrac{w_{11,1/2,j+1/2}+w_{11,1/2,j-1/2}}{h_x} - \dfrac{w_{12,1/2,j+1/2}-w_{12,1/2,j-1/2}}{h_y} \\ -\dfrac{w_{22,1/2,j+1/2}-w_{22,1/2,j-1/2}}{h_y} - \dfrac{w_{12,1/2,j+1/2}+w_{12,1/2,j-1/2}}{h_x}\end{bmatrix} = \begin{bmatrix} -\dfrac{2}{h_x}\sigma_{11,0j} - f_{x0j} \\ -\dfrac{2}{h_x}\sigma_{12,0j} - f_{y0j}\end{bmatrix}.$$

Similar operators are introduced for all necessary boundaries.

An approximation of the boundary conditions specified in terms of stresses in the lower left corner of the computational domain is written as

$$R_h^* \mathbf{w}^h(\mathbf{x}_{00}) \equiv \begin{bmatrix} -2\dfrac{w_{11,1/2,1/2}}{h_x} - 2\dfrac{w_{12,1/2,1/2}}{h_y} \\ -2\dfrac{w_{22,1/2,1/2}}{h_y} - 2\dfrac{w_{12,1/2,1/2}}{h_x} \end{bmatrix} = \begin{bmatrix} -\dfrac{2}{h_x}\sigma_{11,0,1/2} - \dfrac{2}{h_y}\sigma_{12,1/2,0} - f_{x00}^h \\ -\dfrac{2}{h_y}\sigma_{22,1/2,0} - \dfrac{2}{h_x}\sigma_{12,0,1/2} - f_{y00}^h \end{bmatrix}.$$

Finally, as an approximation of the equation of state, we choose

$$[\mathbf{w}^h]_{i+\frac{1}{2},j+\frac{1}{2}} = K[\mathbf{q}^h]_{i+\frac{1}{2},j+\frac{1}{2}}, \text{ where } K = \begin{bmatrix} \lambda+\mu & \lambda & 0 \\ \lambda & \lambda+\mu & 0 \\ 0 & 0 & \mu \end{bmatrix}.$$

As a result, a discrete model of the problem is constructed. It has the same structure as the original conjugate-operator:

$$R_h^* \boldsymbol{\sigma}^h = \mathbf{f}^h,$$
$$\boldsymbol{\sigma}^h = K\boldsymbol{\varepsilon}^h,$$
$$\boldsymbol{\varepsilon}^h = R_h \mathbf{u}^h.$$

From it, just as at the differential level, we obtain an approximation of the Lame equations:

$$R_h^* K R_h \mathbf{u}^h = \mathbf{f}^h.$$

In the case under consideration, it is not difficult to show that the constructed difference scheme approximates the original second-order problem on smooth solutions. Its stability can be established by the method described in [7].

We will solve the resulting problem by the method of minimal residuals (the index h is omitted)

$$\frac{\mathbf{u}^{n+1} - \mathbf{u}^n}{\tau_{n+1}} + R_h^* K R_h \mathbf{u}^n = \mathbf{f}^h,$$

where τ_{n+1} – is the iterative parameter, n – is the iteration number.

4. Simulation results

The constructed difference scheme was tested on a static problem with a solution

$$\boldsymbol{u}(x,y) = \begin{bmatrix} \sin(k_1 x)\sin(l_1 y) \\ \sin(k_2 x)\sin(l_2 y) \end{bmatrix},$$

where k_1, k_2, l_1, l_2 – are parameters. The computational domain is $\Omega = [-\pi,\pi] \times [0,2\pi]$. At the three boundaries of the domain, stress tensors were specified as boundary conditions. The fourth boundary (right side of the rectangle) was fixed by the condition $\mathbf{u} = \mathbf{0}$. The vector of mass forces was also calculated using the exact solution.

The calculation results are presented on Fig. 2, 3. The test was performed for $k_1 = 3$, $k_2 = 2$, $l_1 = 2$, $l_2 = 1$, $N_x = 100$, $N_y = 200$.

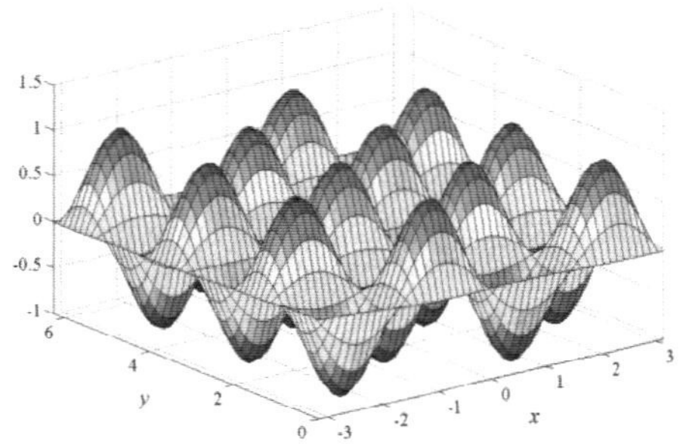

Figure 2. Component u_x of the displacement vector

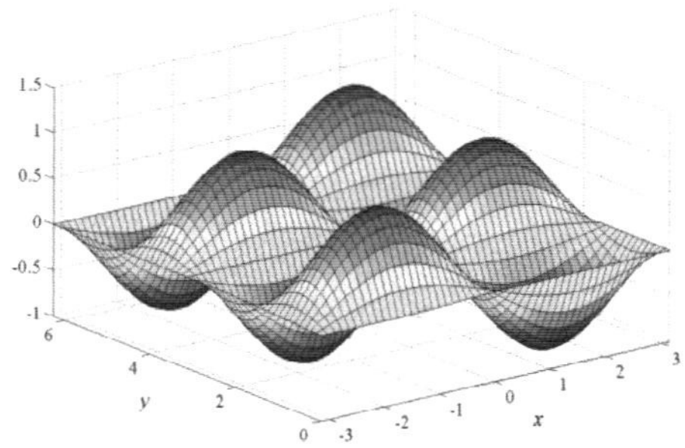

Figure 3. Component u_y of the displacement vector

The obtained test results approximate the exact solution well.

5. Conclusion

A discrete model is constructed for calculating the Lame equation with complex boundary conditions. The model is tested on an analytical solution. A complex boundary condition arises when a microcrack is specified on one of the boundaries. Calculation of microcracks will enable better assessment of the relevance of the simulation and finding out which mechanisms will occur in the case of plasma flow heating in modern plasma and future thermonuclear installations.

6. Acknowledgements

Problem statement have been obtained under support of the Russian Science Foundation (project No 19-19-00272). The reported study (computational experiment) was funded by the RFBR under the research project 18-31-00303. Lazareva G.G. was partly supported by the "RUDN University Program 5-100".

7. References

[1] Vyacheslavov L, Arakcheev A, Burdakov A, Kandaurov I, Kasatov A, Kurkuchekov V, Mekler K, Popov V, Shoshin A, Skovorodin D, Trunev Y, Vasilyev A 2016 Novel electron beam based test facility for observation of dynamics of tungsten erosion under intense ELM-like heat loads *American Institute of Physics* CP1771, 060004.

[2] Arakcheev A S, Apushkinskaya D E, Kandaurov I V, Kasatov A A, Kurkuchekov V V, Lazareva G G, Maksimova A G, Popov V A, Snytnikov A V, Trunev Yu A, Vasilyev A A, Vyacheslavov L N 2018 Two-dimensional numerical simulation of tungsten melting under pulsed electron beam *Fusion Engineering and Design* **132**, pp. 13-17 [3] Nakamura S, Senoh M, Nagahama S, Iwase N, Yamada T, Matsushita T, Kiyoku H and Sugimoto Y 1996 *Japan. J. Appl. Phys.* **35** L74

[3] Arakcheev A S, Arakcheev S A, Kandaurov I V, Kasatov A A, Kurkuchekov V V, Lazareva G G, Maksimova A G, Mashukov V I, Popov V A, Trunev Yu A, Vasilyev A A, Vyacheslavov L N 2019 On the mechanism of surface-parallel cracks formation under pulsed heat loads *Nuclear Materials and Energy* **20** 100677

[4] Landau L D, Lifshitz E M 1986 *Course of Theoretical Physics VII, Theory of elasticity* 196 p.

[5] Konovalov A N, Sorokin S B 1986 *Structure of equation of elasticity theory* 26 p. [in Russian]

[6] Konovalov A N 1995 Numerical methods in static problems of the theory of elasticity *Siberian Mathematical Journal* **XXXVI**, *No. 3, pp. 573–589.* [in Russian]

IOP Conf. Series: Journal of Physics: Conf. Series **1336** (2019) 012016 doi:10.1088/1742-6596/1336/1/012016

[7] Sorokin S B 2015 Justification of a Discrete Analog of the Conjugate-Operator Model of the Heat Conduction Problem *Journal of Applied and Industrial Mathematics*, **9**, No. 1, pp. 119–131
[8] Samarsky AA, Andreev VB 1976 *Difference methods for elliptic equations* 354 p. [in Russian]

IOP Conf. Series: Journal of Physics: Conf. Series **1336** (2019) 012017 doi:10.1088/1742-6596/1336/1/012017

Parallel Realization of the Hybrid Model Code for Numerical Simulation of Plasma Dynamics

M A Boronina, I G Chernykh, E A Genrikh, V A Vshivkov

Institute of Computational Mathematics and Mathematical Geophysics SB RAS, Lavrentiev avenue 6, Novosibirsk, 630090, Russia

E-mail: boronina@ssd.sscc.ru

Abstract. The work is devoted to a parallel realization of a hybrid model for study of plasma dynamics in axially symmetric open magnetic traps. The model is based on MHD approximation for electron component of the plasma and on the kinetic approach for ion component. In the model the particle-in-cells method (PIC) with explicit numerical schemes on staggered grids for the cylindrical coordinate system is applied. The permanent injection of the particles, the long characteristic times, the necessary grid resolution and the conditional stability of the method required development of parallel version of the algorithm. In the parallelization we use the mixed decomposition with static load balancing. We present the parallel algorithm as well as results of the computational experiments on plasma dynamics in regime of diamagnetic confinement in an open magnetic trap.

1. Introduction

The present-day leaders in the field of devices for plasma confinement with purpose of controlled thermonuclear fusion are tokamaks. Experiments on JET tokamak yielded fusion power 16MW and fusion power gain Q around 0.65 [1]. The ratio of the plasma pressure to the magnetic field pressure, β, in tokamaks does not exceed 5-10%. This fact together with the complicated tokamak design, big sizes, high experimental costs brings into consideration open plasma traps. A new concepts of plasma confinement are proposed and needed to be verified. For example, formation of "diamagnetic bubble" allows creating a compact linear open traps for the fusion purposes [2].

Mathematical modelling is the key instrument for the physical effects analysis. The appropriate model for the plasma simulation is based on kinetic Vlasov equation for electrons and ions. However, the differences in masses of electrons and ions make this model inapplicable due to the enormously high computer requirements. Our 2D hybrid model is based on MHD approximation for electron component of the plasma and on the kinetic approach for the ion component [3]. Despite of its economy compared with fully kinetic models the parallel version of the numerical algorithm is natural necessity for the reasonable computational times. The aim of the work is the parallel realization of the hybrid model code for the simulations of the plasma flows interaction as well as testing it on example of beam injection into open magnetic trap. In the paper the algorithm description and the finite-difference schemes are written out and the results of the numerical tests are presented.

Content from this work may be used under the terms of the Creative Commons Attribution 3.0 licence. Any further distribution of this work must maintain attribution to the author(s) and the title of the work, journal citation and DOI.
Published under licence by IOP Publishing Ltd

IOP Conf. Series: Journal of Physics: Conf. Series **1336** (2019) 012017 doi:10.1088/1742-6596/1336/1/012017

2. Problem statement

We consider a cylindrical open trap of radius R and length L with background hydrogen plasma with density n_0 at the initial time moment. Two coaxial round coils on the ends of the trap produce the magnetic field of the mirrors, the magnetic field at the center on the axis $(0, L/2)$ is H_0. In the same point a beam particle source is located. The question is to study the self-consistent beam-plasma interaction.

As we consider an axially symmetric case, we use cylindrical coordinate system in the two dimensional domain and eliminate derivatives $\partial/\partial\varphi$ from the equations.

For the definition of the dimensionless variables we use the characteristic values H_0 for the magnetic field and n_0 for the density. The characteristic time is $t_0 = 1/\omega_{ci}$, where $\omega_{ci} = eH_0/cm_i$ is the ion gyrofrequency. The characteristic length is $L_0 = c/\omega_{pi}$, $\omega_{pi} = \sqrt{4\pi n_0 e^2/m_i}$ is the ion plasma frequency. The characteristic velocity is Alfven velocity $V_A = H_0/\sqrt{4\pi m_i n_0}$.

The ion dynamics can be described with the Vlasov equation, we solve it with the method of characteristics. The equations for the ion trajectories coincide with the equations of the characteristics of kinetic Vlasov equation:

$$\frac{d\vec{r}_i}{dt} = \vec{v}_i, \tag{1}$$

$$\frac{d\vec{v}_i}{dt} = \vec{F}_i, \tag{2}$$

where $\vec{r}_i, \vec{v}_i$ are the ion coordinates and the velocities,

$$\vec{F}_i = \vec{E} + [\vec{v}_i, \vec{H}] - \kappa\vec{j}/n, \tag{3}$$

The force $\vec{F}_i$ includes Lorentz force and the ion-electron friction force with $\kappa = cm_e/eH_0\tau_{ei}$, where τ_{ei} is the characteristic ion-electron collision time [4]. Since we consider the quasi-neutral plasma $n = n_i = n_e$. The current is

$$\vec{j} = n_i\vec{V}_i - n_e\vec{V}_e, \tag{4}$$

where n_e is electron density, $\vec{V}_e$ is electron velocity. The ion velocity $\vec{V}_i$ and the ion density n_i are defined by the ion distribution function $f_i(t, \vec{r}, \vec{v})$:

$$n_i(\vec{r}) = \int f_i(t, \vec{r}, \vec{v})d\vec{v}, \tag{5}$$

$$\vec{V}_i(\vec{r}) = \frac{1}{n_i(\vec{r})}\int f_i(t, \vec{r}, \vec{v})\vec{v}d\vec{v}. \tag{6}$$

The massless electron component is described by the transport equation:

$$\vec{E} + [\vec{V}_e, \vec{H}] - \frac{\nabla p_e}{2n} - \kappa\frac{\vec{j}}{n} = 0. \tag{7}$$

The electron temperature T_e is defined by the heating due to the ion-electron friction, the heat flux, the changes due to the compression/expansion and obeys the equation:

$$n\left(\frac{\partial T_e}{\partial t} + (\vec{V}_e\nabla)T_e\right) = (\gamma - 1)\left(2\kappa\frac{\vec{j}^2}{n} + \kappa_1 div\nabla T_e - p_e div\vec{V}_e\right) \tag{8}$$

with $\kappa_1 = 4\pi e\lambda/H_0 c$, where $p_e = n_e T_e$ is the pressure of the electrons, λ is the thermal conductivity coefficient. The adiabatic index $\gamma = 5/3$. We consider $\kappa = const$, $\kappa_1 = const$.

The electric field $\vec{E}$ and magnetic field $\vec{H}$ follow Maxwell's equations for the case of negligible displacement currents:

$$\frac{\partial \vec{H}}{\partial t} = -rot\vec{E}, \tag{9}$$

$$rot\vec{H} = \vec{j}. \tag{10}$$

At the initial time moment the electric field $\vec{E} = 0$ and the magnetic field is defined from potential of the coils in a doubled domain with zero boundary conditions. On the present stage of the model realization we consider the perturbations of all values at the borders are negligibly small and we assume the particles reflect from the external boundaries.

3. Algorithm description

We use particle-in-cell method on staggered uniform grids to solve the equations (1–10) [5]. The values $H_r(i-1/2,k)$, $H_\varphi(i,k)$, $H_z(i,k-1/2)$, $E_r(i,k-1/2)$, $E_\varphi(i-1/2,k-1/2)$, $E_z(i-1/2,k)$, $V_{er}(i,k-1/2)$, $V_{e\varphi}(i-1/2,k-1/2)$, $V_{ez}(i-1/2,k)$, $V_{ir}(i,k-1/2)$, $V_{i\varphi}(i-1/2,k-1/2)$, $V_{iz}(i-1/2,k)$, $j_r(i,k-1/2)$, $j_\varphi(i-1/2,k-1/2)$, $j_z(i-1/2,k)$, $p_e(i-1/2,k-1/2)$, $T_e(i-1/2,k-1/2$, $n(i-1/2,k-1/2$ are defined in the spatial mesh nodes. The values of particle data $r, z, v_r, v_\varphi, v_z$ are defined on the Lagrangian mesh.

Since in equations (3) (4), (7), (8) the currents appear with a factor as $\vec{j}/n$, we substitute $\vec{U} = \vec{j}/n$ and use the arrays for $\vec{U}$ during the computations. The only equation requiring values $\vec{j}$ is (10) and we use the multiplication $n\vec{U}$ there. The result of the changes is decrease of the number of the operations performed by the processors and increase of the algorithm performance. The definitions $\vec{\mathcal{E}}^* = -[\vec{V}_e, \vec{H}]$, $\vec{\mathcal{E}}_\nabla = -\nabla p_e/2n$, $\vec{\mathcal{E}} = \vec{\mathcal{E}}^* + \vec{\mathcal{E}}_\nabla$ yield $\vec{E} = \vec{\mathcal{E}} + \kappa\vec{U} = \vec{\mathcal{E}}^* + \vec{\mathcal{E}}_\nabla + \kappa\vec{U}$. Here $\mathcal{E}_{\nabla\varphi} = 0$ and we use only two arrays for $\vec{\mathcal{E}}_\nabla$. For equations (3) and (7) rewritten in terms of $\vec{\mathcal{E}}$ we obtain the reduction of the term $k\vec{U}$ and a computation speed-up due to the use of arrays only for $\vec{\mathcal{E}}$. We use the idea and calculate and store the array data of $\vec{\mathcal{E}}^*$ and $\vec{\mathcal{E}}_\nabla$ separately for the calculation of the magnetic field H_φ using equation (9). The additional arrays allow using more compact scheme template and more accurate calculation of the pressure gradient contributions into the magnetic field.

The stationary structure of the diamagnetic bubble implies attainment of the bubble saturation with the particles and further steady plasma flux in the trap, defined by the particle injection and losses. The characteristic times of the system evolution are expected to be $10^4\omega_{ci}^{-1}$, so, for the time unit with $\sim 10^4$ steps it requires $\sim 10^8$ time steps.

Since the solution accuracy is determined by the grid resolution, the bubble boundary layer of the size $\lambda_b \sim 1$ cm has to be described with minimum 10 grid nodes. The bubble sizes 10×100 cm require grids with 100×1000 nodes. The particle-in-cell accuracy is also defined by the number of the model particles in each cell [6], and more accurate results require bigger number of the model beam particles and model background particles. The millions of the injected particles require sufficient processor memory. The grid refinement leads to proportional increase of the number of the background ions. Besides, smaller time step must be taken for finer grids according to the stability condition for the explicit schemes. All these factors necessitate using of a parallel version of the described algorithm for both faster computations and larger data processing.

The domain is divided in equal parts and each subdomain is assigned to one of np_0 groups of processor cores. The particles of the subdomain are distributed evenly among the corresponding group cores [7]. From np processor cores the ones with ranks $0..np_0 - 1$ are denominated as master cores, the exchanges of the auxiliary grid nodes are performed with the master cores. The rest of the cores are denominated as slaves, they receive the same grid data from their masters, process the particles and send the results back. The static core distribution for the particles and the domain decomposition are presented schematically in figure 1.

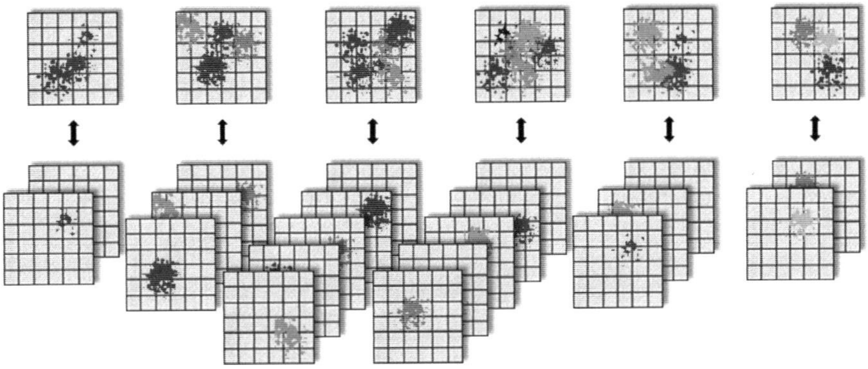

Figure 1. The cores distribution for the mixed decomposition.

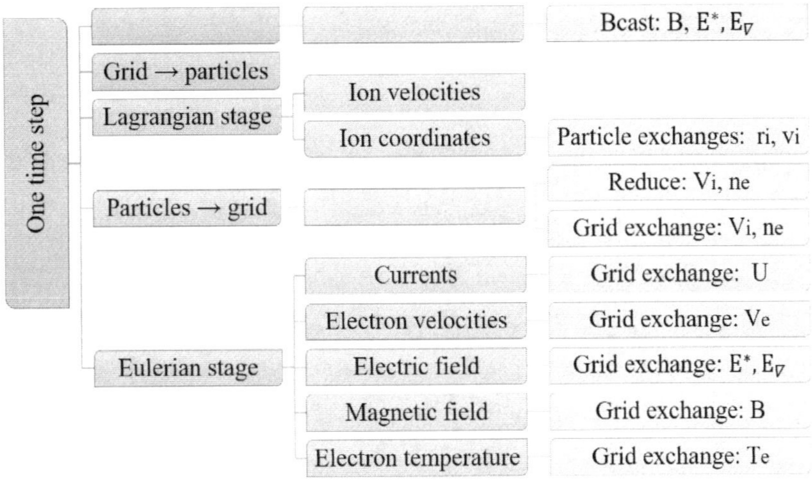

Figure 2. Algorithm block-scheme.

The block-scheme of the algorithm is presented in figure 2. Let us consider one step of the time cycle. It consists from the two stages and switching between them. On the Lagrangian stage the particle velocities and coordinates are calculated. The grid values are calculated on the Eulerian stage.

In the beginning of the cycle the master cores distribute the field values B, E^*, E_∇ to their slaves using function MPI_Bcast.

3.1. Ion motion

We use the scheme [8] for the values r_i, v_i in Cartesian coordinates, and then recalculate them in the cylindrical coordinates:

$$v_{ir}^* = v_{ir}^m + \tau\left(\mathcal{E}_r^{*m} + \mathcal{E}_{\nabla r}^m + v_{i\varphi}^m H_z^m - v_{iz}^m, H_\varphi^m\right),$$

$$v_{i\varphi}^* = v_{i\varphi}^m + \tau\left(\mathcal{E}_\varphi^m + v_{iz}^m H_r^m - v_{ir}^m H_z^m\right)$$

$$v_{iz}^{m+1} = v_{iz}^m + \tau\left(\mathcal{E}_z^{*m} + \mathcal{E}_{\nabla z}^m + v_{\alpha r}^m H_\varphi^m - v_{i\varphi}^m H_r^m\right), \tag{11}$$

$$z_i^{m+1} = z_i^m + \tau v_{iz}^{m+1}, \quad x = r_i^m + \tau v_{ir}^*, \quad y = \tau v_{i\varphi}^*, \quad r_i^{m+1} = \sqrt{x^2 + y^2}, \tag{12}$$

$$\cos\alpha = z/r_i^{m+1}, \quad \sin\alpha = y/r_i^{m+1}$$

$$v_{ir}^{m+1} = v_{ir}^* \cos\alpha + v_{i\varphi}^* \sin\alpha \qquad v_{i\varphi}^{m+1} = v_{i\varphi}^* \cos\alpha - v_{ir}^* \sin\alpha \tag{13}$$

Each processor core computes its own particle velocities and coordinates using bilinear interpolation for the electromagnetic fields. We form buffer arrays for the particles, which move away from their subdomain, to be sent to one of the cores of the neighbour group. The local rank of the receiving core in the neighbour group is defined as $mod(n+1, C_g)$, where C_g is number of cores in group g. This way of particles exchanges yields a uniform particle distribution among the cores in each group.

3.2. Average ion velocities, densities

The switching from the Lagrangian to Eulerian stage implies that each processor core computes the average velocity in cell and density. First, the square of the cell $[r_{a-1/2}, r_{a+1/2}] \times [z_{b-1/2}, z_{b+1/2}]$ is calculated:

$$S_{cell\ a} = \pi r_{a+1/2}^2 - \pi r_{a-1/2}^2 = 2\pi h_r r_a,$$

and then the volume of the cell:

$$V_{cell\ a} = S_{cell\ a}\, h_z = 2\pi h_r h_z r_a.$$

The density values on the spatial grid are

$$n_{i-1/2,k-1/2} = \frac{\sum_j \left(q_j R_{CIC}(\vec{r}_{i-1/2,k-1/2}, \vec{r}_j)\right)}{V_{cell\ i-1/2}}. \tag{14}$$

Here $j = 1..N$ is the number of model particle, q_j is the full charge of the ion j, R_{CIC} is cloud-in-cell shape-function of the particle (bilinear interpolation). Similarly, from the velocities of ions $\vec{v}_j$ the average ion velocity is computed at the nodes of the grid:

$$\vec{V}_{i\ a,b} = \frac{\left(\sum_j q_j \vec{v}_j R_{CIC}(\vec{r}_{a,b}, \vec{r}_j)\right)}{n_{a,b} V_{cell\ a}} \tag{15}$$

Since the density is calculated at points $i - 1/2, k - 1/2$, for the velocity components V_{ir} and V_{iz} in (15) the values $n_{i,k-1/2}, n_{i-1/2,k}$ are densities, calculated as arithmetic average from two corresponding density values.

The masters collect the processed arrays using MPI_Reduce from their group and then exchange with each other with the boundary grid values.

3.3. Currents

The further computations in the time cycle are performed only by the master cores on the spatial grids, so only the grid boundary exchanges are needed.

From Maxwell equation (10) values U_r, U_φ, U_z are determined:

$$U_{r\ i,k-1/2}^{m+1} = -\frac{2}{h_z} \frac{H_{\varphi\ i,k}^m - H_{\varphi\ i,k-1}^m}{(n_{i+1/2,k-1/2} + n_{i-1/2,k-1/2})},$$

$$U_{\varphi\ i-1/2,k-1/2}^{m+1} = \frac{1}{n_{i-1/2,k-1/2}} \left(\frac{H_{r\ i-1/2,k}^m - H_{r\ i-1/2,k-1}^m}{h_z} - \frac{H_{z\ i,k-1/2}^m - H_{z\ i-1,k-1/2}^m}{h_r} \right), \tag{16}$$

$$U_{z\ i-1/2,k}^{m+1} = \frac{2}{h_r} \frac{r_i H_{\varphi\ i,k}^m - r_{i-1} H_{\varphi\ i-1,k}^m}{r_{i-1/2}(n_{i-1/2,k+1/2} + n_{i-1/2,k-1/2})}.$$

IOP Conf. Series: Journal of Physics: Conf. Series **1336** (2019) 012017 doi:10.1088/1742-6596/1336/1/012017

3.4. Electron velocities

The electron velocities are defined from relation (4):

$$\vec{V}_e = \vec{U} - \vec{V}_i. \tag{17}$$

3.5. Electric field

Equation (7) is used to compute the electric field:

$$\mathcal{E}^{m+1}_{\varphi\ i-1/2,k-1/2} = \frac{\left(V^{m+1}_{er\ i,k-1/2} + V^{m+1}_{er\ i-1,k-1/2}\right)\left(H^m_{z\ i,k-1/2} + H^m_{z\ i-1,k-1/2}\right)}{4} -$$
$$- \frac{\left(V^{m+1}_{ez\ i-1/2,k} + V^{m+1}_{ez\ i-1/2,k-1}\right)\left(H^m_{r\ i-1/2,k} + H^m_{r\ i-1/2,k-1}\right)}{4}.$$

The values of $\mathcal{E}^*$ are computed according to the following schemes:

$$\mathcal{E}^*_{r\ i,k-1/2} = \bar{V}^{m+1}_{ez\ i,k-1/2}\frac{H^m_{\varphi\ i,k} + H^m_{\varphi\ i,k-1}}{2} - H^m_{z\ i,k-1/2}\frac{V^{m+1}_{e\varphi\ i+1/2,k-1/2} + V^{m+1}_{e\varphi\ i-1/2,k-1/2}}{2},$$
$$\mathcal{E}^*_{z\ i-1/2,k} = H^m_{r\ i-1/2,k}\frac{V^{m+1}_{e\varphi\ i-1/2,k+1/2} + V^{m+1}_{e\varphi\ i-1/2,k-1/2}}{2} - \bar{V}^{m+1}_{er\ i-1/2,k}\frac{H^m_{\varphi\ i,k} + H^m_{\varphi\ i-1,k}}{2}, \tag{18}$$

where the overline denotes the arithmetic average of the four values in the nearest nodes:

$$\bar{C}_{i,k} = \frac{C_{i+1/2,k+1/2} + C_{i-1/2,k+1/2} + C_{i+1/2,k-1/2} + C_{i-1/2,k-1/2}}{4}.$$

For the value of E_∇ the following schemes are used:

$$\mathcal{E}_{\nabla}{}^{m+1}_{r\ i,k-1/2} = -\frac{n_{i+1/2,k-1/2}T^m_{i+1/2,k-1/2} - n_{i-1/2,k-1/2}T^m_{i-1/2,k-1/2}}{h_r(n_{i+1/2,k-1/2} + n_{i-1/2,k-1/2})},$$
$$\mathcal{E}_{\nabla}{}^{m+1}_{z\ i-1/2,k} = -\frac{n_{i-1/2,k+1/2}T^m_{i-1/2,k+1/2} - n_{i-1/2,k-1/2}T^m_{i-1/2,k-1/2}}{h_z(n_{i-1/2,k+1/2} + n_{i-1/2,k-1/2})}.$$

3.6. Magnetic field

The formulae for the magnetic field are of the following form:

$$H^{m+1}_{r\ i-1/2,k} = H^m_{r\ i-1/2,k} + \frac{\tau}{h_z}\left(\mathcal{E}^{m+1}_{\varphi\ i-1/2,k+1/2} - \mathcal{E}^{m+1}_{\varphi\ i-1/2,k-1/2}\right) + \frac{\tau\kappa}{h_z}\left(U^m_{\varphi\ i-1/2,k+1/2} - U^m_{\varphi\ i-1/2,k-1/2}\right)$$

$$H^{m+1}_{\varphi\ i,k} = H^m_{\varphi\ i,k} + \tau G_{i,k} + \frac{\tau}{h_r}\left(\mathcal{E}^*_{z\ i+1/2,k} - \mathcal{E}^*_{z\ i-1/2,k}\right) + \frac{2\tau\kappa}{h_r}\left(U^m_{z\ i+1/2,k} - U^m_{z\ i-1/2,k}\right) -$$
$$-\frac{\tau}{h_z}\left(\mathcal{E}^*_{r\ i,k+1/2} - \mathcal{E}^*_{r\ i,k-1/2}\right) - \frac{2\tau\kappa}{h_z}\left(U^m_{r\ i,k+1/2} - U^m_{r\ i,k-1/2}\right) \tag{19}$$
$$H^{m+1}_{z\ i,k-1/2} = H^m_{z\ i,k-1/2} - \frac{\tau}{h_r r_i}\left(r_{i+1/2}\mathcal{E}^{m+1}_{\varphi\ i+1/2,k-1/2} - r_{i-1/2}\mathcal{E}^{m+1}_{\varphi\ i-1/2,k-1/2}\right) -$$
$$-\frac{\tau\kappa}{h_r r_i}\left(U^m_{\varphi\ i+1/2,k-1/2} - U^m_{i\varphi\ i-1/2,k-1/2}\right).$$

Here $G_{i,k}$ correspond to $rot\mathcal{E}_\nabla$ and computed from the relation:

$$G_{i,k} = \frac{1}{4h_r h_z \bar{n}_{ik}}\left[\left(T_{i+1/2,k+1/2} - T_{i-1/2,k-1/2}\right)\left(n_{i+1/2,k-1/2} - n_{i-1/2,k+1/2}\right) +\right.$$
$$\left. + \left(T_{i-1/2,k+1/2} - T_{i+1/2,k-1/2}\right)\left(n_{i+1/2,k+1/2} - n_{i-1/2,k-1/2}\right)\right].$$

3.7. Temperature

We use the up-wind scheme for the electron temperature computation:

$$\frac{T^{m+1}_{e\ i-1/2,k-1/2} - T^{m}_{e\ i-1/2,k-1/2}}{\tau} + \frac{V_{er\ i,k-1/2} + V_{er\ i-1,k-1/2}}{2} \Delta_r T_e|_{i-1/2,k-1/2} +$$

$$+ \frac{V_{ez\ i-1/2,k} + V_{ez\ i-1/2,k-1}}{2} \Delta_z T_e|_{i-1/2,k-1/2} = \tag{20}$$

$$= -(\gamma-1) T_{e\ i-1/2,k-1/2} \left[\frac{1}{r_{i-1/2}} \frac{r_i V_{er\ i,k-1/2} - r_{i-1} V_{er\ i-1,k-1/2}}{h_r} + \frac{V_{ez\ i-1/2,k} - V_{ez\ i-1/2,k-1}}{h_z} \right]$$

$$+2\kappa(\gamma-1) \left[\left(U_{r\ i,k-1/2} + U_{r\ i-1,k-1/2} \right)^2 + \left(U_{i\varphi\ i-1/2,k-1/2} \right)^2 + \left(U_{z\ i-1/2,k} + U_{z\ i-1/2,k-1} \right)^2 \right] +$$

$$+ \frac{\kappa_1(\gamma-1)}{n_{e\ i-1/2,k-1/2}} \left[\frac{r_i \left(T_{e\ i+1/2,k-1/2} - T_{e\ i-1/2,k-1/2} \right) - r_{i-1} \left(T_{e\ i-1/2,k-1/2} - T_{e\ i-3/2,k-1/2} \right)}{h_r^2 r_{i-1/2}} + \right.$$

$$\left. + \frac{T_{e\ i-1/2,k+1/2} - 2T_{e\ i-1/2,k-1/2} - T_{e\ i-1/2,k-3/2}}{h_z^2} \right].$$

$$\Delta_r T_e|_{i-1/2,k-1/2} = \begin{cases} \frac{T_{e\ i-1/2,k-1/2} - T_{e\ i-3/2,k-1/2}}{h_r}, & if\ \ V_{er\ i,k-1/2} + V_{er\ i-1,k-1/2} \le 0 \\ \frac{T_{e\ i+1/2,k-1/2} - T_{e\ i-1/2,k-1/2}}{h_r}, & if\ \ V_{er\ i,k-1/2} + V_{er\ i-1,k-1/2} > 0 \end{cases} \tag{21}$$

$$\Delta_z T_e|_{i-1/2,k-1/2} = \begin{cases} \frac{T_{e\ i-1/2,k-1/2} - T_{e\ i-1/2,k-3/2}}{h_r}, & if\ \ V_{ez\ i-1/2,k} + V_{ez\ i-1/2,k-1} \le 0 \\ \frac{T_{e\ i-1/2,k+1/2} - T_{e\ i-1/2,k-1/2}}{h_r}, & if\ \ V_{ez\ i-1/2,k} + V_{ez\ i-1/2,k-1} > 0 \end{cases} \tag{22}$$

For the computations (3.3-3.7) only exchanges with the auxillary grid nodes between the master cores are required after each value computation. The magnetic fields on the boundary $(1,k)$ are computed from the scheme and do not require the exchanges with the right neighbour.

4. Numerical experiments

In the numerical tests we consider the computational domain sizes R=6, L=30. The magnetic field on the axis $H(0, L/2) = 0.1$, the mirror ratio is 3.2. The beam density is 10^2, the injecting ion speed is 0.1, the ion temperature is 0.01. The coefficient of electrical conductivity is 0.2, the coefficient of thermal conductivity is k=0.1. These parameters correspond to the dimensional values $R = 137cm$, $L = 684cm$, $H(0, L/2) = 0.2kG$. The dimensional background ion density is $10^{12}cm^{-3}$, the beam density $10^{14}cm^{-3}$, the average beam speed $4.3 \cdot 10^7 cm/sec$, the ion temperature $10eV$. The normalizing parameters are: $H_0 = 2kG$, $n_0 = 10^{12}cm^{-3}$, $V_A = 4.3 \cdot 10^8 cm/sec$, $L_0 = 23cm$, $t_0 = 5 \cdot 10^{-8} sec$.

In figure 3 and figure 4 the absolute values of the magnetic field at the initial time moment and at t=40 are presented. The simulation results demonstrate formation of the magnetic cavern with high field gradients on its boundary.

IOP Conf. Series: Journal of Physics: Conf. Series **1336** (2019) 012017 doi:10.1088/1742-6596/1336/1/012017

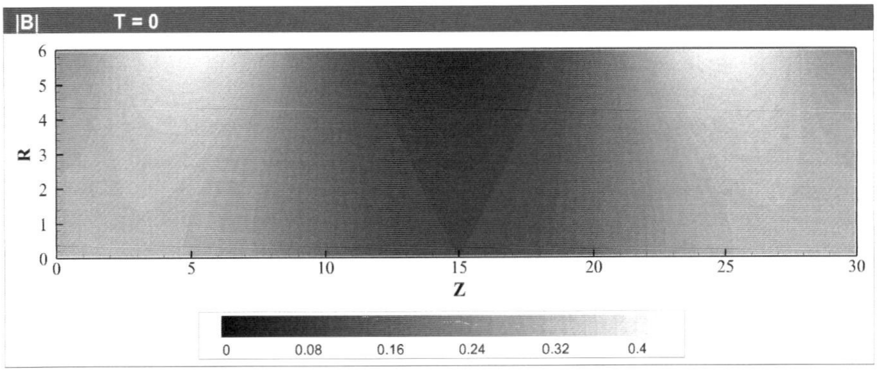

Figure 3. The magnitude of the magnetic field $|B|$ at $t = 0$.

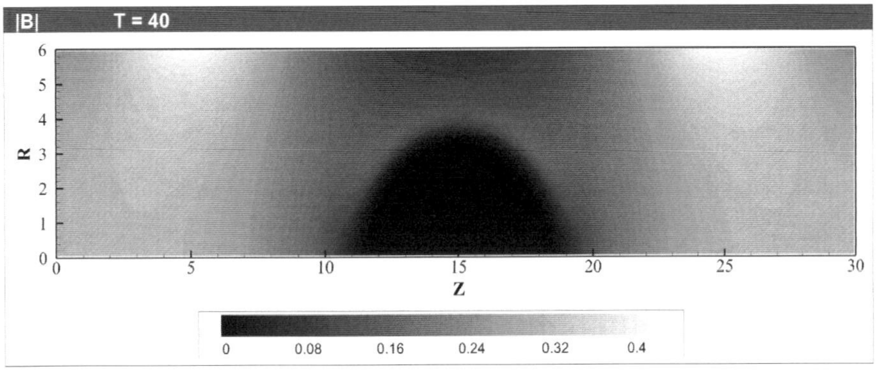

Figure 4. The magnitude of the magnetic field $|B|$ at $t = 40$.

Figure 5 and figure 6 demonstrate the particle velocities (V_z, V_r) for the time moments $t = 10$ and $t = 40$. The beam ions are marked with black colour, the background ions are marked with red. The initially motionless background ions and the injecting beam ions increase their speeds, and later concentrate on the beam front forming background plasma cavern. The decrease of

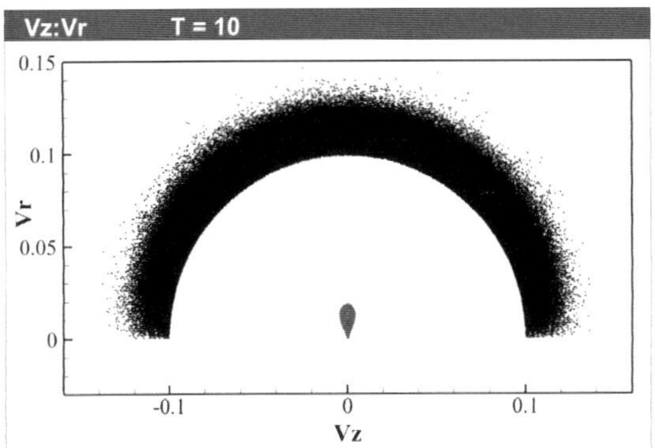

Figure 5. The particle velocities at $t = 10$.

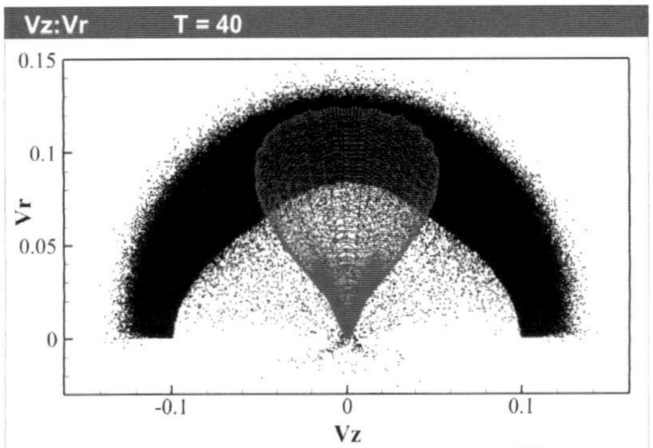

Figure 6. The particle velocities at $t = 40$.

the radial velocities of the beam ions with time and the evolution observed make possible the theoretically expected plasma confinement due to the diamagnetic bubble regime.

For the numerical experiments spatial grid with $N_r = 60$ and $N_z = 300$ nodes and $N_\tau = 10^5$ time steps $\tau = 5 \cdot 10^{-4}$ were used. The number of background ions used $J_b = 7.2 \cdot 10^4$, the total number of ions $J_p = 3.4 \cdot 10^6$. The computations took ~ 9.5 hours.

For the grids $N_r = 30$, $N_z = 150$ and time step $\tau = 8 \cdot 10^{-4}$ from 10 to 72 processor cores in 10..30 groups were used. We denote the core distribution as $np_0 - np$. The two central groups have maximal number of cores, for example, for the case of $10 - 30$ the number is $C_4 = C_5 = 5$. The number C_g falls linearly with the factor δ as the number of group g approaches the domain ends: $C_{g \pm i} = \pm i\delta + C_g$, where c is the number of one or two central groups. For the case $10 - 30$ the number of cores falls according to $\delta = 1$: $C_3 = C_6 = 4$, etc. Another example is distribution $15 - 30$, the central group has $C_7 = 8$ cores and $\delta = 2$: $C_6 = C_8 = 4$, $C_5 = C_9 = 2$. The remote groups have 1 core.

In figure 7 and figure 8 the average numbers of particles $J_c = J_g/C_g$ in a core of a group for the time moments $t = 40$ and $t = 100$ are shown. Here we normalize the groups number g on np_0 and stretch the value along the longitudinal coordinate z to compare the data for different number of groups np_0.

At the beginning of the injection (figure 7) the central cores contain the majority of particles and force the other cores to stand idle increasing the computational time. The cores of

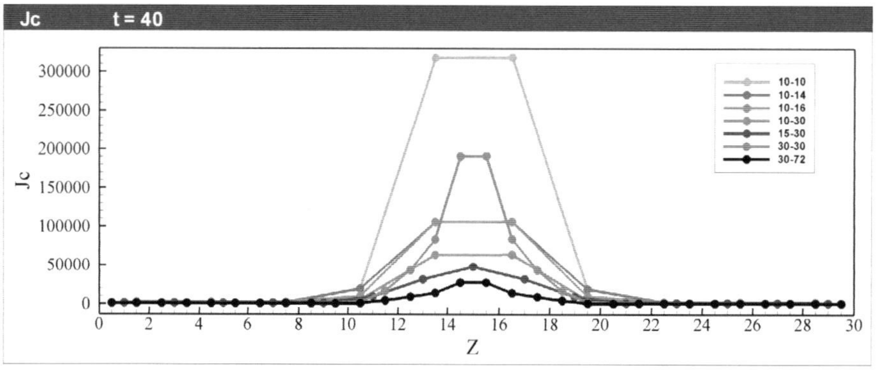

Figure 7. The average number of particles J_c in core at $t = 40$.

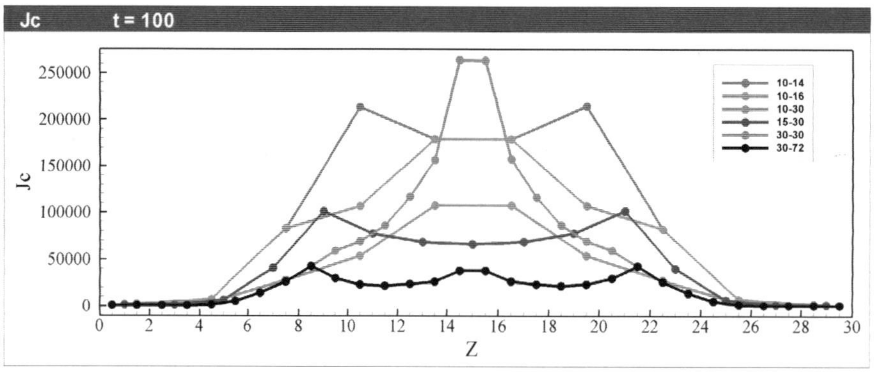

Figure 8. The average number of particles J_c in core at $t = 100$.

Table 1. Computation times, sec.

$np_0 - np$	$10-10$	$10-30$	$15-30$	$30-30$	$30-72$
Ion motion	46423	14261	17652	31585	15338
Currents	52639	11335	12380	32377	4290
Particle exchanges	17	17	24	19	17
MPI_Bcast + MPI_Reduce	7	17	22	6	15
Eulerian stage	164	300	204	40	65
Grid exchanges	54	56	70	46	43
Total time	99455	26071	30549	64238	19898

distribution $15-30$ are loaded almost uniformly at t=80 due to the highest number of central cores. However, further increase of the beam size leads an effective distribution to the one with the underloaded central cores and the overloaded periphery cores (see $10-14$, $15-30$, $30-72$ in figure 8). The core distribution $30-72$ is most effective as it has high number of cores in the center: $C_7 = 8$, $\delta = 1$ and smallest of all considered subdomains with particles. The configuration $10-14$ and $10-16$ have the same number of central cores $C_4 = C_5 = 4$, but higher $\delta = 2$ for $np = 16$ corresponds to an additional core in groups $g = 3$ and $g = 6$. The more uniform core load yields decreasing of the computation times for 3% for $T = 100$. The linear decomposition $np_0 = np$ is the most inefficient case, but bigger number of processors np_0 allows processing smaller domains and, as consequence, smaller number of particles, providing faster computations.

The computation times measured in seconds for the evolution time $T = 1.25 \cdot 10^5\tau$ are presented in Tab. 1. The table demonstrates the strong dependence of the computational time on the particle processing. The calculation of the velocities, coordinates and currents takes $\sim 98\%$ of the total time T_{total}. The largest subdomain with the biggest number of particles in core J_c corresponds to the distribution $10-10$ and the longest computations. The minimal number of particles J_c corresponds to the configuration $30-72$ and yields the best results. Since number of the grid nodes is smaller then the number of particles, the computations on the Eulerian stage take significantly smaller time (less than 1% of T_{total}). The spatial grid size is slowly growing parameter in the simulations, whereas the number of particles is defined by the injection intensity and the evolution time T. In the present simulations the particle exchanges take $\sim 10^{-4}T_{total}$ or less, but for $10^3\omega_{ci}^{-1}$ these times must be taken into account. The times for 2D array exchanges grow with the maximal number of cores in group. A certain distribution within the given np cores may significantly accelerate the computations, for example, for $np = 30$ the most effective configuration in the table corresponds to $np_0 = 10$ and requires about 5 times smaller computation time than for $10-10$. The test simulations demonstrate the mixed decomposition advantages, but a dynamical load balancing may considerably improve it.

The computations were performed on Intel Xeon Phi 7290 processors of Siberian Supercomputer Center cluster (ICM&MG SB RAS, Novosibirsk).

5. Conclusion

A 2D hybrid model realization for study of plasma dynamics in open magnetic systems is presented. The basis of the code is a model with MHD approach for the electron component of the plasma and the kinetic approach for the ions. The parallel PIC code for the cylindrical coordinate system combines domain and particle decomposition. The numerical experiments were performed and the results correspond to the theoretical expectations. The mixed decomposition is powerful method for acceleration of the computation times, and appropriate core distribution may significantly decrease the times. For the problems with highly non-uniform particle distribution in time a dynamic load balancing is required for efficient computations.

Acknowledgements
The work is supported by Russian Science Foundation (project N 19-71-20026).

References
[1] Syme D, Popovichev S, Conroy S, Lengar I and Snoj L 2011 *Nuclear Engineering and Design* **246** 185–190
[2] Beklemishev A 2016 *Physics of Plasmas* **23** 082506
[3] Vshivkova L and Dudnikova G 2017 Hybrid model of particle acceleration on a shock wave front *Numerical Analysis and Its Applications: 6th International Conference, NAA 2016, Revised Selected Papers* pp 737–743
[4] Braginskii S 1965 *Reviews of Plasma Physics* **1** 205–311
[5] Birdsall C and Langdon A 1985 *Plasma physics via computer simulation* (McGraw-Hill)
[6] Lotov K V, Timofeev I V, Mesyats E, Snytnikov A and Vshivkov V 2014 *Physics of Plasmas* **22**
[7] Boronina M and Vshivkov V 2015 *Journal of Plasma Physics* **81** 495810605
[8] Boris J 1970 *Proc. Fourth Conferenceon Numerical Simulations of Plasmas, Washington D.C., USA* 3–67

IOP Conf. Series: Journal of Physics: Conf. Series **1336** (2019) 012018 doi:10.1088/1742-6596/1336/1/012018

On sub-Alfven MHD flows with acceleration in coaxial channels

E V Stepin

Keldysh Institute of Applied Mathematics, Russian Academy of Sciences, Miusskaya sq. 4, 125047 Moscow, Russia
National Research Nuclear University MEPhI, Kashirskoe shosse, 31, 115409 Moscow, Russia

eugene.v.stepin@gmail.com

Abstract. Magnetohydrodynamic (MHD) flows in coaxial channels of plasma accelerators with a longitudinal magnetic field are considered. Sub-Alfven MHD flows with acceleration transonic to the slow magnetosonic speed are studied. Sub-Alfven MHD flows are related to the case when the plasma flow velocity doesn't exceed the Alfven speed corresponding to the longitudinal field along the channel. A narrow coaxial nozzle channel with a constant lower bound is considered. The non-stationary and stationary mathematical MHD models are considered in the quasi-one-dimensional approximation. The corresponding MHD problems are solved numerically. Features of the relaxation process and characteristics of the transonic sub-Alfven flow type with acceleration in the channel are studied. The influence of the longitudinal magnetic field on the steady-state plasma flows are researched. The mechanism of energy transformations is described.

1. Introduction

Mathematical modeling and numerical simulation are a key element in physical process researches and new technique development. Plasma acceleration is an object of interest due to its diverse technical applications. Plasma accelerators based on this effect are used to solve a number of problems facing humanity. In the transport industry, these devices are a component of electric thrusters installed on aerospace and aviation techniques. A spacecraft or an aircraft equipped with a plasma accelerator could achieve values of speed significantly exceeding those in their gas dynamic analogs. In the nuclear energy sector, high energy plasma fluxes formed in a plasma accelerator output could have parameters necessary for the thermonuclear fusion process activation and could be injected into the currently developed magnetic traps.

Depending on a principle of acceleration and plasma parameters, there are various technical realizations of plasma accelerators. In the present paper, plasma accelerators based on plasma acceleration in crossed electric and magnetic fields [1, 2] are considered. In these devices connected to an external source of electric energy, the interaction of the current flowing in plasma with a generated own transverse magnetic field results in plasma acceleration towards the channel output (figure 1).

Content from this work may be used under the terms of the Creative Commons Attribution 3.0 licence. Any further distribution of this work must maintain attribution to the author(s) and the title of the work, journal citation and DOI.
Published under licence by IOP Publishing Ltd

IOP Conf. Series: Journal of Physics: Conf. Series **1336** (2019) 012018 doi:10.1088/1742-6596/1336/1/012018

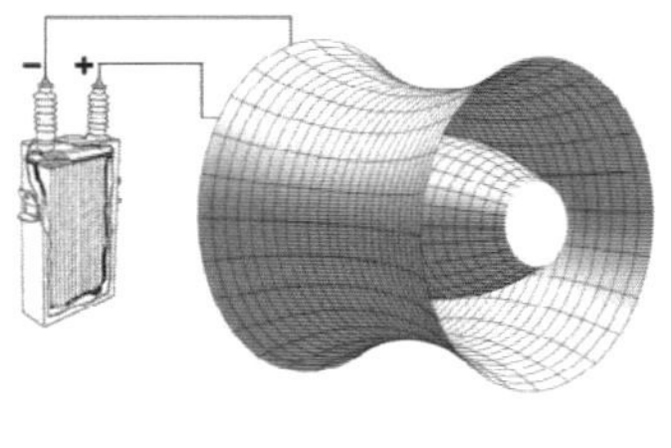

Figure 1a. Schematic view of coaxial plasma accelerator.

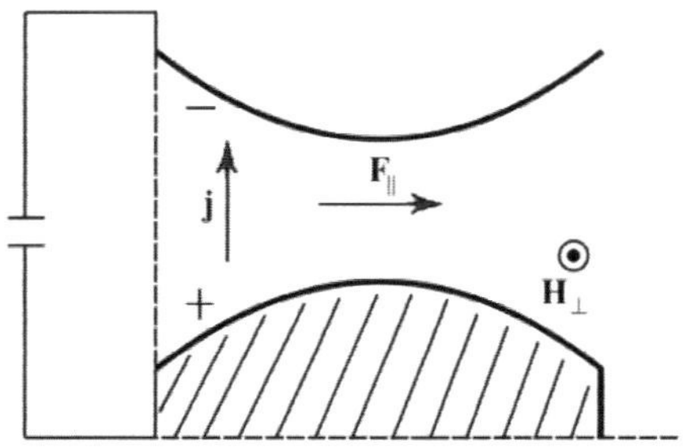

Figure 1b. Scheme of coaxial plasma accelerator in transverse magnetic field.

The quasi-steady-state high-current plasma accelerator (QSPA) with record-high values of the output plasma flow velocity is a prime example of accelerators based on this principle [3]. The development and realization of QSPA were successfully completed because of, inter alia, the mathematical modeling and numerical simulation of plasma fluxes flowing in channels of such accelerators. In these devices, plasma can be considered as a continues medium which behavior is described in terms of magnetic hydrodynamics (MHD) (see, e.g. [4]).

The above-mentioned scheme of the plasma accelerator under consideration can be modified by including in it a longitudinal magnetic field induced by external current-carrying conductors. According to the theory [5], MHD flows in the channels are subdivided into two classes of MHD flows in this case: super-Alfven in which plasma velocity υ exceeds the Alfven speed C_A corresponding to the longitudinal field and sub-Alfven with the opposite relation. In turn, each of the classes is also subdivided into flow types depending on relation of the velocity and one of the magnetosonic speeds – fast magnetosonic C_f or slow magnetosonic C_s. The classification of MHD flows in channels [6] is presented in figure 2.

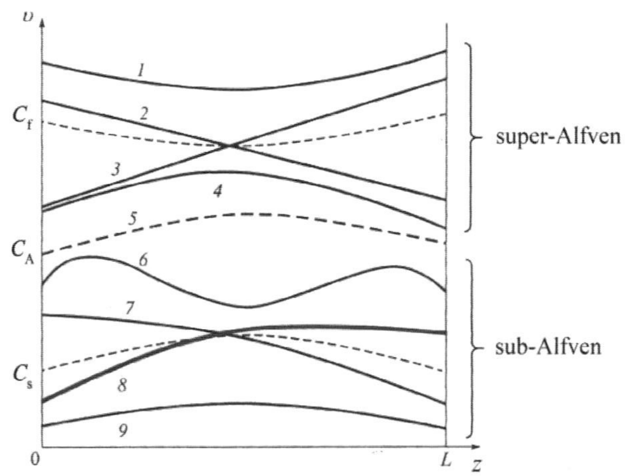

Figure 2. Classification of stationary MHD flows in channels.

The classification was obtained for a narrow nozzle channel with a constant lower bound, but its basic consequences are applied to MHD flows in a channel of arbitrary geometry. Accelerating super-Alfven transonic relative to the fast magnetosonic speed C_f (type 3 on figure 2) and sub-Alfven transonic relative to the slow magnetosonic speed C_s (type 8) are the main object of interest in the theory of plasma accelerators.

IOP Conf. Series: Journal of Physics: Conf. Series **1336** (2019) 012018 doi:10.1088/1742-6596/1336/1/012018

The relevant researches were connected with the study of the super-Alfven accelerating flow type in the transverse magnetic field [7] and in the presence of the longitudinal magnetic field [8, 9].

Features of relaxation process and characteristics of the Alfven type (type 5) in which plasma velocity coincides with the Alfven speed in each point of a channel with a constant lower bound and the near-Alfven types (types 4 and 6) are presented in [10]. The transition of the plasma velocity through the Alfven speed in combined flows with super-Alfven and sub-Alfven regions simultaneously existing in a channel of arbitrary geometry is studied in [11].

Following the previous researches, the present paper is focused on the sub-Alfven transonic (relative to the slow magnetosonic speed) flow (type 8 on figure 2).

2. Mathematical model

In the present paper, the object of modeling is plasma fluxes flowing in a coaxial channel in the presence of a longitudinal magnetic field (figure 3). The nozzle channel is formed by two coaxial electrodes connected to the external source of electric current. The axisymmetric plasma flow considered as a continuous conducting medium with the known thermodynamic parameters enters the channel and completes the formed electric circuit. In this system, the plasma is accelerated by the Ampere force $\mathbf{F}_{\parallel}$ of interaction between the radial electric current $\mathbf{j}$ flowing in it and the transverse magnetic field $\mathbf{H}_{\perp}$ generated by the current flowing in the central electrode. The longitudinal magnetic field $\mathbf{H}_{\parallel}$ additionally induced in the system influence on the plasma flow causing, in particular, its rotation in the channel transverse direction by the force $\mathbf{F}_{\perp}$.

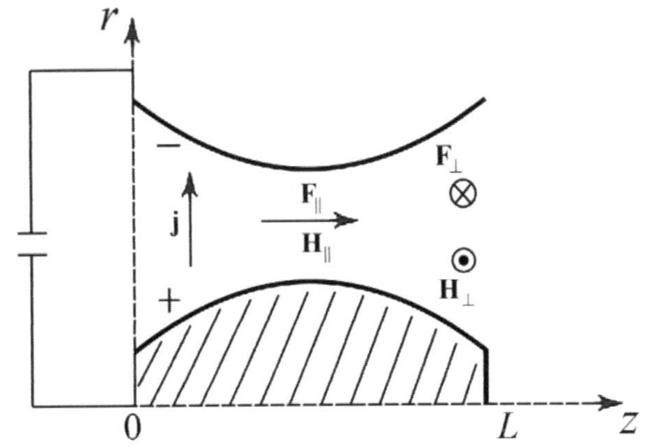

Figure 3. Scheme of coaxial plasma accelerator with longitudinal magnetic field in cylindrical coordinate system (r, φ, z). Cross-section by the plane $\varphi = \text{const}$.

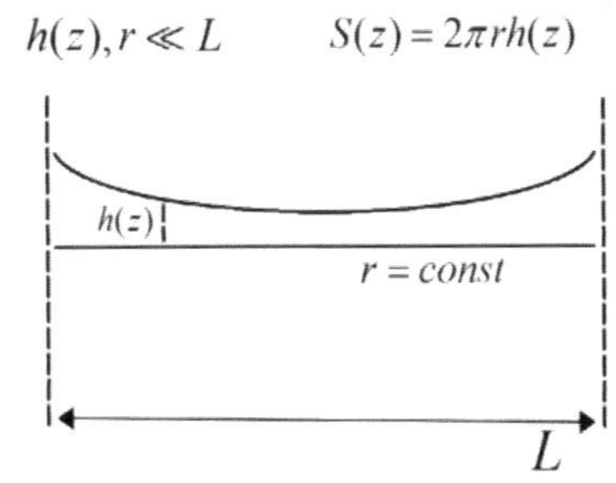

Figure 4. Coaxial narrow nozzle channel with constant lower bound.

Main regularities in parameters and characteristics of the plasma flows under consideration can be obtained in terms of the one-dimensional MHD model in the quasi-one-dimensional approximation (see, e.g. [6]). This approach is applied to narrow channels in which distance between electrodes is much less than its length or thin tubes between close flow trajectories. In this case, the researched parameters are averaged over a channel cross-section and depend only on the longitudinal spatial coordinate z.

In the present paper, a narrow nozzle-type coaxial channel with a constant lower bound is considered (figure 4). Its geometry is determined only by its ring cross-section $S(z)$. The non-stationary axisymmetric $(\partial/\partial\varphi \equiv 0)$ MHD flows in such channels in the quasi-one-dimensional approximation are described by the following system of equations:

$$\begin{cases} \frac{\partial}{\partial t}\rho S + \frac{\partial}{\partial z}\rho \upsilon_z S = 0 \\ \frac{\partial}{\partial t}\rho \upsilon_z S + \frac{\partial}{\partial z}\left(\rho \upsilon_z^2 + p + \frac{H_\varphi^2}{2}\right) S = \left(p + \frac{H_\varphi^2}{2}\right) S' \\ \frac{\partial}{\partial t}\rho \upsilon_\varphi S + \frac{\partial}{\partial z}\left(\rho \upsilon_z \upsilon_\varphi - H_\varphi H_z\right) S = 0 \\ \frac{\partial}{\partial t}\rho \varepsilon S + \frac{\partial}{\partial z}\rho \varepsilon S \upsilon_z + p\frac{\partial}{\partial z}\upsilon_z S = 0 \\ \frac{\partial}{\partial t} H_\varphi S + \frac{\partial}{\partial z}\left(H_\varphi \upsilon_z - H_z \upsilon_\varphi\right) S = 0 \\ \frac{\partial}{\partial t} H_z S = 0,\ \frac{\partial}{\partial z} H_z S = 0, \end{cases} \tag{1}$$

where $p = (\gamma - 1)\rho\varepsilon$.

The last two equations of the system (1) imply that the longitudinal magnetic flux $H_z S$ is constant in time and space, and it can be considered as given. In addition, an initial-boundary value problem about a sub-Alfven transonic MHD flow with acceleration in a narrow nozzle channel with a constant lower bound requires to set four boundary conditions [6]: three at the channel input and one at the channel output.

At the channel input, the role of boundary conditions is played by the known parameters of the accelerator and the plasma entering the channel. Plasma enters the channel with defined thermodynamic parameters in the absence of rotation. The total electric current in the circuit is also known. Turning to the dimensionless variables composed of the aforementioned dimension parameters of the model, these conditions are:

$$\rho = 1,\ \upsilon_\varphi = 0,\ H_\varphi = 1 \tag{2}$$

The considered flow type is characterized by the transition of the plasma flux through the slow magnetosonic speed C_s at the thinnest channel cross-section, and the plasma velocity exceeds it at the channel output. Hence, the boundary condition on the right of the calculation area can be defined as follows:

$$\upsilon_z = C_{sz} + \delta, \tag{3}$$

where δ as well as the magnetic flux $H_z S$ is a parameter of the considered mathematical model. The value of $\delta > 0$ indirectly defines the energy of the plasma flowing into the channel.

Initial conditions of the problem should only provide initial plasma acceleration and therefore can be quite arbitrary.

The MHD problem under consideration with equations (1) and boundary conditions (2), (3) is solved numerically by using the flux-corrected transport (FCT) algorithm with the Boris-Book correction [11]. Its steady-state solutions are obtained in the process of relaxation in time.

Part of the computation results presented in the paper is obtained in the stationary quasi-one-dimensional model based on the first integrals (the laws of conservation) of the system (1). In this case, the computation algorithm is reduced to numerical solving of a single algebraic equation for the plasma density followed from the first integrals. In more detail, the previous researches contain the description of the aforementioned calculation method and its implementation in relation to the research of super-Alfven transonic accelerating [7, 9] and trans-Alfven non-accelerating [11] MHD flows in narrow channels of curvilinear geometry.

IOP Conf. Series: Journal of Physics: Conf. Series **1336** (2019) 012018 doi:10.1088/1742-6596/1336/1/012018

3. Computation results

In this section, the results of numerical simulation of sub-Alfven transonic MHD flows with acceleration (type 8 on figure 2) in narrow coaxial channels are provided.

The coaxial nozzle narrow channel which has a constant lower bound with the following cross-section is considered:

$$S(z) = 2(z - 0.5)^2 + 0.5$$

In the non-stationary initial-boundary value problem, the chosen value of the longitudinal magnetic flux $H_z S = 0.9$ corresponds to a sufficiently strong longitudinal magnetic field. As mentioned before, the parameter δ indirectly defines the energy and temperature of the input plasma, which are arbitrary in the considered problem statement, and is chosen $\delta = 0.345$.

Relaxation process of the MHD flow of this type is presented in figure 5. The initial distributions $(t = 0)$ of the plasma longitudinal velocity υ_z (figure 5a) and azimuthal magnetic field H_φ (figure 5b) are changed in time during numerical solving of the initial-boundary value problem until the steady-state regime of the flow $(t \sim 15)$ is obtained in the calculations.

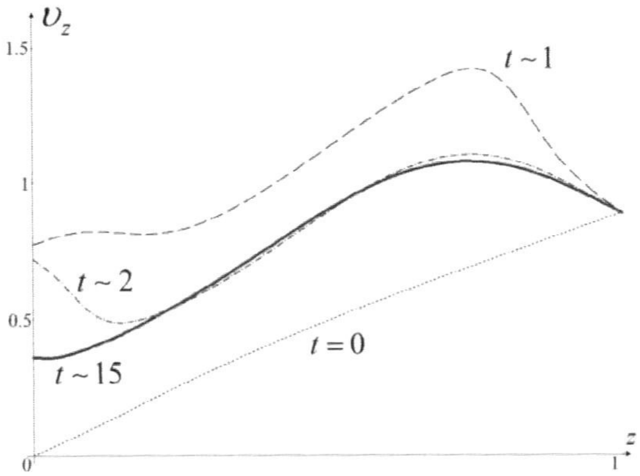

Figure 5a. Relaxation of longitudinal plasma velocity.

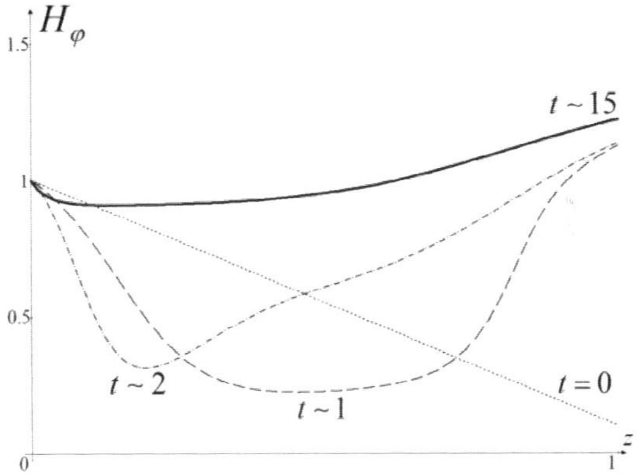

Figure 5b. Relaxation of azimuthal magnetic field.

At the first stage of the relaxation $(0 < t < 1)$, the portrait of the sub-Alfven plasma acceleration along the channel is similar to the super-Alfven one, except to its output section where the boundary condition (3) on the output plasma velocity takes effect. The presented distributions of the plasma parameters attest to the shock wave propagating to the channel enter during the relaxation process. Over time, the influence of the right boundary condition extends to the entire length of the channel, and the distributions become smooth.

The steady-state regime $(t \sim 15)$ is characterized by the plasma acceleration along the channel axis. The velocity distribution isn't monotonic and has a local maximum at the output section of the channel. In the steady-state regime of the flow, the longitudinal plasma velocity υ_z doesn't exceed Alfven speed C_{Az} corresponding to the longitudinal field H_z along the channel length and transits through the slow magnetosonic speed C_{sz} at its center (figure 6).

During the steady-state plasma motion in the channel, generation of an electric current in the flow takes place: the output value of the azimuthal magnetic field intensity H_φ is greater than the input one. The decrease of the plasma density ρ corresponds to the thermal energy consumption (figure 7). In addition, the flow also rotates in the transverse channel direction with the velocity υ_φ.

Summarizing the aforementioned observations, in sub-Alfven transonic MHD flows with acceleration, the accumulated thermal energy of an input plasma transforms into its electromagnetic and kinetic energies of acceleration and rotation. In the super-Alfven transonic accelerating type, on the contrary, electromagnetic energy of a plasma flux isn't generated and also turns into energy of its motion (see, e.g. [9]).

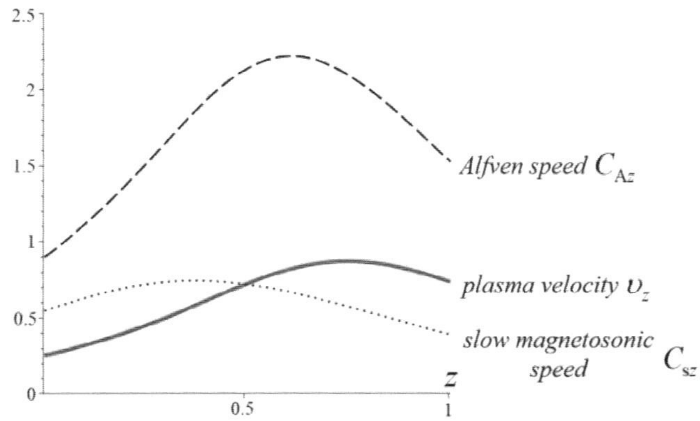

Figure 6. Relation between longitudinal plasma velocity and characteristic magnetic speeds in steady-state sub-Alfven transonic flow type with acceleration.

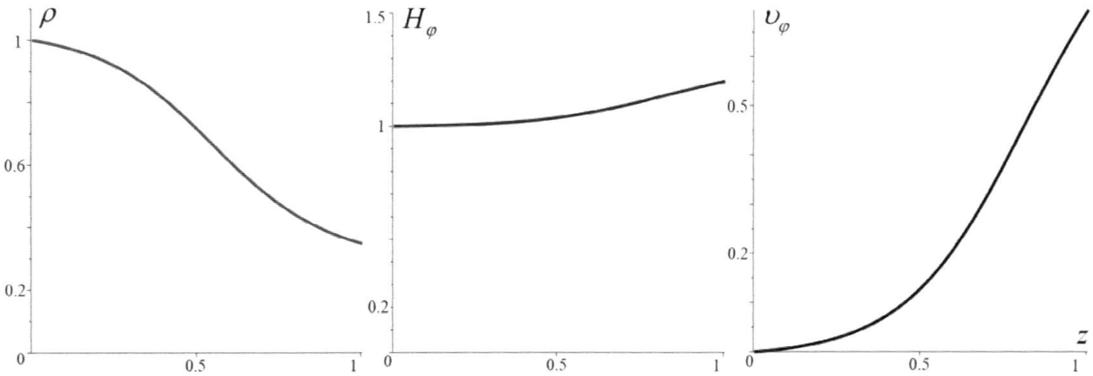

Figure 7. Steady-state distributions of plasma flow parameters.

The final part of the calculations was connected with the study of the longitudinal magnetic field influence on the stationary flows (figure 8). The longitudinal flux increase leads to more intensive thermal energy consumption (a plasma density at the channel output decreases) and growth of acceleration characteristics of the channel (the difference between output and input plasma velocities increases). This is another distinction of the sub-Alfven accelerating mechanism from its super-Alfven analog: in super-Alfven flows, the longitudinal field reduces acceleration properties of the device (see, e.g. [9]). The azimuthal magnetic field of the flow and its rotation aren't directly proportional to the longitudinal field intensity. There is a critical value of the longitudinal flux in which behavior of these parameters is qualitatively changed. The increasing electric current generation corresponding to the increase of the azimuthal field intensity is replaced by its following decrease. This observation is also applied to the flow rotation.

Finally, it is important to note that the sub-Alfven transonic flow type with acceleration exists under any longitudinal magnetic flux $H_z S > 0$.

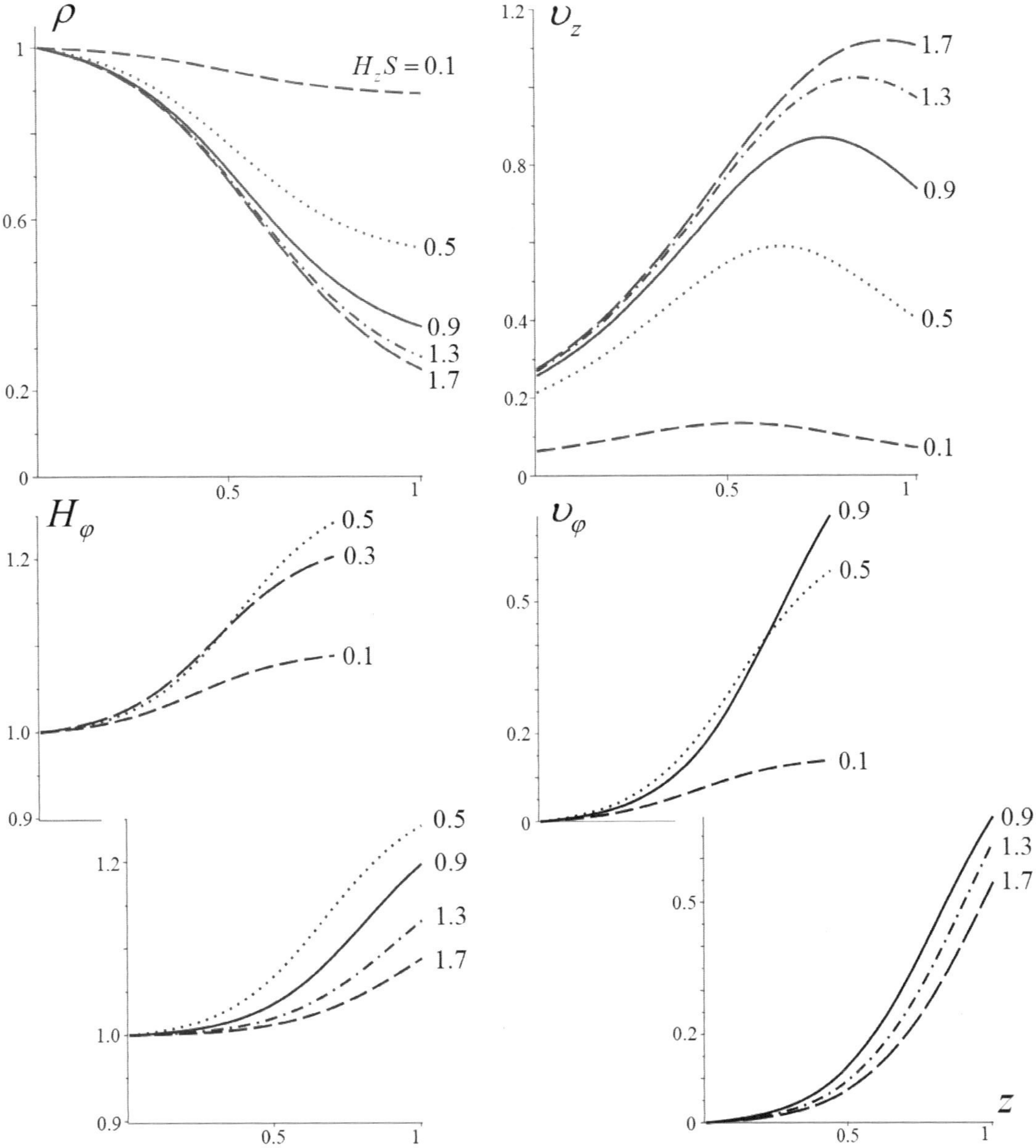

Figure 8. Influence of longitudinal magnetic field on plasma flow parameters.

4. Conclusions

The results of the numerical researches of sub-Alfven transonic accelerating flows in coaxial channels of plasma accelerators with a longitudinal magnetic field are presented. The non-stationary MHD model based on initial-boundary value problems formulated in the quasi-one-dimensional approximation for narrow channels with a constant lower bound and its stationary form are considered.

The following results are highlighted as the most significant:

- The features of the relaxation process and characteristics of the considered flow type are defined;
- The mechanism of energy transformation taking place in the flow is described;
- The influence of the longitudinal field on the steady-state flow parameters is determined.

IOP Conf. Series: Journal of Physics: Conf. Series **1336** (2019) 012018 doi:10.1088/1742-6596/1336/1/012018

References

[1] Morozov A I 1957 *JETP* **5** (2) 215–20
[2] Artsimovich L A, Luk'ianov S Iu, Podgornyi I M, Chuvatin S A 1958 *JETP* **6** (1) 1–5
[3] Volkov Ya F, Kulik N V, Marinin V V, Morozov A I et al 1992 *Sov. J. Plasma Phys.* **18** 718–28
[4] Kulikovskiy A G and Lyubimov G A 1965 *Magnetohydrodynamics* (Boston: Addison-Wesley)
[5] Morozov A I and Solov'ev L S 1980 vol 8 ed by M A Leontovich *Reviews of Plasma Physics* (NY, London: Consultants Bureau) pp 1–103
[6] Brushlinskii K V 2009 *Mathematical and Computational Problems in Magnetohydrodynamics* (Moscow: Binom) [in Russian]
[7] Brushlinskii K V and Styopin E V 2017 *J. Phys.: Conf. Ser.* **788** (1) 012009
[8] Brushlinskii K V and Styopin E V 2017 *J. Phys.: Conf. Ser.* **937** 012007
[9] Brushlinskii K V, Zhdanova N S, Stepin E V 2018 *Comp. Math. and Math. Phys.* **58** (4) 593–603
[10] Styopin E V 2015 *J. Plasma Phys.* **81** 905810309
[11] Stepin E V 2019 *J. Phys.: Conf. Ser.* **1205** (1) 012055
[12] Oran E and Boris J P 1987 *Numerical Simulation of Reactive Flow* (NY, Amsterdam, London: Elsevier)

IOP Conf. Series: Journal of Physics: Conf. Series **1336** (2019) 012019 doi:10.1088/1742-6596/1336/1/012019

Two-temperature advective transonic accretion flows around black holes

Shilpa Sarkar[1,2*], Indranil Chattopadhyay [1]**

[1]Aryabhatta Research Institute of Observational Sciences (ARIES), Manora Peak, Nainital-263002, India

[2]Pt. Ravishankar Shukla University, Great Eastern Road, Amanaka, Raipur, Chhattisgarh 492010, India

E-mail: *shilpa@aries.res.in, **indra@aries.res.in

Abstract. We present here unique transonic two-temperature accretion solutions in presence of radiation processes. Unlike one-temperature flow, the set of equations governing two-temperature flow is not complete, since for the latter, the number of equations is less than the number of flow variables. Consequently, a large number of transonic solutions of the equations of motion exist, for a given set of constants of motion. We invoke the second law of thermodynamics and identified the solution with the maximum entropy as the physically possible solution. In this paper, we compare spherical and rotating accretion two-temperature flows. We also show that the luminosity, as well as radiative efficiency of rotating flows, is higher by an order of magnitude, as compared to spherical flows.

1. Introduction

Accretion is a process by which the gravity of a compact object like white dwarf, neutron star or a black hole (BH), accumulates matter onto it, from the surrounding environment. This mechanism powers the brightest sources in the Universe, which are Active Galactic Nuclei (AGN). Accretion model started with the seminal papers of Hoyle & Lyttleton[1] (1939) and Bondi (1952)[2] who gave the first accretion solutions around a Sun-like star and a compact object respectively. But it was until a decade later, with the discovery of quasars and X-ray binaries, that accretion as a mechanism to explain the huge amount of emission coming from these objects, was realised. In 1964, Salpeter[3], used the Bondi accretion model (available at that time) to explain the luminosities of quasars ($L \sim 10^{12} L_{\odot}$). But, he reported that this flow is too fast to explain it. Shakura & Sunyaev (1973)[4], on the other hand, considered a rotation dominated flow to allow maximum time before the flow enters into the BH. Such a flow has Keplerian azimuthal velocity and negligible radial velocity. The flow is cold with negligible pressure gradient force, geometrically thin and optically thick. These disks were famously known as Shakura-Sunyaev disks (SSD) or Keplerian Disks (KD). The overall spectral profile of a SSD is a modified multi-coloured blackbody. Although this model was successful in explaining the thermal component of X-ray binaries and AGNs, but it could not explain the non-thermal part present in the spectra of these objects. Thorne & Price in 1975[7] argued that the instability present at the inner portion of the accretion disk could puff up the optically thick, radiation pressure dominated region into an optically thin, gas pressure dominated region. Using this assumption, Shapiro, Lightman & Eardley in 1976[6], put forward the first model of

Content from this work may be used under the terms of the Creative Commons Attribution 3.0 licence. Any further distribution of this work must maintain attribution to the author(s) and the title of the work, journal citation and DOI.
Published under licence by IOP Publishing Ltd

two-temperature theory. They assumed that the inner puffed up region consisted of protons and electrons described by two-different temperature distributions and explained the hard X-rays coming from Cygnus X-1. From the above papers, a general conclusion can be drawn that accretion flow should be a combination of Keplerian and sub-Keplerian matter. It has to be transonic, since the matter crosses the horizon at the speed of light, while at the outer boundary it should be subsonic. In 1980, Liang & Thompson (LT80)[8] pointed out this fact and reported that rotating sub-Keplerian flows around BH possess multiple sonic points, due to the presence of gravity and angular momentum. Presence of relativistic effects induces the formation of another sonic point. It was later showed that rotating sub-Keplerian flows passing through the outer sonic point, may undergo shock transition and enter the BH through the inner sonic point (Fukue 1987[9], Chakrabarti 1989[10]). Sub-Keplerian flows also have significant advection and the flow is also hotter than SSD (Abramowicz *et al* . 1988[11], for a brief review see, Bisnovatyi-Kogan & Lovelace 2001[12]). Few works were also focussed to study radiative properties of these systems. Colpi, Marashi & Treves (1984)[13] studied two-temperature version of spherical flows around BHs, where they elaborately discussed the radiative mechanisms while Narayan & Yi (1995) (NY95)[14] considered rotating advection-dominated two-temperature flows in order to obtain the spectra of low accreting BH systems. But in NY95, self-similarity was assumed while obtaining solutions, which is valid only far away from the BH and cannot be assumed throughout the flow. Also electron energy equation was dealt locally, neglecting its advection. In 1996, Nakamura *et al* .[15], were the first to obtain global transonic two-temperature advection dominated accretion solutions. Manmoto *et al* .[16] in 1997, computed the spectra of such flows. Since then, works have been done in two-temperature advection dominated branch of solutions, notable being by Chakrabarti & Titarchuk (1995) [17], Bisnovatyi-Kogan & Lovelace (1997)[18] and Mandal & Chakrabarti (2005)[19]. Few works considered transonicity like Rajesh & Mukhopadhyay (2010)[20] and Dihingia *et al* .(2017)[21].

Flows around BH are trans-relativistic in nature varying from thermally non-relativisitic, adiabatic index (Γ) $\sim$ 5/3, very far away from the central object to mildly-relativistic/relativistic, $\Gamma \rightarrow 4/3$, near the horizon, therefore a fixed Γ equation of state (EoS) is untenable. However, most of the previous works in two-temperature model assumed so. In 2009, Chattopadhyay & Ryu[22], gave an approximate analytical form of EoS for multispecies flow, which is easy to handle and it matches perfectly well[23] with the relativistically perfect EoS[24].

It is worth noting that the number of hydrodynamic equations in one-temperature and two-temperature is same, but there is one more variable in the latter ($T_{\rm e}$=electron and $T_{\rm p}$=ion temperature). Number of unknowns is hence more than the number of equations. Or in other words the *two-temperature model is degenerate*. If we scan the parameter space of two-temperature transonic solutions we can obtain infinite solutions for the same constants of motion. This problem was never been reported and solved for, in literature known so far, although hint of this was given in LT80, where they parametrized $T_{\rm e}$ and $T_{\rm p}$ by a constant ratio. The standard Coulomb interaction between electrons and ions is weak, therefore the energy exchange is not substantial to establish a stronger correlation between $T_{\rm e}$ and $T_{\rm p}$. In absence of any other physical processes dictated by plasma physics, all other parametrization employed by various authors are arbitrary. In Sarkar & Chattopadhyay (2019) (SC19)[25], we gave a detailed methodology to obtain a unique transonic two-temperature accretion solution around BHs, albeit for spherical accretion. In SC19, we derived for the first time, an entropy measure form, close to the horizon. Second law of thermodynamics dictates that nature prefers a solution with maximum entropy. Hence, we select the maximum entropy solution as the unique transonic two-temperature solution for a given set of constants of motion. In this paper, we extend SC19 to rotating, conical accretion flow around BHs. Since, rotating flows possess multiple sonic points, complexity of the problem is increased further. In the methodology section, we will discuss in detail how SC19 is modified in the present paper. We have also used the variable Γ index CR EoS which removed the problem of specifying the adiabatic indices of the species.

The paper is divided into the following sections. In section (2), we will give a brief overview of the basic equations and assumptions used to model the flow. In section (3), we will discuss the methodology to find unique transonic two-temperature solutions, showing some important results in section (4) and would finally conclude in section (5).

2. Basic Equations and Assumptions

We consider here inviscid conical rotating axis-symmetric two-temperature accretion flows around BHs described by Schwarzschild metric. It is to be noted that qualitative feature of the flow is not hampered while neglecting viscosity. It has been observed in viscous, single temperature, rotating accretion flow that angular momentum remains almost constant till $\sim$ few $\times 100$ Schwarzschild radius ($r_{\rm g}$) [26, 27, 28]. Spherical coordinate system has been employed where, r, θ, ϕ have their usual meaning and t is the time coordinate. It is to be noted that throughout the paper, we have used a system of units where, $G = M = c = 1$, such that the unit of length, time and velocity are given as $r_{\rm g} = GM/c^2$, $r_{\rm g}/c = GM/c^3$ and c, respectively, where G = Gravitational constant, M = mass of BH, c = speed of light. In the subsequent sections all the variables are in this unit system unless otherwise mentioned. The system is in steady state and is axis-symmetric. Additionally, all the quantities are averaged in the vertical direction. Therefore, $\partial/\partial t = \partial/\partial\phi = \partial/\partial\theta = 0$.

The energy momentum tensor of the accreting matter is $T^{\mu\nu} = (e+p)u^\mu u^\nu + pg^{\mu\nu} + t_{rad}$, where, e and p are the internal energy density and isotropic gas pressure, all measured in local fluid frame, u^μs are the components of four-velocity and t_{rad} is the radiation tensor. $g^{\mu\nu}$s are the inverse of metric tensor components, non-zero components of which are :

$$g_{tt} = -\left(1 - \frac{2}{r}\right) \; ; \; g_{rr} = \left(1 - \frac{2}{r}\right)^{-1} \; ; \; g_{\theta\theta} = r^2 \; ; \; g_{\phi\phi} = r^2 \sin^2\theta \tag{1}$$

Conservation law for : (1) Energy : $T^{\mu\nu}_{;\nu} = 0$, (2) particle density flux : $(nu^\nu)_{;\nu} = 0$ where, n = particle number density. If we project the four-divergence of the energy-momentum tensor along the space direction, *i.e.* $h^i_\mu T^{\mu\nu}_{;\nu} = 0$, we get the Euler equation, radial component of which is :

$$u^r \frac{du^r}{dr} + \frac{1}{r^2} - (r-3)u^\phi u^\phi + (g^{rr} + u^r u^r)\frac{1}{e+p}\frac{dp}{dr} = 0. \tag{2}$$

The azimuthal component is absent, since the angular momentum is a constant of motion in the absence of viscosity.

On integrating the equation for conservation of particle density flux, we get the form of mass accretion rate, which is another constant of motion :

$$\dot{M} = 4\pi\rho u^r r^2 \cos\theta, \tag{3}$$

where, $\rho = n(m_p + m_e)$ = mass density, m_p and m_e being the mass of proton and electron respectively, θ is the angle which the surface of the flow makes with the normal.

The first law of thermodynamics is $u_\mu T^{\mu\nu}_{;\nu} = 0$ and can be simplified to :

$$u^r \left[\left(\frac{e+p}{\rho}\right)\rho_{,r} - e_{,r}\right] = \Delta Q, \tag{4}$$

where, $\Delta Q = Q^+ - Q^-$. Q^+ is the rate of heating present in the flow and Q^- gives the cooling rate. Generalised Bernoulli constant, a constant of motion, is obtained by integrating Eq.(2) with the help of Eq.(4) and is given by,

$$E = -hu_t \exp(X_f), \tag{5}$$

where, $h = (e+p)/\rho$, is the specific enthalpy and $X_f = \int \frac{\Delta Q_{\rm p}+\Delta Q_{\rm e}}{\rho h u^r} dr$. E is constant even in the presence of dissipation. In case of no dissipation, $X_f = 0$ and $E \rightarrow \mathcal{E} = -hu_t$, which is the canonical form of relativistic Bernoulli constant [26].

To close the equations of motion, we use CR EoS [22], which for electron-proton plasma, in geometric units, is given by,

$$e = n_{\rm e} m_{\rm e} \left(f_{\rm e} + \frac{f_{\rm p}}{\eta} \right) = \frac{\rho f}{\tilde{K}}, \tag{6}$$

where, $\tilde{K} = 1 + 1/\eta$ and $\eta = m_{\rm e}/m_{\rm p}$. Here, $f_i = 1 + \Theta_i \left(\frac{9\Theta_i+3}{3\Theta_i+2} \right)$ and $\Theta_i = \frac{kT_i}{m_i c^2}$ is the non-dimensional conjugate of the dimensional temperature (T_i), where k = Boltzmann constant and i ≡electron or proton. Polytropic index and adiabatic index are defined as, $N_i = \frac{df_i}{d\Theta_i}$ and $\Gamma_i = 1 + \frac{1}{N_i}$. If we rewrite the first law of thermodynamics Eq.(4), separately for protons and electrons using Eq.(6), we get the differential equation for their temperatures :

$$\frac{d\Theta_{\rm p}}{dr} = -\frac{\Theta_{\rm p}}{N_{\rm p}} \left(\mathcal{A} + \frac{1}{v(1-v^2)} \frac{dv}{dr} \right) - \mathbb{P}\eta \quad \text{and} \quad \frac{d\Theta_{\rm e}}{dr} = -\frac{\Theta_{\rm e}}{N_{\rm e}} \left(\mathcal{A} + \frac{1}{v(1-v^2)} \frac{dv}{dr} \right) - \mathbb{E}, \tag{7}$$

where, $\mathcal{A} = -\frac{2r-3}{r(r-2)}$, $\mathbb{P} = \frac{\Delta Q_{\rm p} \tilde{K}}{\rho u^r N_{\rm p}}$, $\mathbb{E} = \frac{\Delta Q_{\rm e} \tilde{K}}{\rho u^r N_{\rm e}}$, These equations are not independent, but are coupled by a Coulomb coupling term. If we simplify Eq.(2), using Eqs.(3),(6-7), we obtain:

$$\frac{dv}{dr} = \frac{\mathcal{N}}{\mathcal{D}}, \tag{8}$$

where, $v = \sqrt{u_r u^r/(1+u_r u^r)}$, $\mathcal{N} = -\frac{1}{r(r-2)} + \frac{\lambda^2 \gamma_\phi^2 (r-3)}{r^4} + a^2 \mathcal{A} + \frac{\Gamma_{\rm p} N_{\rm p} \mathbb{P} + \Gamma_{\rm e} N_{\rm e} \mathbb{E}}{h\tilde{K}} - \frac{\Delta Q}{\rho h u^r}$ and $\mathcal{D} = \frac{v}{1-v^2} \left(1 - \frac{a^2}{v^2} \right)$. The sound speed is defined as, $a = \sqrt{\mathcal{G}/h\tilde{K}}$, where $\mathcal{G} = \frac{\Gamma_{\rm p}\Theta_{\rm p}}{\eta} + \Gamma_{\rm e}\Theta_{\rm e}$.

2.1. Sonic point conditions and shock conditions :

The mathematical form of Eq.(8) suggests that there exists a point where the numerator vanishes. For smooth continuous flow, the denominator also has to vanish. Thus, dv/dr has a 0/0 form and this point is called the sonic point of the flow. The sonic point conditions are : $\mathcal{N} = 0$ and $\mathcal{D} = 0$. At the sonic point we need to employ L'Hospital rule in order to find $(dv/dr)_c$.

The relativistic shock conditions or the Rankine-Hugoniot conditions [29] are : (1) Conservation of mass flux : $[\dot{M}] = 0$, (2) energy flux : $[E] = 0$ and (3) momentum flux : $[(e+p)u^r u^r + pg^{rr}] = 0$, where, the square brackets denote the difference of the quantities across the shock.

2.2. Radiative processes :

- Equipartition of magnetic energy density demands that magnetic field lines reconnect such that magnetic field is not frozen into the plasma [30, 31]. The heat dissipated[32] in such a way ($Q_{\rm B}$) is assumed to be absorbed by protons and electrons in equal portions ($\delta = 0.5$).
- Coulomb coupling ($Q_{\rm ep}$) cause energy exchange between electrons and protons [33].
- Inverse bremsstrahlung ($Q_{\rm ib}$) acts as a cooling mechanism for protons [34, 35].
- Electron is cooled via three processes : bremsstrahlung ($Q_{\rm br}$)[36], synchrotron ($Q_{\rm syn}$)[37] and inverse-Comptonization ($Q_{\rm ic}$)[37].
- If the temperature of the electrons is less than the energy of the interacting photons, then electrons can gain heat via Compton heating ($Q_{\rm comp}$).

Thus to conclude we have, for protons : $Q_{\rm p}^+ = \delta Q_{\rm B}$ and $Q_{\rm p}^- = Q_{\rm ep} + Q_{\rm ib}$, and for electrons : $Q_{\rm e}^+ = (1-\delta)Q_{\rm B} + Q_{\rm ep} + Q_{\rm comp}$ and $Q_{\rm e}^- = Q_{\rm br} + Q_{\rm syn} + Q_{\rm ic}$. In addition, all the relativistic transformations have been considered to compute the actual energy loss.

2.3. Entropy accretion rate expression :

To obtain the expression for entropy, one needs to integrate the first law of thermodynamics by switching off explicit heating and cooling terms . Due to the presence of Coulomb interaction term, Eq.(4) is not analytically integrable. However, close to the BH horizon, gravity overpowers all other interactions. Therefore, at $r = r_{\rm in}$ (asymptotically close to the horizon), $Q_{\rm ep}$ can be neglected. Thus we obtain,

$$\frac{d\Theta_{\rm p}}{dr} = \frac{\Theta_{\rm p}}{N_{\rm p}}\frac{1}{n_{\rm p}}\frac{dn_{\rm p}}{dr} \quad \text{and} \quad \frac{d\Theta_{\rm e}}{dr} = \frac{\Theta_{\rm e}}{N_{\rm e}}\frac{1}{n_{\rm e}}\frac{dn_{\rm e}}{dr} \tag{9}$$

Hence, we can integrate equation (9) at $r_{\rm in}$ to obtain :

$$n_{\rm ein} = \kappa_1 \exp\left(\frac{f_{\rm ein}-1}{\Theta_{\rm ein}}\right)\Theta_{\rm ein}^{\frac{3}{2}}(3\Theta_{\rm ein}+2)^{\frac{3}{2}} \text{ and } n_{\rm pin} = \kappa_2 \exp\left(\frac{f_{\rm pin}-1}{\Theta_{\rm pin}}\right)\Theta_{\rm pin}^{\frac{3}{2}}(3\Theta_{\rm pin}+2)^{\frac{3}{2}} \tag{10}$$

where, κ_1 and κ_2 are the integration constants which are measure of entropy. Subscript 'in' indicates quantities measured just outside the horizon. Due to charge neutrality in the flow $n_{\rm ein} = n_{\rm pin} = n_{\rm in}$. Therefore we can write,

$$n_{\rm in} = \sqrt{n_{\rm ein}n_{\rm pin}} = \kappa\sqrt{\exp\left(\frac{f_{\rm ein}-1}{\Theta_{\rm ein}}\right)\exp\left(\frac{f_{\rm pin}-1}{\Theta_{\rm pin}}\right)\Theta_{\rm ein}^{\frac{3}{2}}\Theta_{\rm pin}^{\frac{3}{2}}(3\Theta_{\rm ein}+2)^{\frac{3}{2}}(3\Theta_{\rm pin}+2)^{\frac{3}{2}}}, \tag{11}$$

where, $\kappa = \sqrt{\kappa_1\kappa_2}$. Thus, the expression of entropy accretion rate can be written as,

$$\dot{\mathcal{M}}_{\rm in} = \frac{\dot{M}}{4\pi\kappa(m_{\rm e}+m_{\rm p})}$$
$$= \sqrt{\exp\left(\frac{f_{\rm ein}-1}{\Theta_{\rm ein}}\right)\exp\left(\frac{f_{\rm pin}-1}{\Theta_{\rm pin}}\right)\Theta_{\rm ein}^{\frac{3}{2}}\Theta_{\rm pin}^{\frac{3}{2}}(3\Theta_{\rm ein}+2)^{\frac{3}{2}}(3\Theta_{\rm pin}+2)^{\frac{3}{2}}}u^r r^2 cos(\theta) \tag{12}$$

In the next section we will briefly discuss the significance of this formula and its use as a tool to remove degeneracy in two-temperature problem.

3. Methodology for obtaining unique transonic solutions

3.1. Finding of sonic points and obtaining a transonic solution :

This is the first step to find a general two-temperature accretion solution. Since the system is dissipative, sonic points are not known a priori. In order to compute the location of the sonic point/points, for a given set of constants of motion (E, $\dot{M}$ and λ), we start, with integration from $r_{\rm in} = 2.001$. At $r_{\rm in}$, $X_f = 0$ and from Eq.(5), $E \simeq \mathcal{E}$. Simplifying this, we obtain an analytical form for $v_{\rm in}$ (which is a function of $T_{\rm p\ in}$, $T_{\rm e\ in}$). We assume a value of $T_{\rm p\ in}$ and iterate $T_{\rm e\ in}$, while simultaneously calculating $v_{\rm in}$, until both the sonic point conditions (see, section 2.1) are satisfied. But as discussed before, rotating flows can possess multiple sonic points. So we keep $T_{\rm p\ in}$ same and reduce/increase $T_{\rm e\ in}$ by a large factor, and iterate it again until the sonic point conditions are satisfied. However, multiple sonic points form only for a limited range of constants of motion. Once we get the sonic point, we integrate Eqs.(7) and (8) inwards and outwards from the sonic point, using $4^{\rm th}$ order Runga-Kutta method to obtain transonic solutions. At the sonic point, $dv/dr = (dv/dr)_c$ (see section 2.1).

Fig.1(a1-a4), shows the Mach number ($M = v/a$) plot vs $\log r$ plot for 'spherical' flows ($\lambda = 0$) and in panel (b1-b6) solutions are for 'rotating' conical flows ($\lambda = 2.5$) around BH. Parameters used are $E = 1.002$, $\dot{M} = 0.05\dot{M}_{\rm edd}$, $M_{\rm BH} = 10M_{\odot}$, $\theta = 60°$. Thus, the constants of motion are same in all the cases, but $T_{\rm p\ in}$ is varied to get different solutions. The dotted solutions in every case are mathematical transonic outflows. Solid lines denote the global accretion solutions since it connects BH horizon from infinity.

We can see that in case of spherical flows the global solution always passes through a single sonic point (a1-a4), while in case of rotating flows, the flow might pass through one or more sonic points (b1-b6). In (b1-b2), global solution passes through the outer sonic point only, while in (b6) it passes through the inner sonic point. In both the cases only one physical sonic point is present. In (b3-b6) solutions possess multiple sonic points, out of which in (b4), we see that the flow harbours shock. So the global solution first passes through the outer sonic point, becomes supersonic and then after going through a shock transition, jumps to the subsonic branch and enter the BH supersonically after passing through the inner sonic point.

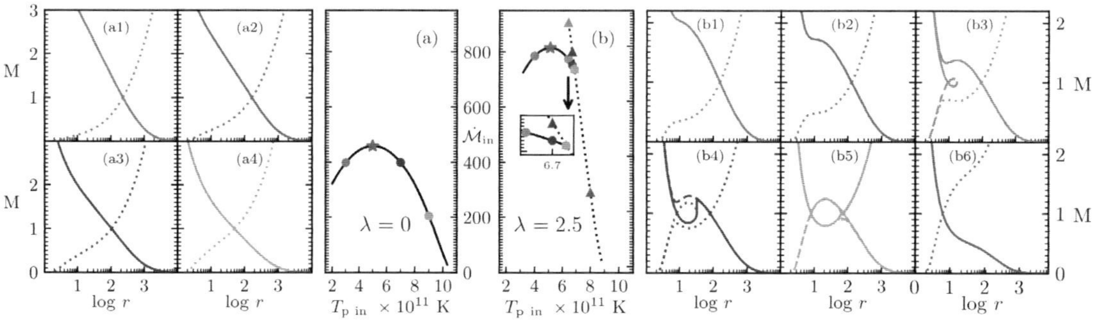

Figure 1: Left : (a) $\dot{\mathcal{M}}_{\rm in}$ vs $T_{\rm p\ in}$ plot for spherical flows. Single sonic point exist for every $T_{\rm p\ in}$. Panels (a1-a4) M vs log r plot for various $T'_{\rm p\ in}s$ marked with solid coloured dots in panel (a). The values of $T_{\rm p\ in}$ are : (a1) $3.0 \times 10^{11} K$, (a2) $4.0 \times 10^{11} K$, (a3) $5.0 \times 10^{11} K$, (a4) $6.0 \times 10^{11} K$. Right : (b) Shows $\dot{\mathcal{M}}_{\rm in}$ vs $T_{\rm p\ in}$ plot for rotating conical flows ($\lambda = 2.5$). Solid black curve is for the solutions passing through outer sonic point, while dotted black curve is for solutions passing through inner sonic points. Panels (b1-b4): Shows solutions for various $T'_{\rm p\ in}s$ marked in panel (b), value of which are: (b1) $4.0 \times 10^{11} K$, (b2) $5.1 \times 10^{11} K$, (b3) $6.4 \times 10^{11} K$, (b4) $6.7 \times 10^{11} K$, (b5) $6.86 \times 10^{11} K$, (b6) $8.0 \times 10^{11} K$. Entropy maximises at a certain $T_{\rm p\ in}$ in both the cases (green star). Parameters used are $E = 1.002$, $\dot{M} = 0.05\dot{M}_{\rm edd}$, $M_{\rm BH} = 10M_{\odot}$, $\theta = 60°$.

3.2. Selection of a unique solution :

A solution should be unique for a given set of constants of motion, but in Fig.1(a1-a4) and (b1-b6), it was noticed that a change in $T_{\rm p\ in}$ gave solutions with completely new topology and sonic point properties. In Fig.1, we plotted the case of spherical flows, in order to contrast for the complexity that arises due to the presence of angular momentum. Any wrong choice of solution would give us a completely different information and hence a wrong spectrum. So it is necessary to select the correct solution. The only way to deal with this degeneracy is to have a *measure of entropy* of the system. This measure of entropy is plotted in panel (a) and (b) against the corresponding $T'_{\rm p\ in}s$, using the entropy form given by Eq.(12).

In case of spherical flows, which possess single sonic points, there is a clear entropy maxima at $T_{\rm p\ in} = 5 \times 10^{11}$ K (green star). By employing the second law of thermodynamics, we select this solution (panel a2). Thus, we are able to obtain a unique transonic two-temperature solution for the given set of constants of motion [25].

In case of rotating flows, global solutions may pass through one or more sonic points. So the system is complicated unlike spherical flows. Dotted black curve in panel (b) is for solutions passing through inner sonic points while solid black curve is for solutions passing through outer sonic points. In a certain range of $T_{\rm p\ in}$ both inner and outer sonic points exist (solution b3-b5).

We can see that the entropy maximises in the outer sonic point branch of solutions (green star). Since flow starts from infinity, it would select the maximum entropy solution. Then, it could pass through an inner sonic point only if the shock conditions are satisfied (see section 2.1). For the present case, $T_{\rm p\ in} = 5.1 \times 10^{11} K$ possess maximum entropy and the global solution passes through outer sonic point (solution b2). In the sections to follow, we use this method of maximising entropy to obtain global advective transonic two-temperature accretion solutions.

4. Results

4.1. General transonic two-temperature solution :

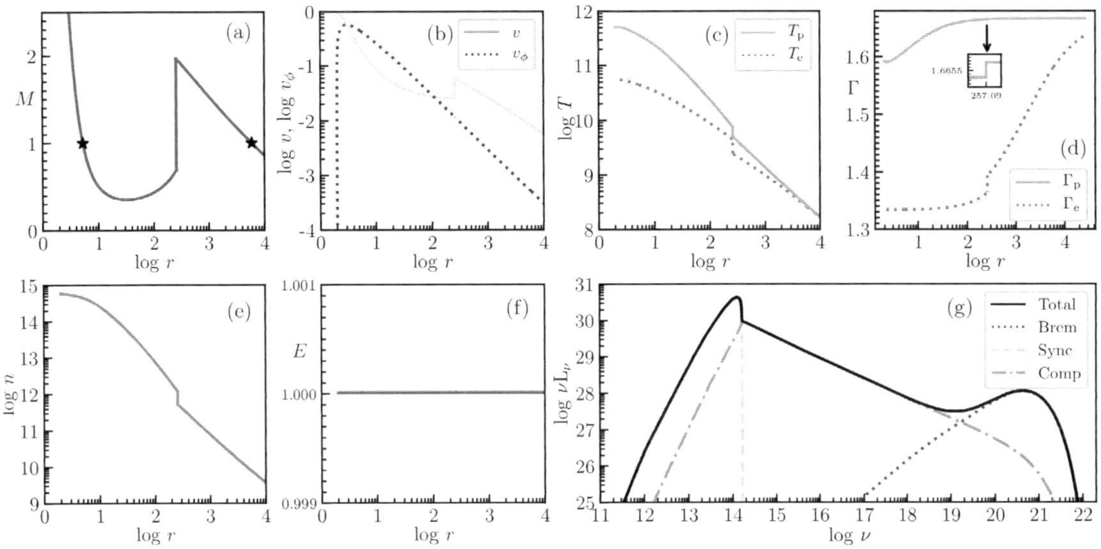

Figure 2: We plot (a) Mach number (M), (b) v (solid, cyan) and v_ϕ (dotted, blue), (c) $T_{\rm p}$ (solid, orange), $T_{\rm e}$ (dotted, red), (d) $\Gamma_{\rm p}$ (solid, orange), $\Gamma_{\rm e}$ (dotted, red), (e) log(n) (solid, magenta), (f) E (solid, brown), with respect to log r. In panel (g) we plot the total spectrum (solid black) along with contributions coming from bremsstrahlung emission (dotted violet), synchrotron (dashed yellow) and inverse Comptonization (dotted-dashed grey). The set of flow parameters used are $E = 1.00001$, $\lambda = 3.0$, $\dot{M} = 0.001\dot{M}_{\rm edd}$, $M_{\rm BH} = 10M_\odot$ and $\theta = 60^\circ$

In Fig.(2), we plot a typical two-temperature transonic advective accretion disk solution with its corresponding flow variables. The parameters used are, $E = 1.00001$, $\lambda = 3.0$, $\dot{M} = 0.001\dot{M}_{\rm edd}$, $M_{\rm BH} = 10M_\odot$ and $\theta = 60^\circ$. In panel (a) we plot the global solution, which passes through multiple sonic points (green solid line). First the solution passes through an outer sonic point $r_{\rm co} = 5935.59r_{\rm g}$ and then through an inner sonic point $r_{\rm ci} = 6.29r_{\rm g}$ (denoted by black stars) through a shock transition at $r_{\rm shock} = 257.09r_{\rm g}$. The infall speed of the supersonic matter after passing through the outer sonic point, is slowed down, due to the twin effect of centrifugal force and thermal pressure. This slowed down matter acts as a barrier to the matter coming from behind. This causes the formation of a shock. After the shock, matter becomes subsonic but it ultimately falls into the BH supersonically after passing through the inner sonic point. The radial three-velocity (v) in co-rotating frame and flow velocity in the azimuthal direction (v_ϕ) is plotted in panel (b) (solid cyan and dotted blue respectively). The fate of v and v_ϕ near the horizon is quite different, one approaching c while the other going to 0, since near the BH, gravity is so strong that matter only has a radial velocity component and negligible azimuthal

IOP Conf. Series: Journal of Physics: Conf. Series **1336** (2019) 012019 doi:10.1088/1742-6596/1336/1/012019

component. T_p (solid, orange) and T_e (dotted, red) as a function of r is plotted in panel (c). At very large r, $T_p \simeq T_e$. But as the matter flows inwards, cooling processes in electrons starts to dominate. Thus, protons and electrons settle down into two-different temperatures. They are however, coupled by a Coulomb coupling term which acts as an energy exchange term between the protons and electrons. But the term is weak, unlike in one-temperature case, where it is infinite. Panel (d), shows that gamma index of both protons (solid, orange) and electrons (dotted, red) varies with the flow. This justifies our use of CR EoS. Γ_p and $\Gamma_e \sim 1.66$ at very large distances, suggesting that both of them are thermally non-relativistic. When the flow nears the BH, Γ_e becomes mildly-relativistic and then relativistic with $\Gamma_e \sim 1.33$ near the horizon, though Γ_p never becomes truly relativistic. In (e), we have plotted number density (in units of cm^{-3}) vs log r. Since accretion is an example of convergent flow, the number density increases with decrease in radius. In panel (f), we prove that the generalised Bernoulli constant is a constant of motion throughout the flow, even in the presence of dissipation. In panel (g), we plot the spectrum for the accretion flow. It is plotted by summing up all the emissions coming from each radius. Spectrum of each emission process is also plotted separately. Bremsstrahlung is shown in dotted violet line, synchrotron by dashed yellow line and inverse-Comptonization by dotted-dashed grey line.

4.2. Luminosity and Efficiency of Bondi flows as compared to rotating flows :

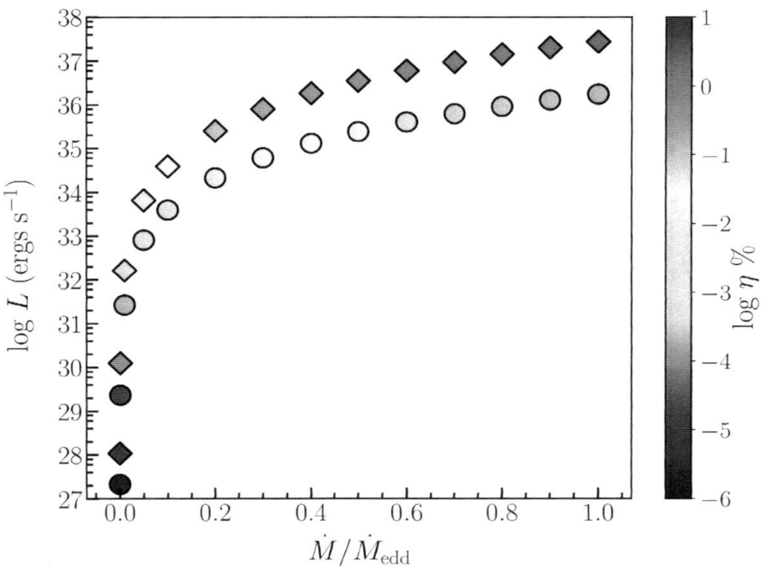

Figure 3: Plot shows the variation of luminosity with $\dot{M}$ which is normalised with respect to Eddington rate ($\dot{M}_{\rm edd}$). The color bar shows the change in efficiency (in logarithmic scale). The circled points are for spherical flows while diamond shaped points are for flows having angular momentum $\lambda = 2.4$. The other parameters used here are, $E = 1.001$, $M_{\rm BH} = 10M_{\odot}$ and $\theta = 60°$

In Fig.3, we have compared the change in luminosity with increase in accretion rate ($\dot{M}$), which is in terms of Eddington rate ($\dot{M}_{\rm edd}$). Circled points are for spherically symmetric flows whereas diamond shaped points are for flows having angular momentum $\lambda = 2.4$. For both the flows parameters used are $E = 1.001$, $M_{\rm BH} = 10M_{\odot}$ and $\theta = 60°$. We can see that as the supplied matter is increased the luminosity increases, irrespective of the type of flow. It is apparent from the plot that Bondi flows are less luminous flows, about an order of magnitude lower than rotating flows. The presence of angular momentum slows down the matter giving it more time to radiate. The color bar plots the efficiency of the system, which is given by the formula, $\tilde{\eta} = L/(\dot{M}c^2)$. The efficiency also have a similar trend, with the increase in accretion rate efficiency increases but flows with angular momentum have a higher efficiency than spherical flows, at a given an accretion rate.

IOP Conf. Series: Journal of Physics: Conf. Series **1336** (2019) 012019 doi:10.1088/1742-6596/1336/1/012019

5. Conclusions

In this paper we present our results in case of two-temperature rotating flows, that is flows having some angular momentum. Pure GR treatment was made, which helped us, model the strong gravity of BHs. In these flows, Coulomb coupling is never too strong, to make the flow attain a single temperature. We have also presented results of spherical flows in few cases in order to contrast the differences of theses flows with flows having angular momentum.

Our focus in this paper is to obtain unique transonic solutions in two-temperature model. We, in SC19, have reported that two-temperature solutions are degenerate in nature. Infinite transonic solutions exist for the same constants of motion. We solved the problem with the help of the entropy measure (Eq.12) which was obtained from the first principles. In this paper we first demonstrated the use of this entropy form in case of spherical flows, since spherical flows are simple and possess single sonic points. We could see a clear maxima at a certain solution. We applied the same methodology in case of rotating flows as well. But these systems are complex. Still the entropy measure form of ours worked and we again selected the solution with maximum entropy. This helped us in selecting a unique solution from the infinite solutions obtained, for a given set of constants of motion, in case of rotating conical flows. This is the first time, to the best of our knowledge, that such work has been done. This methodology to remove the degeneracy in two-temperature flows is required since any wrong choice of solution would give us a wrong solution and hence an overall wrong information of the system.

In our previous paper SC19, although we have worked on spherical transonic two-temperature flows but we did not consider any GR and doppler effects. We also ignored any photon trapping effects by BH. This led us to high luminosity values in case of spherical flows. In this paper we took these corrections into account and compared the luminosities of these flows with rotating flows. We saw that at $1\dot{M}_{\rm edd}$, Bondi flows have efficiencies $\simeq 0.14\%$, while rotating flows have efficiencies $\simeq 2.12\%$. Rotating flows were always an order of magnitude more luminous than spherical flows. Similar trend was also observed in efficiency.

References

[1] Hoyle, F., and Lyttleton, R. A. 1939, *Proc. Cam. Phil. Soc.*, **35**, 405
[2] Bondi, H. 1952, *MNRAS*, **112**, 195.
[3] Salpeter, E. E., 1964, *ApJ*, **140**, 796.
[4] Shakura N. I., Sunyaev R. A., 1973, *A&A*, **24**, 337.
[5] Thorne K. S., Price R. H., 1975, *ApJ*, **195**, L101
[6] Shapiro, S. L., Lightman, A. P., & Eardley, D. M., 1976, *ApJ*, **204**, 187
[7] Thorne K. S., Price R. H., 1975, *ApJ*, **195**, L101
[8] Liang E. P. T., Thompson K. A., 1980, *ApJ*, **240**, L271
[9] Fukue J., 1987, *PASJ*, **39**, 309
[10] Chakrabarti S. K., 1989, *ApJ*, **347**, 365
[11] Abramowicz, M. A., Czerny, B., Lasota, J. P., & Szuszkiewicz, E. 1988, *ApJ*, **332**, 646
[12] Bisnovatyi-Kogan, G. S., Lovelace, R.V.E., 2001, *New Astron. Rev.*, **45**, 663
[13] Colpi M., Maraschi L., Treves A., 1984, *ApJ*, **280**, 319
[14] Narayan R., Yi I., 1995, *ApJ*, **452**, 710
[15] Nakamura K. E., Kusunose M., Matsumoto R., Kato S., 1996, *PASJ*, **48**, 761
[16] Manmoto T., Mineshige S., Kusunose M., 1997, *ApJ*, **489**, 791
[17] Chakrabarti, S. K., Titarchuk, L. G., 1995, *ApJ*, **455**, 623
[18] Bisnovatyi-Kogan, G. S., Lovelace, R.V.E., 1997, *ApJ*, **486**, L43
[19] Mandal S., Chakrabarti S. K., 2005, *A&A*, **434**, 839
[20] Rajesh S. R., Mukhopadhyay B., 2010, *MNRAS*, **402**, 961
[21] Dihingia I. K., Das S., Mandal S., 2017, *MNRAS*, **475**, 2164
[22] Chattopadhyay I., Ryu D., 2009, *ApJ*, **694**, 492
[23] Vyas, M. K., Kumar, R., Mandal, S., Chattopadhyay, I., 2015, *MNRAS*, **453**, 2992
[24] Chandrasekhar, S., 1938, An Introduction to the Study of Stellar Structure, Dover, New York
[25] Sarkar, S., Chattopadhyay, I., 2019, *Int. J. Mod. Phys. D*, **28**, 1950037
[26] Chattopadhyay I., Chakrabarti S. K., 2011, *Int. J. Mod. Phys. D*, **20**, 1597

[27] Kumar R., Chattopadhyay I., 2014, *MNRAS*, **443**, 3444
[28] Chattopadhyay I., Kumar R., 2016, *MNRAS*, **459**, 3792
[29] Taub, A. H., 1948, *Phys. Rev.*, **74**, 328
[30] Schwartzman, V. F., 1971, *Soviet Astr. AJ*, **15**, 377.
[31] Bisnovatyi-Kogan, G. S., & Ruzmaikin, A. A., 1974, Ap&SS, **28**, 45
[32] Ipser J. P., Price R. H., 1982, *ApJ*, **255**, 654.
[33] Stepney, S., & Guilbert, P. W. 1983, *MNRAS*, **204**, 1269
[34] Boldt E., Serlemitsos P., 1969, *ApJ*, **157**, 557.
[35] Jones, F. C., 1971, *ApJ*, **169**, 503.
[36] Novikov I. D., Thorne, K. S., 1973, in Dewitt B. S., Dewitt C., eds, Black Holes. Gordon and Breach, New York, p. 343.
[37] Wardziński G., Zdziarski A., 2000, *MNRAS*, **314**, 183

IOP Conf. Series: Journal of Physics: Conf. Series **1336** (2019) 012020 doi:10.1088/1742-6596/1336/1/012020

Hybrid numerical model of the plasma flow dynamics in open magnetic systems

V.A. Vshivkov, M.A. Boronina, E.A. Genrikh, G.I. Dudnikova, L.V. Vshivkova, A.M. Sudakov

The Institute of Computational Mathematics and Mathematical Geophysics
SB RAS, Novosibirsk
E-mail: vsh@ssd.sscc.ru

Abstract. The report provides a brief overview of the numerical models used to solve plasma physics problems. The necessity of using hybrid models is shown, and a hybrid model that can be used in computer modeling of plasma flows in open magnetic systems is considered in more detail. The features of using hybrid models in two-dimensional cylindrical coordinates are given. To solve the equations of particle motion, the Boris algorithm is considered and its improvement is proposed based on the analytical solution of equations at a time step. We present a number of results of numerical simulation of the interaction of plasma flows obtained on the basis of a hybrid model. The calculations were performed in relation to the conditions of laboratory experiments with laser plasma at the KI-1 facility of the ILF SB RAS and the conditions of the diamagnetic regime at the CAT facility of the INP SB RAS.

1. Introduction

The numerical models used to solve plasma physics problems can be divided into three groups: hydrodynamic, kinetic, and hybrid. The system of equations of the two-fluid MHD model is obtained from kinetic equations when the local distribution function is slightly different from the Maxwell one. The system of simple plasma transport equations contains the equations of continuity, motion and conservation of thermal energy for electrons and ions [1]

$$
\begin{gathered}
\frac{\partial n_e}{\partial t} + \nabla \cdot (n_e \boldsymbol{V}_e) = 0; \\
\frac{\partial n_i}{\partial t} + \nabla \cdot (n_i \boldsymbol{V}_i) = 0; \\
m_e n_e \frac{d_e \boldsymbol{V}_e}{dt} = -\nabla p_e - \frac{\partial \boldsymbol{\pi}_{e\beta}}{\partial x_\beta} - e n_e \left(\boldsymbol{E} + \frac{1}{c} \boldsymbol{V}_e \times \boldsymbol{H}\right) + \boldsymbol{R}; \\
m_i n_i \frac{d_i \boldsymbol{V}_i}{dt} = -\nabla p_i - \frac{\partial \boldsymbol{\pi}_{i\beta}}{\partial x_\beta} + Z e n_i \left(\boldsymbol{E} + \frac{1}{c} \boldsymbol{V}_i \times \boldsymbol{H}\right) - \boldsymbol{R}; \\
\frac{3}{2} n_e \frac{d_e T_e}{dt} + p_e \nabla \cdot \boldsymbol{V}_e = -\nabla \cdot \boldsymbol{q}_e - \boldsymbol{\pi}_{e\beta} \cdot \frac{\partial \boldsymbol{V}_e}{\partial x_\beta} + Q_e; \\
\frac{3}{2} n_i \frac{d_i T_i}{dt} + p_i \nabla \cdot \boldsymbol{V}_i = -\nabla \cdot \boldsymbol{q}_i - \boldsymbol{\pi}_{i\beta} \cdot \frac{\partial \boldsymbol{V}_i}{\partial x_\beta} + Q_i,
\end{gathered}
\tag{1}
$$

Content from this work may be used under the terms of the Creative Commons Attribution 3.0 licence. Any further distribution of this work must maintain attribution to the author(s) and the title of the work, journal citation and DOI.
Published under licence by IOP Publishing Ltd

where

$$p_e = n_e T_e; \qquad p_i = n_i T_i;$$

$$\frac{d_e}{dt} = \frac{\partial}{\partial t} + (\boldsymbol{V}_e \nabla); \qquad \frac{d_i}{dt} = \frac{\partial}{\partial t} + (\boldsymbol{V}_i \nabla).$$

This system of equations is supplemented by the Maxwell system of equations for electromagnetic fields.

When modeling a rarefied plasma, the distribution functions are nonequilibrium, so it is necessary to solve the kinetic equations. To study the dynamics of a collisionless plasma, the mathematical model consisting of the Vlasov equations for each plasma component (electrons and ions of different elements) and Maxwell equations for electromagnetic fields is best suited [2]

$$\begin{gathered}
\frac{\partial f_\alpha}{\partial t} + \boldsymbol{v}\frac{\partial f_\alpha}{\partial \boldsymbol{r}} + \frac{q_\alpha}{m_\alpha}\left(\boldsymbol{E} + \frac{1}{c}\boldsymbol{v} \times \boldsymbol{H}\right)\frac{\partial f_\alpha}{\partial \boldsymbol{v}} = 0, \\
\nabla \times \boldsymbol{H} = \frac{4\pi}{c}\boldsymbol{j} + \frac{1}{c}\frac{\partial \boldsymbol{E}}{\partial t} = \frac{4\pi}{c}\sum_\alpha q_\alpha \int f_\alpha \, \boldsymbol{v} d\boldsymbol{v} + \frac{1}{c}\frac{\partial \boldsymbol{E}}{\partial t}, \\
\nabla \times \boldsymbol{E} = -\frac{1}{c}\frac{\partial \boldsymbol{H}}{\partial t}, \\
\nabla \cdot \boldsymbol{E} = 4\pi\rho = 4\pi \sum_\alpha q_\alpha \int f_\alpha \, d\boldsymbol{v}, \\
\nabla \cdot \boldsymbol{H} = 0.
\end{gathered} \tag{2}$$

Here α is the sort of particles (ions or electrons), q_α is the particle charge.

For the numerical solution of problems using this model, the method of particles in cells is most effective due to its universality for a wide range of physical parameters [3, 4]. There are a number of 2- and 3-dimensional electromagnetic kinetic PIC codes (KARAT, VORPAL, OOPIC, MANDOR, etc.), which are used, in particular, to solve the problems of the interaction of a laser pulse or electron beam with a plasma. However, the application of the particle method in three-dimensional problems requires large computational resources - memory and computer speed. This is due to the fact that in the particle in cells method in the solution area a grid is introduced (the grid step determines the accuracy of the solution), in each cell of which a sufficiently large number of model particles (up to 1000 particles) is placed. But an even greater limitation on the effectiveness of the numerical model is imposed by the different scales of physical phenomena determined by the behavior of light (electrons) and heavy (ions) plasma particles. In particular, the time step in the numerical model is selected from the conditions of accuracy and stability of motion of the lightest particles modeling the electronic component of the plasma. If the effects under study are determined by the movement of ionic components, then a significant limitation when using the model will be associated with a long counting time.

To reduce the requirements for the speed and memory of a computer, in comparison with completely kinetic models, combined (hybrid) models are used [5-7]. Their feature is that one of the plasma components (ions or electrons) is described kinetically, and the other is considered in the hydrodynamic approximation. Hydrodynamic equations for various plasma components can be obtained from the corresponding kinetic equations, provided that the particle distribution function is equilibrium or close to this state. For example, for an electronic component, these conditions can be a sufficiently large magnetic field or a high plasma density. For each specific problem, an analysis of the possibility of using the hydrodynamic approximation for any plasma component is necessary. In the course of calculations, it is necessary to ensure that the plasma parameters do not go beyond the applicability of the hydrodynamic approximation.

The simplest hybrid electrostatic model can be considered under conditions when the electron temperature is much higher than the ion temperature ($T_e \gg T_i$) [8, 9]. Such conditions are often realized in laboratory experiments and solar wind plasma. In this model, only ionic component motion is considered, and the electron density is described by the Boltzmann distribution $n(\boldsymbol{r},t) = n_0 \exp(e\varphi(\boldsymbol{r},t)/T_e)$, where $\varphi(\boldsymbol{r},t)$ is the electric field potential, n_0 is undisturbed plasma density. The system of equations has the form

$$\frac{\partial f}{\partial t} + \boldsymbol{u}\frac{\partial f}{\partial \boldsymbol{r}} - \frac{e}{m_i}\cdot\frac{\partial \varphi}{\partial \boldsymbol{r}}\cdot\frac{\partial f}{\partial \boldsymbol{u}} = 0,$$
$$\Delta\varphi = 4\pi e\left[n_0 \exp\left(\frac{e\varphi}{T_e}\right) - \int_{-\infty}^{\infty} f d\boldsymbol{u}\right], \tag{3}$$

where $f = f(\boldsymbol{r},\boldsymbol{u},t)$ is the ion distribution function, $\boldsymbol{u}$ is the ion velocity. In this formulation in [6, 7], the problems of the propagation of large-amplitude waves and the decay of an arbitrary discontinuity in the density of ions in a nonisothermal plasma were solved.

One of the first works in which the hybrid model is described is article [5]. It presents hybrid numerical algorithms for modeling low-frequency ($\omega \ll \omega_{pe}, \omega_{ce}$) electromagnetic and electrostatic phenomena in a magnetized plasma. Maxwell's equations are solved within a small transverse bias current (Darwin model) and quasineutrality. Numerical models treat electrons as a massless liquid, and ions as particles. The numerical stability of the algorithms is studied analytically and confirmed by computer experiments.

2. Hybrid model

In this paper, we consider the following hybrid model. In it, the ionic components are described using kinetic equations, and the hydrodynamic type equations are used for the electronic component. In this model, it is assumed that the plasma is quasineutral and bias currents $\partial E / \partial t$ are neglected.

Equations for ionic components

$$\frac{\partial f_\alpha}{\partial t} + \boldsymbol{v}\frac{\partial f_\alpha}{\partial \boldsymbol{r}} + \frac{\boldsymbol{F}_\alpha}{m_\alpha}\frac{\partial f_\alpha}{\partial \boldsymbol{v}} = 0,$$
$$\boldsymbol{F}_\alpha = q_\alpha\left(\boldsymbol{E} + \frac{1}{c}\boldsymbol{v}\times\boldsymbol{H}\right) + \boldsymbol{R}_\alpha. \tag{4}$$

where the index α denotes the sort of ions.

Hydrodynamic equations for electrons

$$m_e\left(\frac{\partial \boldsymbol{V}_e}{\partial t} + (\boldsymbol{V}_e\cdot\nabla)\boldsymbol{V}_e\right) = -e\left(\boldsymbol{E} + \frac{1}{c}\boldsymbol{V}_e\times\boldsymbol{H}\right) - \frac{\nabla p_e}{n_e} + \boldsymbol{R}_e,$$
$$n_e\left(\frac{\partial T_e}{\partial t} + (\boldsymbol{V}_e\cdot\nabla)T_e\right) + (\gamma-1)p_e\nabla\cdot\boldsymbol{V}_e = (\gamma-1)(Q_e - \nabla\cdot\boldsymbol{q}_e). \tag{5}$$

The quasineutrality condition has the form $n_e = \sum_\alpha Z_\alpha n_\alpha = n$, where $n_\alpha = \int f_\alpha(\boldsymbol{r},\boldsymbol{v},t)\,d\boldsymbol{v}$. The friction force is calculated by the formulas

$$\boldsymbol{R}_e = \frac{e}{\sigma}\boldsymbol{j}, \qquad \boldsymbol{R}_\alpha = -\frac{Z_\alpha e}{\sigma}\boldsymbol{j}, \qquad \boldsymbol{R}_e + \sum_\alpha \boldsymbol{R}_\alpha = 0\,.$$

Electron heating Q_e and heat flux due to thermal conductivity $\boldsymbol{q}_e$ are calculated by the formulas

$$Q_e = \frac{j^2}{\sigma}, \qquad \boldsymbol{q}_e = -\kappa\nabla T_e,$$

where σ is the plasma conductivity, κ is the thermal conductivity coefficient, $p_e = n_e T_e$ is the electron pressure [1].

A system of Maxwell equations without bias currents is added to these equations

$$\begin{aligned} \nabla \times \boldsymbol{H} &= \frac{4\pi}{c}\boldsymbol{j}, \\ \nabla \times \boldsymbol{E} &= -\frac{1}{c}\frac{\partial \boldsymbol{H}}{\partial t}, \\ \nabla \cdot \boldsymbol{H} &= 0, \\ \nabla \cdot \boldsymbol{E} &= 0. \end{aligned} \tag{6}$$

In the absence of energy flows across the boundaries of the solution domain, the law of conservation of total energy is satisfied for this system

$$\sum_{\alpha} \frac{m_\alpha}{2} \int f_\alpha(\boldsymbol{r}, \boldsymbol{v}, t)\ v^2 d\boldsymbol{v} d\boldsymbol{r} + \int \left(n_e \frac{m_e V_e^2}{2} + \frac{H^2}{8\pi} + \frac{n_e T_e}{\gamma - 1} \right) d\boldsymbol{r} = \text{const.}$$

The kinetic equation for the ion component is solved using the particle-in-cell method, in which the ion component of the plasma is replaced by a set of model particles. The equations of motion for model particle with number j are obtained from the characteristics of the kinetic equation

$$\begin{aligned} \frac{d\boldsymbol{r}_j}{dt} &= \boldsymbol{v}_j, \\ \frac{d\boldsymbol{v}_j}{dt} &= \frac{q_j}{m_j}\left(\boldsymbol{E} + \frac{1}{c}\boldsymbol{v}_j \times \boldsymbol{H} \right) + \frac{1}{m_j}\boldsymbol{R}_\alpha, \end{aligned} \tag{7}$$

and the hydrodynamic equations for electrons are solved on a grid, in the same way as the Maxwell equations. This model (4-7) is used to solve the problem of the interaction of an ion beam with a background plasma in the magnetic field of an open plasma trap.

Open plasma traps have a cylindrical geometry, and cylindrical coordinates (r, φ, z) can be entered into them. Assuming that all functions are independent of the angular coordinate φ, we pass from the three-dimensional problem to the two-dimensional one, where all functions depend on only two spatial variables (r, z). The transition to two-dimensional coordinates significantly saves computer memory and the number of arithmetic operations. However, one cannot take into account the possibility of the development of a number of angular instabilities and the design features of the trap.

From the point of view of implementing the particle-in-cell method, the following features appear in the algorithm:

1) Model particles have different masses (charges) depending on the distance to the axis.
2) The cell volume is proportional to the distance to the axis.
3) Features of setting boundary conditions on the axis for charge density and current appear.
4) In the algorithm for calculating the coordinates and velocities of particles, it is necessary to use the method proposed in [10], which consists in a local transition to Cartesian coordinates, the calculation of new coordinates and particle velocities, and the reverse transition to cylindrical ones.

3. Particle pusher

In the particle-in-cell method, Boris's algorithm is usually used to calculate new velocities [10]

$$\begin{aligned} \frac{\boldsymbol{v}_1 - \boldsymbol{v}^{n-1/2}}{\tau/2} &= \frac{q}{m}\boldsymbol{E}^n, \\ \frac{\boldsymbol{v}_2 - \boldsymbol{v}_1}{\tau} &= \frac{q}{mc}\frac{\boldsymbol{v}_2 + \boldsymbol{v}_1}{2} \times \boldsymbol{H}^n, \end{aligned} \tag{8}$$

$$\frac{\widetilde{\boldsymbol{v}} - \boldsymbol{v}_2}{\tau/2} = \frac{q}{m}\boldsymbol{E}^n.$$

At each time step, the coordinate calculation algorithm is divided into three stages. In the first and third, the particle velocity changes only under the influence of an electric field, and in the second - only under the influence of a magnetic field. The circuit has a second order of approximation in time. Despite the formally implicit form, the system of three equations for the three velocity components is easily solved analytically. The advantage of the second stage is that when it is used, the velocity modulus does not change, which corresponds to the law of motion of a charged particle in a magnetic field. Moreover, kinetic energy is retained even if the magnetic field or time step is very large.

Let us consider separately the second stage of the method (motion in a magnetic field). If the magnetic field is directed along the z axis, $\boldsymbol{H} = (0, 0, H_z)$, and the vector $\boldsymbol{v}_1 = (1,0,0)$, then the vector $\boldsymbol{v}_2$ will lie in the (x, y) plane and have the following coordinates

$$\boldsymbol{v}_2 = \left(\frac{4 - \omega^2\tau^2}{4 + \omega^2\tau^2}, -\frac{4\omega\tau}{4 + \omega^2\tau^2}, 0\right), \qquad \omega = \frac{qH_z}{mc}. \tag{9}$$

The equation of the second stage in a differential form can be solved analytically. This solution with the same initial data has the form $(t = \tau)$

$$\boldsymbol{v}_2 = (\cos\omega\,\tau, -\sin\omega\tau\,, 0). \tag{10}$$

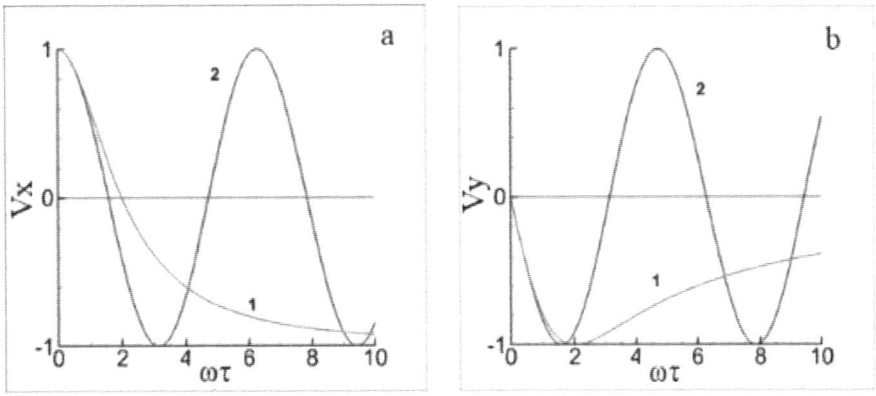

Figure 1. The dependence of the components of the velocity vector a - V_x and b - V_y on $\omega\tau$, obtained by formulas (9) - curve 1 and (10) - curve 2.

Figure 1 shows the solutions of the second stage of the Boris scheme using difference schemes and analytical formulas. For small τH_z $(\omega\tau < 0.5)$, the solutions differ little from each other, and for large, the character of these curves changes. The exact solution corresponds to cyclotron rotation of the ion in a magnetic field, while in the numerical solution the sign of the velocity vector is reversed. It should be noted that for any values of τH_z the length of the velocity vector remains equal to 1, and the scheme is stable. When a particle enters a region with high magnetic field values, then its motion is considered incorrect.

Since when calculating the velocities during the time step τ, the electric and magnetic fields remain constant, it is possible to include electric fields in the analytical solution. Suppose again that the magnetic field is directed along the z axis, $\boldsymbol{H} = (0, 0, H_z)$, and the velocity and electric field vectors lie in the (x, y) plane, $\boldsymbol{v}_1 = (v_x^1, v_y^1, 0)$, $\boldsymbol{E} = (E_x, E_y, 0)$. Then the solution will look like

$$v_x^2 = \left(v_x^1 - \frac{cE_y}{H_z}\right)\cos\omega\,\tau + \left(v_y^1 + \frac{cE_x}{H_z}\right)\sin\omega\tau + \frac{cE_y}{H_z}, \tag{11}$$

$$v_y^2 = \left(v_y^1 + \frac{cE_x}{H_z}\right)\cos\omega\,\tau - \left(v_x^1 - \frac{cE_y}{H_z}\right)\sin\omega\tau - \frac{cE_x}{H_z}.$$

In formulas (11), the velocity separation into two parts is clearly visible: the drift velocity in crossed fields and the Larmor rotation velocity. In the three-dimensional case, for an arbitrary direction of all vectors, the algorithm for calculating the speed at step τ has the following form:

1) First, the projections of the velocity and electric field vectors on the direction of the magnetic field are calculated

$$\boldsymbol{v}_{\parallel}^1 = (\boldsymbol{v}_1 \cdot \boldsymbol{H})\frac{\boldsymbol{H}}{H^2}, \qquad \boldsymbol{E}_{\parallel} = (\boldsymbol{E} \cdot \boldsymbol{H})\frac{\boldsymbol{H}}{H^2}.$$

2) Subtracting from the velocity and electric field vectors the parts parallel to the magnetic field, we obtain the parts of these vectors perpendicular to the magnetic field

$$\boldsymbol{v}_{\perp}^1 = \boldsymbol{v}_1 - \boldsymbol{v}_{\parallel}^1, \qquad \boldsymbol{E}_{\perp} = \boldsymbol{E} - \boldsymbol{E}_{\parallel}.$$

3) The drift velocity is calculated

$$\boldsymbol{v}_{dr} = \frac{c}{H^2}\boldsymbol{E}_{\perp} \times \boldsymbol{H}.$$

4) Next, using the analytical formulas, we find new velocity components perpendicular to the magnetic field

$$v_{\perp x}^2 = \left(v_{\perp x}^1 - v_{dr,x}\right)\cos\omega\,\tau + \frac{1}{H}\left(v_{\perp y}^1 H_z - v_{\perp z}^1 H_y + cE_x\right)\sin\omega\tau + v_{dr,x},$$
$$v_{\perp y}^2 = \left(v_{\perp y}^1 - v_{dr,y}\right)\cos\omega\,\tau + \frac{1}{H}\left(v_{\perp z}^1 H_x - v_{\perp x}^1 H_z + cE_y\right)\sin\omega\tau + v_{dr,y},$$
$$v_{\perp z}^2 = \left(v_{\perp z}^1 - v_{dr,z}\right)\cos\omega\,\tau + \frac{1}{H}\left(v_{\perp x}^1 H_y - v_{\perp y}^1 H_x + cE_z\right)\sin\omega\tau + v_{dr,z},$$

where

$$\omega = \frac{q}{mc}H = \frac{q}{mc}\sqrt{H_x^2 + H_y^2 + H_z^2}.$$

5) The addition to the speed due to the longitudinal electric field and the new longitudinal speed are calculated

$$\boldsymbol{v}_{\parallel}^2 = \boldsymbol{v}_{\parallel}^1 + \tau\frac{q}{m}\boldsymbol{E}_{\parallel}.$$

6) Adding new longitudinal and lateral speeds, we obtain the full speeds

$$\boldsymbol{v}_2 = \boldsymbol{v}_{\parallel}^2 + \boldsymbol{v}_{\perp}^2.$$

An important problem in modeling open traps is the motion of model particles in the region of magnetic plugs. It is known that when particles move along an alternating magnetic field, they retain a magnetic moment associated with kinetic energy transverse to the magnetic field [11]. Since at each time step the particle moves in a constant field, it may not retain the magnetic moment and, therefore, may not be reflected from magnetic plugs. To verify this assumption, a test calculation of the motion of a group of particles in the vicinity of a magnetic plug with the same initial velocity was performed. Figure 2 shows the trajectories of three model particles moving in the direction of the z axis.

IOP Conf. Series: Journal of Physics: Conf. Series **1336** (2019) 012020 doi:10.1088/1742-6596/1336/1/012020

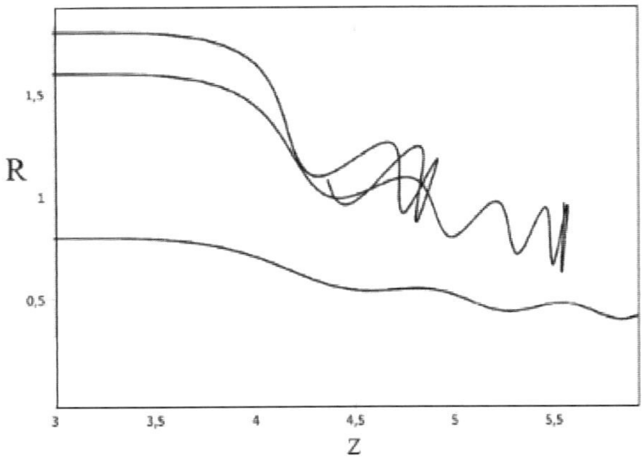

Figure 2. Particle trajectories in the field of magnetic plugs.

As can be seen from the figure, a particle located closer to the axis, flew to the border, and the other two were reflected from the border. Thus, the algorithm used allows one to correctly describe the plasma motion in open magnetic traps.

4. Calculation results

Let us present the results of numerical simulation of the interaction of plasma flows obtained on the basis of a hybrid model.The calculations were performed in relation to the conditions of laboratory experiments with laser plasma at the KI-1 stand [12] and the conditions of the diamagnetic regime at the SAT facility [13].

1. The case of a constant magnetic field. In this formulation, the mechanisms of collisionless interaction of plasma flows and the nature of perturbations generated by a dense plasma piston cloud were studied [2, 6, 14]. Figure 3 shows the magnetic field lines at successive times that characterize the change in the magnetic structure during expansion of a dense plasma cloud in a low-density magnetized background plasma. This process is accompanied by the formation of a magnetic cavity, the size of which depends on the kinetic energy of the cloud and the magnetic field strength.

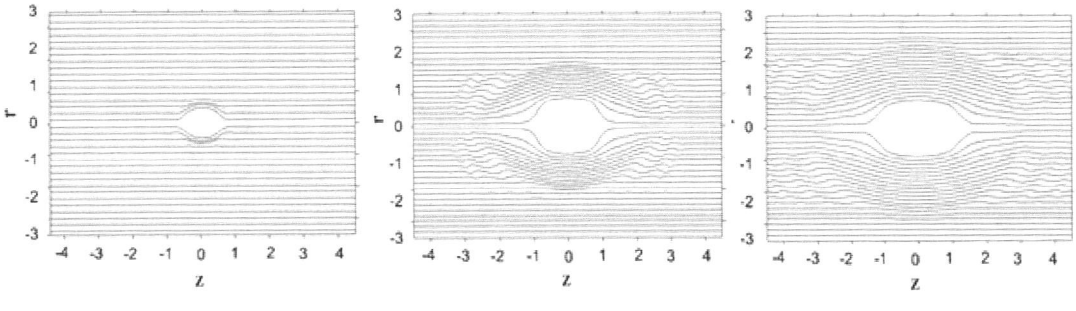

Figure 3. The magnetic field lines at successive times t = 0.15, 0.5, 1.5 [2].

The evolution of the background plasma density is shown on Figure 4. From this figure it is seen that the formation of a region of reduced density (density cavity) is accompanied by the generation of disturbances that propagate in the form of shock waves. Their structure depends on the angle of propagation with respect to the direction of the magnetic field and on the Alfven-Mach number of the plasma cloud [2].

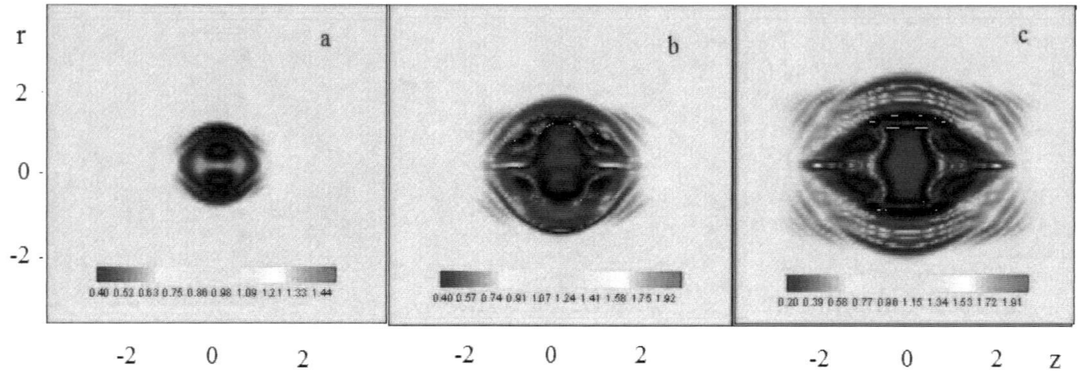

Figure 4. Temporal evolution of the background plasma density at successive instants of time t = 0.15, 0.5, 1.5 [2].

2. Inhomogeneous magnetic field. In this formulation, the problem of ion beam injection into the magnetic system of an open plasma trap is solved [13]. The initial configuration of the magnetic field and the direction of the incoming beam are shown on Figure 5.

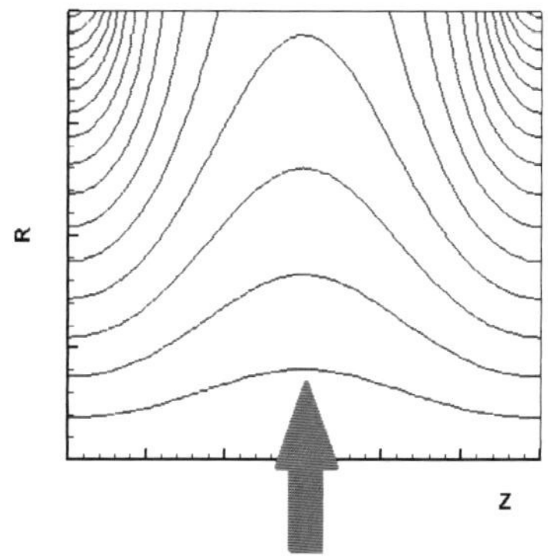

Figure 5. The configuration of the magnetic system of the open trap

Features of this task compared to the previous one are:

1) in the presence of large gradients of the magnetic field of the mirror cell, which requires a small time step.

2) the need for continuous injection of an ion beam that has a directed velocity and a Maxwell velocity distribution (temperature).

3) the need to determine the pressure of the ionic component of the plasma, for the calculation of which an algorithm was used based on the determination of temperature as the dispersion of the particle distribution function.

Here are some simulation results that characterize the possibility of heating and confining the plasma in the diamagnetic mode of an open plasma trap. Presented on Figure 6 the distribution of the beam ions and the background plasma on the (r, z) plane at successive times gives a clear idea of the dynamics of plasma flows in the beam-plasma system under consideration. The movement of the ion beam is accompanied by the formation of a cavity of the density of the background plasma and the displacement of the magnetic field (Figure 7).

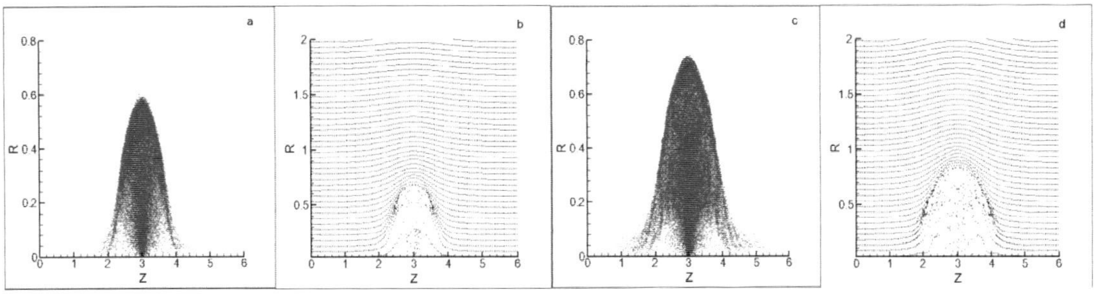

Figure 6. Distribution of beam ions (a, c) and background plasma (b, d) on the (r, z) plane at successive times t = 20 (a, b) and t = 35 (c, d).

In Figure 7 (upper) shows the distribution of the magnetic field pressure at a time corresponding to the formation of its quasistationary structure. The pressure of the magnetic field inside the cavity is 10% of its initial value. At the boundary of the cavity, layers of increased magnetic field pressure and ion pressure (Figure 7, lower) are formed, which have comparable values. The results obtained correspond to the available theoretical estimates of plasma dynamics in the diamagnetic regime of the mirror cell and can be used in planning experiments on heating and plasma confinement using the CAT facility (INP SB RAS) [13].

In all the figures presented in the work, physical variables are presented in a dimensionless form. The spatial dimensions are normalized to the dispersion length $L = c/\omega_{0i}$, the time to $t_0 = \omega_{iH}^{-1}$. For the conditions of experiments on CAT facility [13], B_0 =0.2 T, $n_0 = 10^{12} cm^{-3}$, $L = 22\ cm$, $t_0 = 5 \cdot 10^{-8} s$.

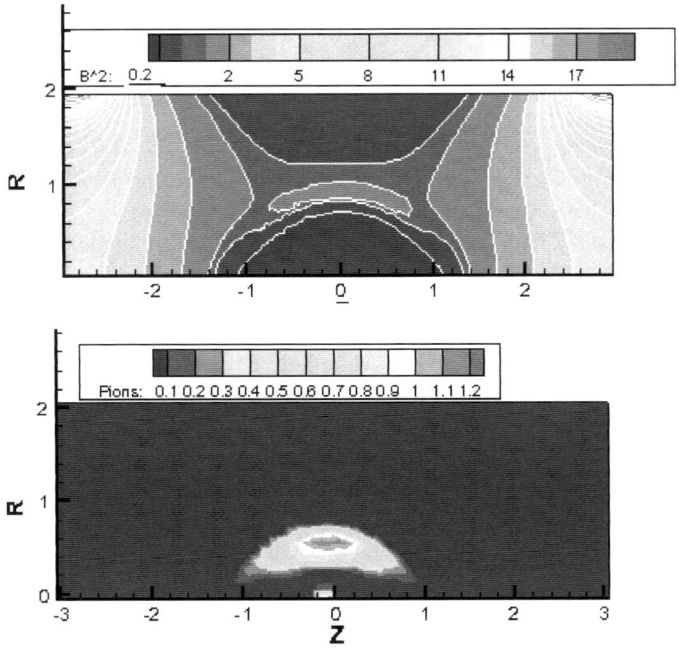

Figure 7. Distribution of magnetic field pressure (upper) and ion pressure (t = 15).

5. Conclusion

To solve the problem of plasma dynamics in open magnetic systems, a hybrid model has been created that takes into account the basic laws of physical processes and anomalous dissipation mechanisms. A modification of the Boris algorithm for solving the equations of particle motion is proposed. Two examples of using a hybrid model in cylindrical coordinates are considered. In both cases, the movement of the ion beam is accompanied by the formation of a cavity of the density of the background plasma and the displacement of the magnetic field. The above results give a clear idea of the dynamics of plasma flows in the beam-plasma systems under consideration.

This work was supported by the Russian Science Foundation, project 19-71-20026.

References

[1] Braginsky S I 1963 *Reviews of Plasma Physics* vol 1 pp 183-272 (in Russian)
[2] Berezin Yu A, Dudnikova G I, Liseykina T V and Fedoruk M P 2017 *Modeling of unsteady plasma processes* (Novosibirsk: Novosibirsk State University Press) p 359 (in Russian)
[3] Berezin Yu A and Vshivkov V A 1980 *Particle method in the dynamics of rarefied plasma* (Novosibirsk: Nauka) p 94 (in Russian)
[4] Birdsall Ch K and Langdon A B 1985 *Plasma Physics via Computer Simulation* (McGraw-Hill Book Company) p 479
[5] Byers J A, Cohen B I, Condit W C and Hanson J D 1978 *J Comput Phys* **27** (3) pp 363-96
[6] Berezin Yu A, Dudnikova G I, Fedoruk M P and Vshivkov V A 1998 *J. Comp. Math.: Fluid Dynamics* **10** pp 117-26
[7] Vshivkova L and Dudnikova G 2017 *Lecture Notes in Computer Science* vol 10187 LNCS pp 737-43
[8] Malkov M A, Sagdeev R Z, Dudnikova G I, Liseykina T V, Diamond P H, Papadopoulos K, Liu C-S and Su J-J 2016 *Physics of Plasmas* **23** (4) doi.org/10.1063/1.4945649
[9] Berezin Yu A and Vshivkov V A 1977 Numerical methods in plasma physics (Moskow Nauka) pp 150-53 (in Russian)
[10] Boris J P 1970 *Fourth Conference on Numerical Simulation of Plasmas* (Washington) pp 3-67
[11] Krall N A and Trivelpiece A W 1973 *Principles of plasma physics* (McGraw-Hill Book Company) p 686
[12] Zakharov Yu P, Ponomarenko A G, Tishchenko V N, Antonov V M, Melekhov A V, Posukh V G, Prokopov P A and Terekhin V A 2016 *Quantum Electronics* **46** pp 399–405
[13] Bagryansky P A et al 2016 *AIP Conf. Proc.* 1771 030015 https://doi.org/10.1063/1.4964171
[14] Dudnikova G I, Orishich A M, Ponomarenko A G, Vshivkov V A and Zakharov Yu P 1990 *Plasma Astrophysics* ESA SP-311 p 301

IOP Conf. Series: Journal of Physics: Conf. Series **1336** (2019) 012021 doi:10.1088/1742-6596/1336/1/012021

Parallel Algorithm for Supercomputing Simulation of Dust-Gaseous Gravitating Systems Using Particle-In-Cell and SPH Methods

N.V. Snytnikov[1], **O.P. Stoyanovskaya**[2,3], **T.A. Glushko**[3]
[1] Institute of Computational Mathematics and Mathematical Geophysics SB RAS
[2] Lavrentyev Institute of Hydrodynamics SB RAS
[3] Novosibirsk State University
E-mail: nik@ssd.sscc.ru

Abstract.
We present a supercomputer numerical model for simulating 3D dynamics of dust-gas gravitating circumstellar disk. The model combines SPH method for solving gas dynamical equations, particle-in-cell (PIC) method for solving Vlasov equation, and convolution FFT-based grid method for solving Poisson equation. The approach is based on 3D domain decomposition, sorting of particles with regard to grid cell, and transferring necessary amount of particles between subdomains. The algorithm is aimed to be run on supercomputers with distributed memory and on hybrid supercomputers.

1. Introduction

One of the important problems of modern computational astrophysics is to develop supercomputing models for simulation of planets formation in circumstellar dust-gas disks. A lot of observational data regarding formation of exoplanets appeared in the last two decades [1]. This data requires theoretical interpretations and explanations with the help of numerical experiments, conducted with high resolution and for large timescale to track fragmentation of disk into clumps and their subsequent migration.

In this regard meshless method of smoothed particle hydrodynamics [2] (SPH) has some advantages over grid methods: it allows to treat rapid changes in density on a very small space scale. On the other hand a traditional implementation of gravitational solver used in conjunction with SPH makes use of tree-code algorithms [3], and it is more complicated and less scalable in comparison with grid-based Poisson solvers.

In this paper we describe numerical model and developed parallel algorithm which combines particle-in-cell method for simulating collisionless dust component, SPH method for simulating 3D dynamics of two-phase dust-gas mixture, and grid FFT-based method for calculating gravitational potential. This numerical model was previously used for simulating thin 2D gas disk [4], while present work is focused on its 3D version.

2. Numerical model

In the developed software we use two approaches for simulating dynamics of dust. First one is based on the solving Vlasov equation using particle-in cell method, while the second approach

Content from this work may be used under the terms of the Creative Commons Attribution 3.0 licence. Any further distribution of this work must maintain attribution to the author(s) and the title of the work, journal citation and DOI.
Published under licence by IOP Publishing Ltd

employs gas-dynamics equations solved with SPH with pressure term equal to zero. Depending on the particular numerical experiment we can use either PIC method or SPH.

2.1. Gravitating Dust Component

Vlasov equation (known also as Boltzmann collisionless equation) consists of distribution function $f = f(t, \mathbf{r}, \mathbf{u})$, which is dependent on time t, space coordinate $\mathbf{r}$ and velocity $\mathbf{u}$, with the initial distribution $f^0(\mathbf{r}, \mathbf{u})$:

$$\frac{\partial f}{\partial t} + \mathbf{u}\frac{\partial f}{\partial \mathbf{r}} - \nabla\Phi(t, \mathbf{r})\frac{\partial f}{\partial \mathbf{u}} = 0, \quad f(0, \mathbf{r}, \mathbf{u}) = f^0(\mathbf{r}, \mathbf{u})$$

Dynamics of dust is governed by self-consistent gravitational force with potential $\Phi = \Phi(t, \mathbf{r})$ with boundary conditions imposed on isolated system:

$$\Delta\Phi(t, \mathbf{r}) = 4\pi G\rho(t, \mathbf{r}), \quad \Phi(t, \mathbf{r})|_{|\mathbf{r}|\to\infty} = 0,$$

where G – gravitational constant.

Density $\rho = \rho(t, \mathbf{r})$ is calculated by distribution function f:

$$\rho(t, \mathbf{r}) = \int\limits_{\mathbf{u}} f(t, \mathbf{r}, \mathbf{u})d\mathbf{u}.$$

Solving Vlasov equation is made on a uniform Cartesian grid with a standard implementation of particle-in-cell method [5, 6].

Gravitational potential is calculated with convolution method [5]. Instead of solving Dirichlet problem for Poisson equation in infinite domain:

$$\begin{aligned} \Delta\Phi(\mathbf{x}) &= \rho(\mathbf{x}), \\ \Phi(\mathbf{x})|_{\mathbf{x}\to\infty} &= 0, \end{aligned} \tag{1}$$

we then calculate the Poisson integral:

$$\Phi(\mathbf{x_0}) = -\int \frac{\rho(\mathbf{x})d\mathbf{x}}{|\mathbf{x_0} - \mathbf{x}|}. \tag{2}$$

Such approach allows to use fast Fourier transform and accomplish the calculations in $O(M^3 \log M)$ operations (where M is a number of grid nodes in one dimension).

2.2. SPH method for simulating dynamics of dust-gas mixture

We consider three dimensional two-phase dust-gas mixture where dust and gas interact with each other by friction force and by gravitational force:

$$\frac{\partial \rho_{\mathrm{g}}}{\partial t} + \nabla(\rho_{\mathrm{g}} v) = 0, \quad \frac{\partial v}{\partial t} + (v \cdot \nabla)v = -\frac{\nabla P}{\rho_{\mathrm{g}}} - \frac{K(v-u)}{\rho_{\mathrm{g}}} - \frac{\Phi}{\rho_{\mathrm{g}}}, \tag{3}$$

$$\frac{\partial \rho_{\mathrm{d}}}{\partial t} + \nabla(\rho_{\mathrm{d}} u) = 0, \quad \frac{\partial u}{\partial t} + (u \cdot \nabla)u = -\frac{K(v-u)}{\rho_{\mathrm{d}}} - \frac{\Phi}{\rho_{\mathrm{d}}}, \tag{4}$$

where ρ_{g} and ρ_{d} – density of gas and dust, v and u – velocities of gas and dust, P – pressure , $K(v-u)$ – friction force between dust and gas:

$$K = \frac{\rho_{\mathrm{d}}}{t_{stop}}. \tag{5}$$

IOP Conf. Series: Journal of Physics: Conf. Series **1336** (2019) 012021 doi:10.1088/1742-6596/1336/1/012021

Equation (4) means that dynamics of dust is modeled as a continuous medium with pressure term set to zero.

For solving this equation we use SPH implementation of [7] in Cartesian grid with the following smoothing kernel $W_{ab}(q)$ for particles a and b:

$$W_{ab}(q) = \frac{1}{\pi h^3} \begin{cases} 1 - \frac{3}{2}q^2 + \frac{3}{4}q^3, & \text{if } 0 \leq q \leq 1, \\ \frac{1}{4}(2-q)^3, & \text{if } 1 < q \leq 2, \\ 0, & \text{otherwise} \end{cases} \tag{6}$$

where $q = \dfrac{|\bar{r}_a - \bar{r}_b|}{h}$, and h is a smoothing radius.

Method of calculating friction force is available in details in paper [8].

Testing of the method implementation is done using three dimensional shock tube problem in a rectangular domain with periodic boundary conditions, where shock wave propagates along x axis. Such condition allows to compare solution of 3D problem with solution of 1D problem.

We use an equation for internal energy of gas e:

$$\rho_{\mathrm{g}} \left(\frac{\partial e}{\partial t} + v \nabla e \right) = -p \nabla v, \tag{7}$$

which is tied with the pressure using the following equation of state, where γ is an adiabatic exponent:

$$p = \rho_{\mathrm{g}} e (\gamma - 1). \tag{8}$$

Computational domain is set up to the rectangle $[0,1] \times [0,0.1] \times [0,0.1]$, and is divided in 2 parts by plane $x = 0.5$. Initial parameters in left and right parts are set to:

$$[\rho_l, p_l, v_l, e_l] = [1, 1, 0, 2.5],$$
$$[\rho_r, p_r, v_r, e_r] = [1, 0.8, 0, 2],$$
$$\gamma_l = \gamma_r = 7/5.$$

Relation between mass of dust and mass of gas is set to $\varepsilon = 1$.

Periodical boundary conditions are set for y and z, providing propagating of shock along x axis.

$$\begin{aligned} v|_{y=0} &= v|_{y=0.1}, \quad v|_{z=0} = v|_{z=0.1}, \\ u|_{y=0} &= u|_{y=0.1}, \quad u|_{z=0} = u|_{z=0.1}, \\ \rho_{\mathrm{g}}|_{y=0} &= \rho_{\mathrm{g}}|_{y=0.1}, \quad \rho_{\mathrm{g}}|_{z=0} = \rho_{\mathrm{g}}|_{z=0.1}, \\ \rho_{\mathrm{d}}|_{y=0} &= \rho_{\mathrm{d}}|_{y=0.1}, \quad \rho_{\mathrm{d}}|_{z=0} = \rho_{\mathrm{d}}|_{z=0.1}. \end{aligned}$$

We used constant smoothing radius $h = 0.01$, and time step $\tau = 0.001$, satisfying Courant condition $CFL = 0.1$.

Figure 1 demonstrates results of the obtained numerical solution at time $t = 0.1$ for coefficient of friction force $K = 500$. For this test we used sequential (non-parallel) version of the program with moderate number of particles $N = 2 \times 300 \times 32 \times 32$. It can be seen that solution is able to reproduce main features of reference solution, but for better precision it is required to use significantly more particles.

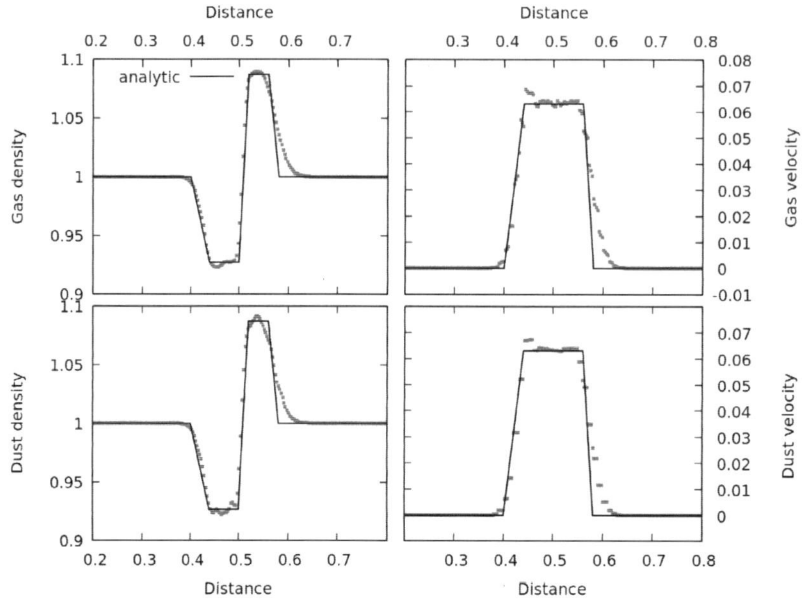

Figure 1. Solution of 3D problem of shock tube for time $t = 0.1$ and $t_{stop} = 0.002$, $K = 500$, $\varepsilon = 1$. Projection on x axis is shown. Upper pictures show gas functions, lower pictures show dust functions. Density is shown on the left, velocity is shown on the right. Red dots mark solution obtained with SPH method; black line marks reference solution. Number of particles is $N = 2 \times 300 \times 32 \times 32$, $h = 0.01$ and $\tau = 0.001$.

3. Parallel Algorithm

The developed parallel algorithm is essentially the same for both SPH and PIC particles, mainly dealing with calculating new positions of particles and transferring them among subdomains. Parallel implementation of 3D FFT-based gravitational solver uses a standard method of data transposing [9, 10].

Initial 3D computational domain is divided into K subdomains of the same size $S_1, S_2, .., S_K$ by x axis. Group of processors G_k is assigned to each subdomain S_k. Number of processors in group G_k is equal to P_k, $P_k \geq 1$. Total number of processors is $P = \sum P_k$.

For each subdomain we assign a grid, which is used for calculating gravitational potential and for calculating friction force between gas and dust phases. Number of nodes (cells) in each grid is chosen in such a way that all grid functions can be stored in RAM of one computational device and keep enough memory for particles (coordinates, velocities and gasdynamical functions). Typically it is equal to $0.5 \div 2$ million of nodes (128^3). Number of processors in group G_k is chosen to provide approximately the same amount of particles on the same processor and it may change during a numerical experiment in case if more particles will be accumulated in some particular subdomains.

The general scheme of the algorithm for 3D PIC method was suggested earlier in [11]. Its main features adopted for solving both SPH and PIC are the following.

- We use sorting of the particles (both SPH and PIC) with regard to the cell of the grid. This sorting is done on each time step and it allows to search efficiently for all neighbour particles.
- To optimize searching for neighbour in one cell (in case if number of SPH particles is big) the construction of K-d trees can be used inside each cell.

- The main computational efforts are concentrated in the calculating gas-dynamical functions in SPH method. Calculating gravitational potential is the least time-consuming part of the algorithm.
- The amount of computations required for processing one PIC or SPH particle (calculating its coordinates and gas parameters) is much higher than transferring this particle from RAM of one processor to another. It means that it is reasonable to have a moderate number of communications (transferring of particles) between subdomains for each step.
- In order to satisfy Courant condition for a single timestep SPH particle can move between nearest cells only. Because of this fact the number of particles which needed to be transfered between subdomains doesn't exceed 0.1 - 1% of the total number of SPH-particles.
- Processing of SPH particles requires that neighbour particles are located on the same processor. This suggests to use so-called "ghost" zones or overlapping zones between subdomain: when the information about particles located near the boundaries of subdomains is distributed on both subdomains.

The results of preliminary proof-of-concept numerical experiments with number of SPH particles $N_{SPH} = 10^8$, and PIC particles $N_{PIC} = 10^8$ on $2048 \times 256 \times 256$ grid nodes, and using 128 CPU-cores of Intel Xeon X5570(at Lomonosov Supercomputer Center), showed the calculation of a single timestep is about 20 seconds, with less than half of computations spent on communications.

4. Conclusions

We developed parallel algorithm for numerical model which combines SPH method for solving 3D gas-dynamics equations, PIC method for solving Vlasov equation and grid FFT-based method for calculating gravitational potential. It is based on domain decomposition and dynamic load balancing of particles and group of processors re-assigned to the subdomains depending on the particular density distribution.

First tests have shown that algorithm is promising for conducting numerical experiments with a big number of SPH and PIC particles on supercomputers with distributed memory. It could be employed for the simulation of non-stationary astrophysics problems, which require series of computations with tens of thousands of time steps.

Acknowledgements. Nikolay Snytnikov is supported by budget project 0315-2019-0009 for ICMMG SB RAS. Numerical experiments have been conducted using Siberian Supercomputer Center (Novosibirsk) and Lomonosov supercomputer in MSU.

References

[1] A.W. Howard. Observed properties of extrasolar planets. Science. V.340. P.572-576. (2013)
[2] Gingold R.A., Monaghan J.J. Smoothed particle hydrodynamics - Theory and application to non-spherical stars. Monthly Notices of the Royal Astronomical Society. Vol. 181. P. 375-389. (1977)
[3] Barnes J.E., Hut P. A Hierarchical O(NlogN) Force-Calculation Algorithm. Nature. 1986. Vol. 324, P. 446-449. (1977)
[4] O. P. Stoyanovskaya, N. V. Snytnikov, V. N. Snytnikov. Modeling circumstellar disc fragmentation and episodic protostellar accretion with smoothed particle hydrodynamics in cell. Astronomy and Computing, Volume 21, p. 1-14. (2017)
[5] Hockney, R.W. and Eastwood, J.W. Computer Simulation Using Particles. McGraw-Hill. New York (1981)
[6] Berezin Yu.A. Vshivkov V.A. Particle-in-cell method in the plasma dynamics. Nauka. (1980)
[7] J. Monaghan, A. Kocharyan. SPH simulation of multi-phase flow. Computer Physics Communications, Volume 87, Issue 1-2, p. 225-235. (1995)
[8] Gas-monodisperse dust mixtures in smoothed particle hydrodynamics: computing of stiff non-linear drag. Second Workshop on Numerical Modeling in MHD and Plasma Physics: Methods, Tools, and Outcomes. (2019) (submitted)
[9] O. Ayala, L.P. Wang. Parallel implementation and scalability analysis of 3D Fast Fourier Transform using 2D domain decomposition. Parallel Computing. Vol.39. P. 58-77. (2013)

IOP Conf. Series: Journal of Physics: Conf. Series **1336** (2019) 012021 doi:10.1088/1742-6596/1336/1/012021

[10] M.Frigo and S.G.Johnson. FFTW software. http://www.fftw.org
[11] 2. N Snytnikov. Scalable parallel algorithm for simulation of 3D dynamics of gravitating systems using particle-in-cell method. CEUR Workshops Proceedings Vol. 1576, Proceedings of the 10th Annual International Scientific Conference on Parallel Computing Technologies Arkhangelsk, Russia, p. 298-307. (2016)

IOP Conf. Series: Journal of Physics: Conf. Series **1336** (2019) 012022 doi:10.1088/1742-6596/1336/1/012022

3D numerical modeling of the plasma beam expansion using the MHD-kinetic approach

Lyudmila Vshivkova[1], **Konstantin Vshivkov**[2], **Galina Dudnikova**[1]

[1]Institute of Computational Mathematics and Mathematical Geophysics SB RAS, Lavrentiev ave., 6, Novosibirsk, 630090, Russia

[2]Institute of Laser Physics SB RAS, Lavrentiev ave., 15b, Novosibirsk, 630090, Russia

E-mail: lyudmila.vshivkova@sscc.ru

Abstract. In the current work a three-dimensional numerical model of a plasma beam expansion in the background plasma based on the kinetic description of an ion plasma component is proposed. To describe electrons the magneto-hydrodynamics equations are used. The Vlasov kinetic equations have been solved using the particle-in-cell (PIC) method. The finite-difference schemes have been utilized to solve the Maxwell system of equations and equations for electrons. The series of computations have been carried out to study collisionless mechanisms of the plasma flow interaction and a structure of the generated waves dealing with the conditions of laboratory experiments on the facility of ILP SB RAS.

1. Introduction

The 3D numerical modeling of the interaction of high-speed collisionless plasma flows is considered in the paper. Frequently, one can find these processes requiring the study in the astrophysics. The examples of such phenomena are, for instance, the Solar flares, the interaction of the Solar wind with the Earth's magnetosphere, the Supernova explosions, etc. The interaction processes of collisionless plasma flows are studied in different experiments, such as the barium cloud experiments in the Earth's magnetosphere tail (AMPTE) [1] and the experiments on the KI-1 facility (ILP SB RAS) [2]. KI-1 is the only experiment facility in Russia where such processes are studied. Another similar facility LAPD (LArge Plasma Device) [3] is in the University of California (UCLA).

Before, the processes of the plasma flow interaction had been modeled in the 2D axisymmetric geometry (r, z), mostly [4]. At the same time some important physical processes, such as 3D instabilities, had been neglected. Moreover, it was impossible to take into account the constructional facility characteristics. However, the 3D computations, comparing with those in the 2D case, require more computational resources (such as a computer memory and speed), therefore, numerical algorithms have to admit a problem solution with large space and time steps. In particular, it is necessary to use the hybrid models where an ion component of plasma is described by the kinetic approach and an electron one is considered as a fluid.

In the current paper a mathematical model and a 3D numerical algorithm based on the particle in cell (PIC) method are proposed and, also, numerical computation results are given.

Content from this work may be used under the terms of the Creative Commons Attribution 3.0 licence. Any further distribution of this work must maintain attribution to the author(s) and the title of the work, journal citation and DOI.
Published under licence by IOP Publishing Ltd

2. Numerical model

The 3D problem is considered in the following statement. The considered computational domain in the Cartesian coordinates (X, Y, Z) has the parallelepiped shape and contains the stationary plasma background of a density n_0. In the domain, there is a uniform magnetic field $\vec{B} = (0, 0, B_z)$. The plasma flow is modeled by a plasma beam expanding in the cross to the magnetic field direction. At the initial time the beam is at the left domain boundary $X = 0$ and has the shape of a sphere of a limited radius r_0. Plasma density in the sphere is spreaded uniformly. The velocities of beam particles are distributed depending on angle θ and an initial velocity on the beam front. The angle θ is the angle between a radius vector r of a separate particle, which comes out from an injection point, and the axis passing through the beam center and the injection point (Fig.1).

With time the plasma beam starts to expand in the cross direction to $\vec{B}$ and to interact with the magnetic field as well as with the background plasma. In the problem statement

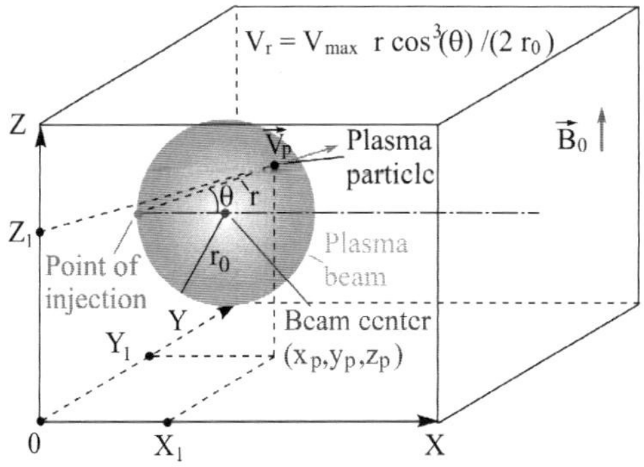

Figure 1. Schematic of the calculation model.

the boundary conditions are the same as the initial conditions on the domain boundary (the boundary conditions of the first kind). To model the process of the plasma and the magnetic field interaction, the 3D hybrid model based on the 2D numerical model described in [4], [5] is used. The considered physical process of the interaction of the collisionless plasma and the magnetic field is described by the Vlasov's kinetic equation for ions and by the magnetic hydrodynamics (MHD) equations for electrons.

The equation for ions is

$$\frac{\partial f_i}{\partial t} + \vec{v}\frac{\partial f_i}{\partial \vec{r}} + \vec{F}_i \frac{1}{m_i}\frac{\partial f_i}{\partial \vec{v}} = 0, \quad \vec{F}_i = e\left(\vec{E} + \frac{1}{c}\vec{v} \times \vec{B}\right) - \vec{R}_e, \tag{1}$$

where f_i is the ion distribution function, m_i is an ion mass, $\vec{r}$ and $\vec{v}$ are a coordinate vector, a velocity vector, $\vec{F}_i$ is the Lorentz force. The ion-electron friction force is $\vec{R}_e = m_e \kappa \left(\vec{V}_i - \vec{V}_e\right)$, m_e is an electron mass, κ is the collision frequency. The average velocities of ions and electrons are $\vec{V}_i$ and $\vec{V}_e$, $\vec{E}$ and $\vec{B}$ are electric and magnetic field intensities. The plasma density n and the average ion velocity $\vec{V}_i$ are computed from the distribution function f_i. In our case, we consider the hydrogen ions only.

For electrons, the MHD equations given below are used. The motion equation for the electron component is written as

$$0 = -en\left(\vec{E} + \frac{1}{c}\vec{V}_e \times \vec{B}\right) - \nabla P_e + n\vec{R}_e, \tag{2}$$

here P_e is an electron pressure ($P_e = nT_e$) and T_e is an electron temperature. Here the assumption $m_e/m_i \ll 1$ is used.

The equation for the electron temperature T_e is

$$n\left(\frac{\partial T_e}{\partial t} + \left(\vec{V}_e \cdot \nabla\right) T_e\right) - (\gamma - 1) P_e \nabla \cdot \vec{V}_e = (\gamma - 1)\left(Q_e - \nabla \cdot q_e\right), \tag{3}$$

here γ is the adiabatic index, Q_e is the electron heating and q_e is due to the plasma thermal conductivity. $Q_e = \vec{J}^2/\sigma$, $q_e = -\chi \nabla T_e$, where $\vec{J}$ are the plasma current, σ is the plasma conductivity and χ is the thermal conductivity coefficient.

To determine the magnetic and electric fields, the Maxwell's equations are used:

$$\begin{gathered} \nabla \times \vec{B} = \frac{4\pi}{c}\vec{J}, \\ \nabla \times \vec{E} = -\frac{1}{c}\frac{\partial \vec{B}}{\partial t}, \\ \nabla \vec{E} = 0, \quad \nabla \vec{B} = 0. \end{gathered} \tag{4}$$

In the paper, two important assumptions have been made. Firstly, it is the quasineutrality assumption, when the electron charge density is equal to the ion charge density $n_i = n_e = n$. Secondly, the displacement current is neglected in Maxwell equations. The displacement current is neglected as the process frequencies $\Omega \ll \omega_{pe} = \sqrt{4\pi n e^2/m_e}$ and $|\vec{V}_i|$, $|\vec{V}_e| \ll c$.

Finally, the sum of the friction forces has to be zero. If there is the friction force then there has to be the electron heating as the energy has to transfer into heating of the electron component. In this model, there are no influxes of energy through the boundaries, then the conservation of energy law has to be maintained. In the general form the energy conservation law is as follows

$$\sum_\alpha \frac{m_\alpha}{2} \int f_\alpha(\vec{r}, \vec{v}, t) v^2 d\vec{v} d\vec{r} + \int \left(n_e \frac{m_e V_e^2}{2} + \frac{B^2}{8\pi} + \frac{n_e T_e}{\gamma - 1}\right) d\vec{r} = const, \tag{5}$$

where α is an ion particle sort (in the paper only one sort of hydrogen ions is used).

When solving the problems of the plasma physics, the energy conservation law is one of the criteria of the solution validity as many processes in plasma are instable. The difficulty of the 3D modeling is in high volume of processed data and large computation time. It leads to the necessity to debug the algorithm, essentially, in order to have opportunity to get high-quality results with the relatively large time and space steps.

3. Algorithms and solution methods

To solution the kinetic Vlasov's equation (1), PIC method is used [4], [6], [7]. The motion equations for model particles are characteristics of the Vlasov's equation.

$$\begin{gathered} \frac{d\vec{r}_j}{dt} = \vec{v}_j, \\ \frac{d\vec{v}_j}{dt} = \frac{q_j}{m_j}\left(\vec{E} + \frac{1}{c}\vec{v}_j \times \vec{B}\right) - \vec{R}_e, \end{gathered} \tag{6}$$

where j is a number of particle, $\vec{r}_j$ and $\vec{v}_j$ are a coordinate vector and a velocity vector, q_j is a particle charge, m_j is a particle mass.

IOP Conf. Series: Journal of Physics: Conf. Series **1336** (2019) 012022 doi:10.1088/1742-6596/1336/1/012022

The plasma density n and the average velocity $\vec{V_i}$ are calculated by using model particles.

$$n = \frac{1}{r}\sum_j q_j R\left(r - r_j\right) R\left(z - z_j\right), \quad \vec{V_i} = \frac{1}{rn}\sum_j q_j \vec{v}_j R\left(r - r_j\right) R\left(z - z_j\right). \tag{7}$$

The accuracy of PIC method depends on the number particle per cell. The form factor of the PIC method is used, when the particles are "smearing" into two cells:

$$\overline{R}(x) = \begin{cases} 0, & |x| > h; \\ \frac{1}{h}\left(1 - \frac{x}{h}\right), & |x| \leqslant h. \end{cases}$$

The equation for the electron temperature and Maxwell equation are solved in the usual way on a uniform rectangular grid by using a finite-difference schemes first order accuracy in time and space $O(\tau, h)$, with specification of the boundary and the initial conditions [4], [6]. In the model, the law of energy conservation (5) is used to control the accuracy of the problem solutions.

The scheme of the program is shown in Fig. 2. Let us consider in more detail the individual blocks of the program.

Block 0: Setting up the initial data. The initial magnetic field intensity $\vec{B}_0$ and the initial particle velocity $\vec{v}_0$ are set on the fractional time step $m - 1/2$. Electric field intensity $\vec{E}_0$, the particle coordinate vectors $\vec{r}_0$ and the initial density n_0 are set on the time step m. In this block, the initial temperature T_0 and the number of particles N_{part} are also set.

Block 1: Calculation of the magnetic field intensity $\vec{B}^m$. Since, at the initial moment of time, the electric field intensity $\vec{E}$, is set on the time step m, then from the Faraday law it possible to calculate magnetic field $\vec{B}$ also at the same time step m.

Block 2: Calculation of velocity and coordinate of particles. This is the so-called Lagrange stage. At this stage, from the motion particles equations (6) the velocities and the coordinates are computed on the time - shifted grids $\left(\vec{v}^{\,m+1/2}, \vec{r}^{\,m+1}\right)$. To calculate the electric field intensity $\vec{E}$ and the magnetic field intensity $\vec{B}$ to the point of the particle position the linear interpolation is produced. At this stage, an implicit scheme is obtained for velocities, which is transformed into a system of linear algebraic equations at each time step. The scheme has a second order of accuracy. The plasma density n at the time step $m + 1$ and the average particle velocities $\vec{V}$ at the fractional time step $m + 1/2$ are calculated by the found coordinates $\vec{r}^{\,m+1}$ and the velocities $\vec{v}^{\,m+1/2}$ of model particles, respectively.

Block 3: Calculation of the magnetic field intensity $\vec{B}^{\,m+1/2}$. The magnetic field intensity is calculated from Faradays law.

Block 4: Calculation of the currents $\vec{J}^{\,m+1/2}$ and the average electron velocities $\vec{V_e}^{\,m+1/2}$. The currents, at the fractional time step, are found from the Maxwell equations, the intensity on the desired time layer $m+1/2$ is already calculated in block 3. The average electron velocities $\vec{V_e}$ on the layer $m + 1/2$ are determined from the equation $\vec{J}^{\,m+1/2} = n^{m+1/2}\left(\vec{V}_i^{\,m+1/2} - \vec{V}_e^{\,m+1/2}\right)$.

Due to neglect of the displacement current term, the electric field intensity $\vec{E}$ has to be determined from the MHD equation for the plasma electron component. The electric field intensity $\vec{E}$ is calculated on the time step $m + 1$.

Block 6: Calculation of the temperature and the energy. In this block the electron temperature is calculated at the time layer T_e^{m+1} from the equation for temperature (3). Next, the kinetic energy of the beam w_{cloud}, the background particles energy w_{bg}, the magnetic field energy w_{mag} and the total energy w_{mag} also are calculated.

Block 7: Recording data. The data are recorded for further processing.

Checks the completion condition of the computational. Computational is terminated when preset time step is reached. If the preset step is not yet reached then the computational is continued from block 1 and the magnetic field intensity is started to calculate at the next time step.

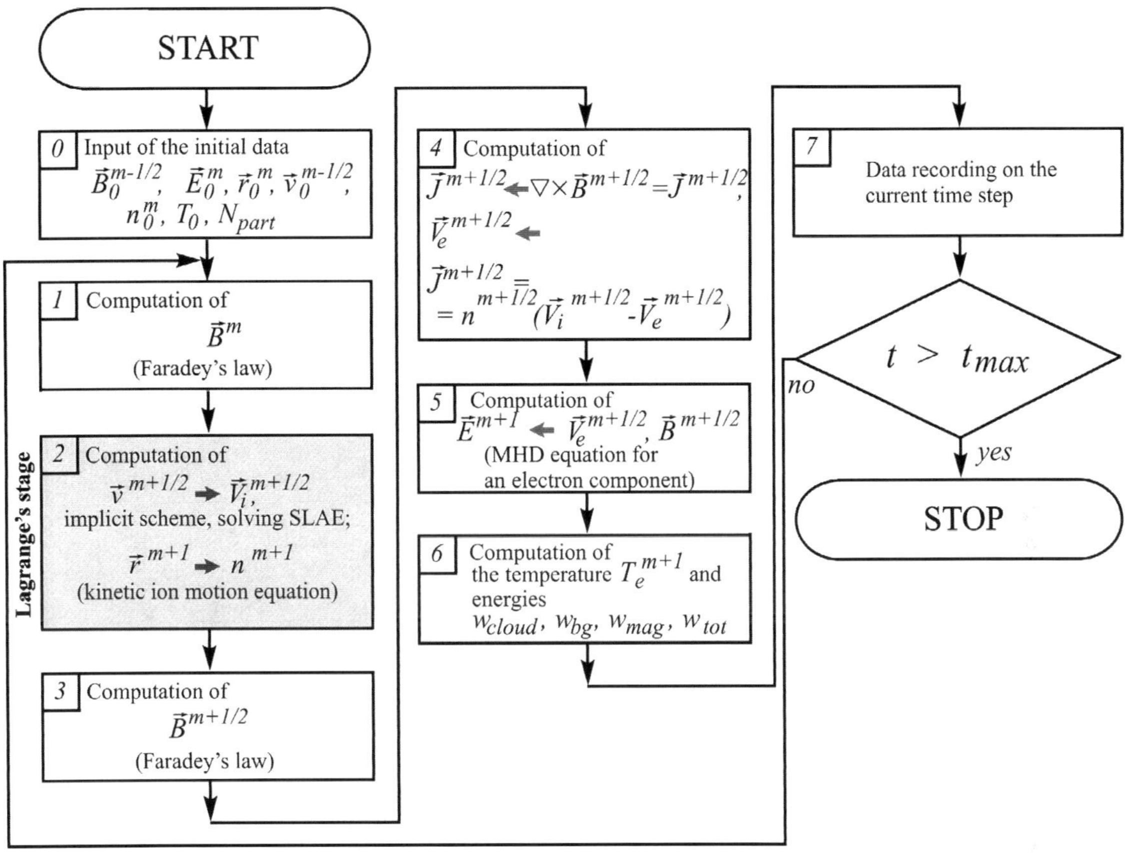

Figure 2. Flowchart of the 3D hybrid code. The main steps of the computational are presented.

4. Results and discussion

To study the dynamics of the interaction of a plasma beam with a uniform magnetic field, numerical experiments were carried out with characteristic parameters close to the parameters of KI-1 facility (ILP SB RAS) [2], [9]. The aim of the numerical experiments was to demonstrate the possibility of numerical simulation of the processes observed at the KI-1 facility based on the presented 3D hybrid numerical model.

The characteristic parameters of the problem are given in Table 1. For the given Mach number $M_A > 1.8$, the deceleration radius $\tilde{R}$ is equal to the gas-dynamic radius R_{gas} and is calculated by the follow formula:

$$\tilde{R} = R_{gas} = \left(\frac{3M}{4\pi n_* m_i}\right)^{1/3}, \tag{8}$$

where M is the mass of a spherical plasma cloud, n_* is a density of the background plasma. Mach number $M_A = V_{max}/V_A$, where again V_{max} is a initial particle velocity at the beam front

IOP Conf. Series: Journal of Physics: Conf. Series **1336** (2019) 012022 doi:10.1088/1742-6596/1336/1/012022

and $V_A = B_0/\sqrt{4\pi n_* m_i}$ is Alfvén velocity. It should be noted that the formula (8) is derived under the assumption that the spherical cloud is expanding. To estimate the deceleration radius of the beam according to (8), it is assumed that the beam is a part of some large spherical plasma cloud, whose mass is eight times larger then the mass of the beam. The delta parameter is calculated as follows: $\delta = (R_{gas}/R_L)^2$. For the effective interaction of the beam particles with the magnetic field, it is necessary that the condition $\delta > 1$ be satisfied. Also, important characteristic dimensions are larmor radius $R_L = V_{max}/\omega_{iH}$ and the length of c/ω_{pi}, where $\omega_{iH} = eB_0/(m_i c)$, ω_{pi} is the ion plasma frequency. Numerical parameters are given in Table 2.

Table 1. Characteristic parameters of the problem

name	designation	dimension	value
Alfvén velocity	V_A	$[cm/s]$	3.2×10^6
deseleration radius by field	R_B	$[cm]$	100.00
deseleration radius by backgr.	R_{gas}	$[cm]$	42.00
larmor radius	R_L	$[cm]$	27.00
delta parameter	δ	$[-]$	2.50
Mach number	M_A	$[-]$	6.50
characteristic length	c/ω_{pi}	$[cm]$	4.16

Table 2. Numerical parameters of the problem

name	designation	dimension	value
number of cells by X, Y, Z	$im \times lm \times km$	$[-]$	$32 \times 32 \times 32$
size of calc. region by X, Y, Z	$xm \times ym \times zm$	$[cm]$	$115 \times 115 \times 115$
number of background particles in cell	lp	$[-]$	27.00
number of beam particles	jmc	$[-]$	0.8×10^6

Numerical experiments were carried out when the number of physical particles N_{H^+} and the initial radius of the cloud r_0 were varied. The beam plasma as well as background plasma consist of the hydrogen ions. At the initial time, the beam is located at the left boundary of calculation domain $X = 0$ and has the spherical shape of radius r_0.

Consider the dynamics of the plasma beam density when Mach number $M_A = 6.5$ and the initial beam radius $r_0 = 10[cm]$. The velocities of the model particles are distributed as follows:

$$V_r = V_{max} r \cos^3(\theta)/(2r_0),$$

where V_r is the velocity of an individual particle directed to the the positive direction of the axes along the straight line connecting the injection point and the particle position point (radius vector of the particle). The initial beam velocity is V_{max}, r is the module of particle radius vector, r_0 is the initial radius of the cloud, θ is the angle between the radius vector of the particle and the axis, passing through the injection point and the center of the beam (Fig.1).

Fig. 3 (a – c) shows the plasma density beam at the XZ plane at consistent moments of time $4[\mu s]$, $6[\mu s]$ and $8[\mu s]$. At the initial moment of time, the plasma beam begins to expand

IOP Conf. Series: Journal of Physics: Conf. Series **1336** (2019) 012022 doi:10.1088/1742-6596/1336/1/012022

in the transverse direction to the magnetic field, transferring its energy to the field and the background plasma (Fig. 3 (a – c)). In the process of expansion, the beam is decelerated by the magnetic field (Fig. 3 (b, c)). After reaching the deceleration radius $\tilde{R}$, the reverse process begins: the magnetic field begins to act on the beam, compressing it in the opposite direction (Fig. 3 (c)). As the plasma beam expands, a current layer forms at the beam boundary and also an ion concentration appears on its front (Fig. 3 (a – c)).

Fig. 4 (a – c) illustrates the density of a plasma beam at the YX plane at consistent moments of time ($4[\mu s]$, $6[\mu s]$ and $8\ [\mu s]$). It can be seen from the figure that a dense current layer is formed (Fig. 4 (a)) with pronounced inhomogeneities (Fig. 4 (b, c)), which indicate a developing instability of the current layer.

Figures 5 (a – c) and 6 (a – c) show the background plasma density at the XZ and YX planes at successive times ($4[\mu s]$, $6[\mu s]$, and $8\ [\mu s]$). It can be seen that the particles of the background plasma are displaced from the region where the beam is located. This also indicates the displacement of the magnetic field and the formation of a magnetic cavity – a region of almost zero magnetic field. At the boundary of the magnetic cavity, a higher ion density concentration which has asymmetric shape is noticeable 6 (a – c).

The distributions of the ion density of the beam and the background plasma presented at the figures 3 – 6 provide a visual representation of the nature of the plasma flows and their interaction for the Mach numbers $M_A \gg 1$.

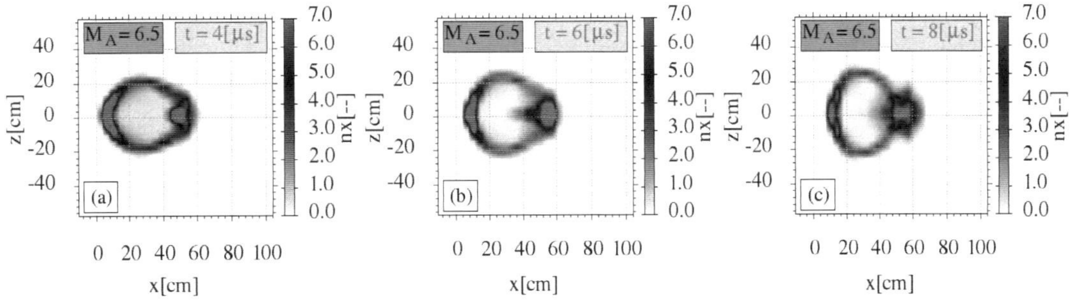

Figure 3. Beam density at successive times in the XZ plane.

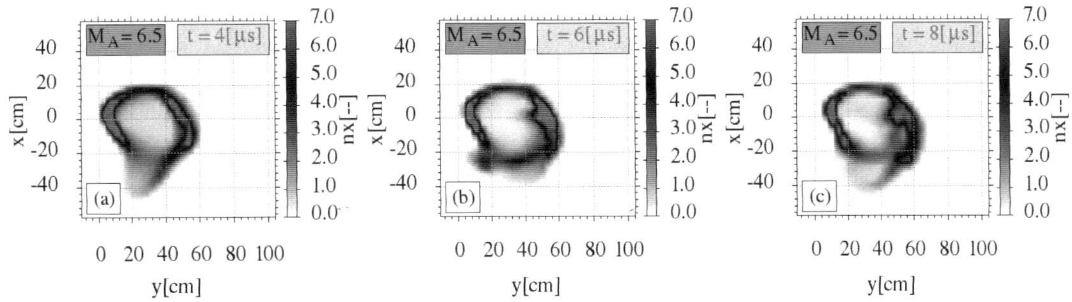

Figure 4. Beam density at successive times in the YX plane.

The figures 7 and 8 show the phase planes for the particles of the beam and the background particles along the rays that are drawn from the injection point and belong to the XZ plane. Phase planes were constructed for the rays drawn at different angles (45° and 86°), measured

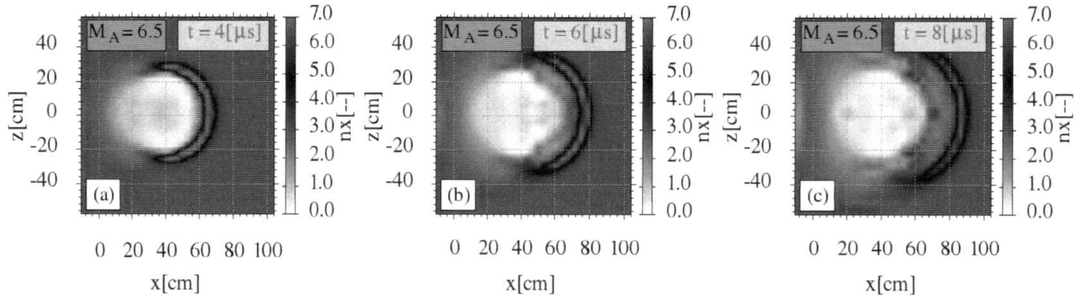

Figure 5. Background density at successive times in the XZ plane.

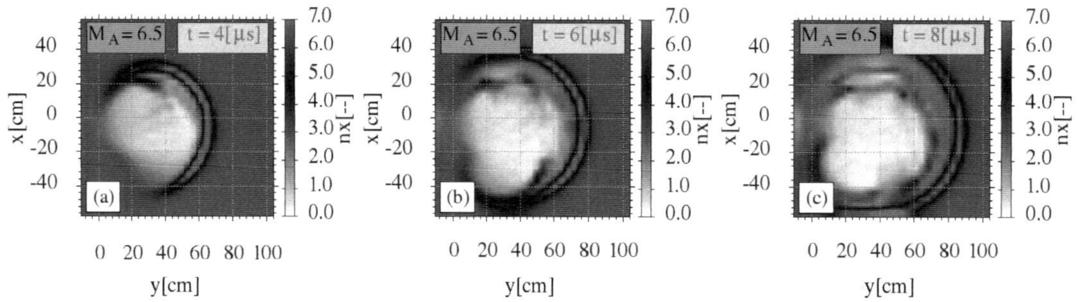

Figure 6. Background density at successive times at the YX plane.

from the Z axis. From the figures 7 (a, b) it can be seen that the particles of the beam that move at the angles 45° and 86°, begin to wrap around the deceleration radius $\tilde{R}$. The irregular pattern of the distribution of the beam and background particles on the phase plane (Fig. 7 (c) and 8 (c)), talks about developing instability. For the particles flying at an angle of 86° (Fig. 7 (c) and 8 (c)), the irregular pattern is observed to a greater extent, since the expansion occurs strictly across the magnetic field.

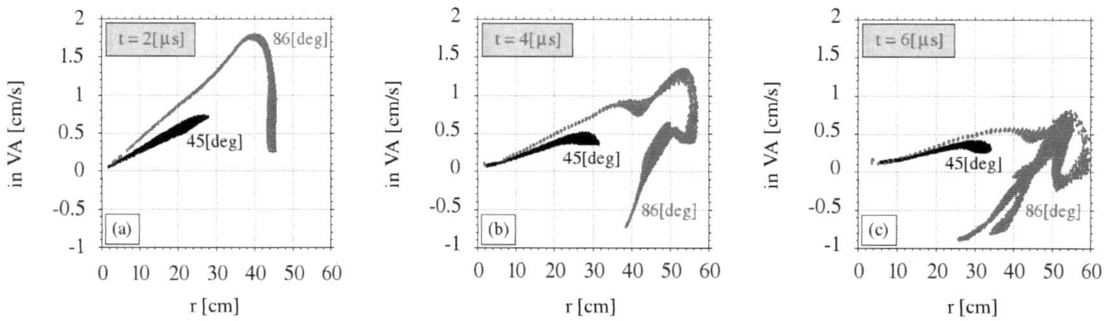

Figure 7. The phase planes of the beam particles along the rays r, lying at angles 45° and 86°, at successive times.

5. Conclusion

Thus, the 3D numerical model allowing study the dynamics of nonstationary nonlinear processes of the high-speed (super-Alfvénic) plasma flow interaction in the constant magnetic field

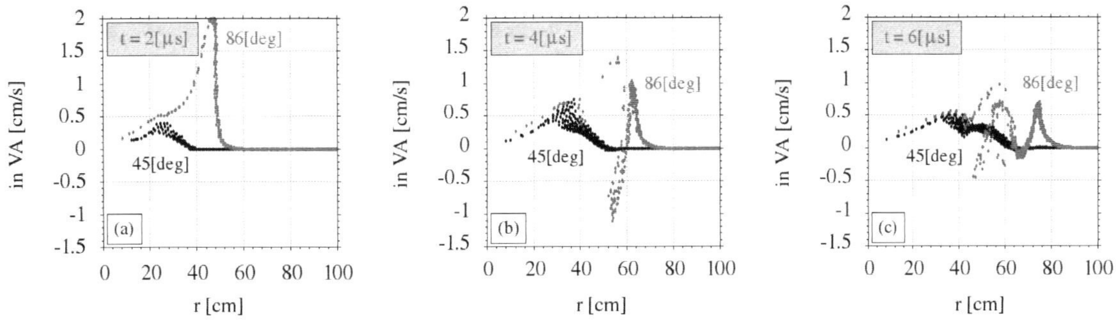

Figure 8. Phase planes of background particles along the rays r, lying at angles 45° and 86°, at successive times.

has been developed. When computing, there have been observed such effects as the current layer formation, its instability and the magnetic cavity generation. The familiar phenomena are observed in the laboratory experiments taking place on the KI-1 facility (ILP SB RAS) [9], [10].

The model and the algorithm have been developed in the frame of the Russian Foundation for Basic Research of grant N 18-29-21025 and the budget project ICMMG SB RAS N 0315-2019-0009. The computations have been carried out in the frame of the budget project of ILP SB RAS N 0307-2016-0002 using the Siberian Supercomputer Center SB RAS (SSCC SB RAS) facility.

References

[1] Bernhardt P A, Rousser-Dupre R A, Pongratz M V, et al 1987 *J. Geophys. Res.* **92** (A6) 5777
[2] Zakharov Yu P, Antonov V M, Boyarintsev E L, et al 2006 *J. Plasma phys.* **32** (3) 207
[3] Gekelman W, et al 1991 *J. Review of Scientific Instruments* **62** (12) 2875
[4] Berezin Yu A, Dudnikova G I, Liseikina T V and Fedoruk M P 2018 *Novosibirsk: NSU*
[5] Vshivkov V A, Boronina M A, Genrikh E A, Dudnikova G I, Vshivkova L V and Sudakov A M 2019 *J. Phys.: Conf. Series* (these proceedings)
[6] Berezin Yu A and Vshivkov V A 1980 *,,Nauka"*
[7] Hockney R W and Eastwood J W 1987 *Moscow ,,Mir"*
[8] Berezin Yu A, Vshivkov V A, Dudnikova G I, Zakharov Yu P, Orishich A M, et al 1988 *Preprint ITAM SB AS USSR Novosibirsk* (13)
[9] Vchivkov K V, Nakashima H, Zakharov Yu P, Esaki T, et al 2003 *Japanese J. of appl. phys.* **42** (1) 6590
[10] Zakharov Yu P 2003 *IEEE Trans. PS.* (6) 1243

IOP Conf. Series: Journal of Physics: Conf. Series **1336** (2019) 012023 doi:10.1088/1742-6596/1336/1/012023

The cuFFT code for N-body simulation

Vladimir Prigarin[1,2], **Dmitry Karavaev**[1], **Viktor Protasov**[1,2]

[1] Institute of Computational Mathematics and Mathematical Geophysics SB RAS, 6-th Lavrentyev avenue, Novosibirsk, Russian Federation

[2] Novosibirsk State Technical University, 20 Prospekt Karla Marksa, Novosibirsk, Russia

E-mail: vovkaprigarin@gmail.com

Abstract. The complexity of astrophysical processes lies in the joint consideration of components of various nature. For example, in the collision problem of galaxies, the three-dimensional dynamics of an interstellar gas and a stellar component is considered. The modeling of these components can be based on completely different classes of numerical methods. One possible solution to this problem is to use the Eulerian-Lagrangian approach, in which physical quantities are concentrated at material points, which is typical for the SPH (Smoothed Particle Hydrodynamics) method, and the forces are calculated on an adaptive grid attached to a system of material points. This approach uniformly takes into account both the dynamics of a continuous medium and discrete particles, and also eliminates a number of drawbacks inherent in the original method. The calculation of gravitational interaction is carried out by solving the Poisson equation for the gravitational potential. In this case, all particles are projected onto the computational grid and the potential values in each cell are already calculated on it. The solution of the Poisson equation for the gravitational potential is performed using the fast Fourier transform. The article describes the new cuFFT code Virtual Planetarium for modeling astrophysical objects based on the SPH method, supplemented by the Godunov method for calculating pressure and momentum flows between particles, and the fast Fourier transform method for solving the Poisson equation for the gravitational potential. The paper describes the rationale for the transition to such a computational model, kinetic and hydrodynamic approaches are described in detail. Simulation of the collapse of an isothermal gas cloud is performed. Method to reproduce the evolution of instabilities in form of two density arms is realized.

1. Introduction

Modern theoretical astrophysics has practically exhausted the possibilities of analytical and semi-analytical solutions that can be obtained under the assumptions of spatial symmetry of the processes under consideration. A computational experiment has become main tool of theoretical astrophysics. The complexity of astrophysical processes can be explained by the joint consideration of components of various nature. To describe such a process, it is necessary to solve the equations of hydrodynamics to describe gas motion and to solve kinetic equation for recording the stars motion. Solution methods are based on completely different classes of numerical methods. One of the trends for resolving such problems is the use of Lagrangian-Eulerian methods (Arbitrary Lagrangian-Eulerian - ALE in foreign literature), which are simultaneously suitable both for solving hydrodynamic equations and for describing particles motion. Although it is worth noting that there are studies on the modification of mathematical models to describe various components by a single class of hyperbolic equations [1]. We will

Content from this work may be used under the terms of the Creative Commons Attribution 3.0 licence. Any further distribution of this work must maintain attribution to the author(s) and the title of the work, journal citation and DOI.
Published under licence by IOP Publishing Ltd

focus on the development of the Lagrangian-Eulerian approach.

In the last decade, several codes have been created in the field of computational astrophysics, successfully and at the same time in various ways, implementing the Lagrangian-Eulerian approach. Among them are AREPO [2], BETHE-HYDRO [3], GIZMO [4] codes. One way or another, all codes are based on representing the solution in the form of a set of material points with physical characteristics placed in them, which is typical for the method of smoothed particles [5] (Smooth Particles Hydrodynamics - SPH in foreign literature). Next, different mechanism is used for the method of interaction with neighboring points. In particular, it is a way to build grids. Let us dwell on the codes description in details.

The AREPO code is based on moving mesh technology which in turn is based on the Voronoi and Delaunay triangulations with Lloyd regularization. This approach allows you to adapt the grid to the solution. The classical Godunov method is used as the main for solving the hydrodynamic equations. This is due to the fact that it is rather difficult to build a higher-order scheme on a moving mesh. To solve the Poisson equation, an approach is used based on writing the equation for the total mechanical energy into the equation for the sum of all types of energy (internal, kinetic and potential). Such an equation for the total energy on the right side has a time derivative of the potential and its gradient, which is calculated using the Poisson integral by particle-grid methods. An individual time step is used for time integration. This approach with all the advantages is quite difficult in terms of computational costs. And the question about of the quality of solution in the areas described by less detailed grid cells remains open. However, the AREPO code is one of the most used in the World at the moment.

The BETHE-HYDRO code is based on the ALE approach combining the advantages of both the Eulerian and Lagrangian approaches. The hydrodynamic equations are formulated in a Lagrangian non-conservative form and are solved on an unstructured grid. The numerical method is based on the operator approach, which allows one to construct consistent schemes for approximating the gradient and divergence operators. To solve the Poisson equation in a one-dimensional formulation, the sweep method (or the Thomas method in foreign literature) is used. In the two-dimensional formulation of the Poisson equation, it is solved using the conjugate gradient method. Then, the potential is corrected to preserve the total energy (the sum of the kinetic, internal and potential energies) of the system. It is worth noting that it is still not possible to save the full energy of the system completely. But the error in the collapse problem is about 10-2 percent that is negligible. Unfortunately, the code was not developed for the three-dimensional case.

In the GIZMO program code, a new meshless approach to solving the equations of gravitational gas dynamics is developed and implemented. The approach is based on a combination of classical grid methods and the SPH method. The method consists in using the equations of gas dynamics in Euler coordinates, which, using the Galerkin variational principle, are multiplied by trial functions. Feature of these functions is the fact they are not tied to the computational grid, but to individual particles, similar in nature to SPH particles. To determine the values at the boundaries of the region, we apply solution of the Riemann problem using the Godunov method.

An analysis of these codes shows that the main computational problem is the construction of grid (or its analogue) to describe the interaction between particles. At the same time, the ideology of the kinetic approach based on the description of the motion of material points is preserved in all methods. In this concept, computational model is developed and described in the paper. To organize the calculations, we will use a uniform cubic grid in which a cloud of particles will be placed. This approach makes it quite simple to organize the Eulerian stage of calculations.

The kinetic computational model and the numerical methods that is used to solve them are described in second section. The third section is devoted to discussion points that is a natural extension of the models and implementations described in the paper. The fourth section is devoted to the results of computational experiments.

2. The Numerical Model

We write the equations of gravitational hydrodynamics with zero pressure in the form of mass (1) and angular momentum (2) conservation laws in Euler coordinates:

$$\frac{\partial \rho}{\partial t} + \nabla \cdot (\rho \vec{u}) = 0, \tag{1}$$

$$\frac{\partial \rho \vec{u}}{\partial t} + \nabla \cdot (\rho \vec{u} \times \vec{u}) = -\rho \nabla \Phi, \tag{2}$$

supplemented by the Poisson equation for the gravitational potential (3):

$$\Delta \Phi = 4\pi G \rho, \tag{3}$$

where ρ is the density, $\vec{u}$ is the velocity vector, Φ is the gravitational potential, G is the gravitational constant.

We will consider the dynamics of an arbitrary number of material points (particle). The mass conservation law of particle is obtained naturally when integrating the equation (1) in a small neighborhood of a point at which other particles do not fall. We write the second Newton's law for an arbitrary number of particle using equation (2) in the Lagrangian formulation:

$$\rho \frac{d\vec{u}}{dt} = -\rho \nabla \Phi. \tag{4}$$

In equation (4), it is necessary to reduce the density function, which is nonzero in a small neighborhood of the point. As a result, we obtain the equations for the particles motion:

$$\frac{d\vec{u}}{dt} = -\nabla \Phi, \quad \frac{d\vec{x}}{dt} = \vec{u}. \tag{5}$$

To solve equations (5), we need to calculate the potential gradient. To do this, we introduce a uniform cubic Eulerian grid with a step h in each direction and project each particle onto the corresponding cell. It means each particle will contribute to the mass of the corresponding cell (i, j, k). Thus, in each cell, we can calculate the density by the equation (6):

$$\rho_{i,j,k} = \frac{\sum_l M_l}{h^3}, \tag{6}$$

where M_l is the mass of the particle falling into the cell (i, j, k). Knowing the density in each cell of the computational domain using the method based on the fast Fourier transform, we can find the gravitational potential. The study and description of the method for solving the Poisson equation is described in [6, 7] in details. Since the function of the gravitational potential is a smooth function, we will use the central differences to find the potential gradient.

To solve equations (5) for a particle in time step, we will use a two-layer scheme with stepping, where velocities are calculated by the equation (7):

$$u_x^{m+1/2} = u_x^m - \frac{\tau}{2h} \left(\Phi_{i+1,j,k} - \Phi_{i-1,j,k} \right), \tag{7}$$

$$u_y^{m+1/2} = u_y^m - \frac{\tau}{2h} \left(\Phi_{i,j+1,k} - \Phi_{i,j-1,k} \right),$$

$$u_z^{m+1/2} = u_z^m - \frac{\tau}{2h}\left(\Phi_{i,j,k+1} - \Phi_{i,j,k-1}\right),$$

Coordinates are calculated using formula (8):

$$\vec{x}^{m+1} = \vec{x}^m + \tau \vec{u}^{m+1/2}. \quad (8)$$

Such a scheme corresponds to the Euler Lagrangian approach used by the team of authors earlier [1, 6, 7]. Choice of appropriate time step is an important condition for numerical scheme stability. To do this, we use the Courant-Friedrichs-Levy condition

3. The Discussion

In this section, we describe discussion points related to the constructed computational model and to the software implementation. These discussion points are not yet included in the code, but can be used when developing the code.

(i) In the introduction, it was said about using only uniform cubic grid. This is done only to simplify the description of the computational model and its use for the needs of the Novosibirsk planetarium. Of course, the use of material points allows us to introduce adaptive meshes (Adaptive Mesh Refinement - AMR in a foreign literature) in a fairly simple and natural way, applying the appropriate numerical methods to restore the gravitational potential. In the future, we plan to develop our approach in this direction.

(ii) In the second section, we used the pressureless equation of state, which depends only on the density function. For the projection of particles on an Eulerian grid, it is sufficient to use the masses conservation law to determine the density. However, to describe the ideal gas, it is necessary to use thermodynamic variable – entropy, and projection method similar to the projection of material points velocities onto a cell.

(iii) The simplest projection of material points onto computational grid was also used. To eliminate computational noise, it is necessary to use smoothing kernels similar to SPH and particle-in-cell methods. In the future, if necessary, the question of choosing a core will be considered separately.

(iv) The formulas for recalculating the velocities and coordinates of material points are written in vector form explicitly. That fact explains usage of vector SSE extensions built into Intel and AMD processors of modern personal computers. The developed program code is oriented for such computing devices. Also, most of personal computers are equipped with NVIDIA graphics accelerators, for which special libraries have been created to perform fast Fourier transform calculation that is a part of the CUDA technology. It shall be noted, that the developed code works using cuFFT.

4. The Numerical Simulation

As a model problem, we consider the collapse of an isothermal cloud in dimensionless variables. To do this, we introduce an auxiliary computational grid in a cubic region of domain size. At the initial moment of time, in a sphere of unit radius we uniformly distribute particles of unit total mass with a model temperature and Keplerian rotation speed. In our model problem, we will study the behavior of gas particles ensemble at various values of velocity dispersion. The problem of studying the fragmentation of rotating cold cloud is especially interesting in the context of the evolution process of elliptical galaxies [8]. Analysis of the simulation results shows (see fig. 1) that instabilities evolve into two density arms form when increasing velocity dispersion up to the same initial density distribution. In the present study, we do not plan to analyze evolution of such instabilities similar to [7]. The ability to reproduce such disturbances by the developed computational model is the main result of computational experiments. In the future, we will consider evolution of instabilities in many bodies system in the context of modeling planetary systems and galaxies [9] in details.

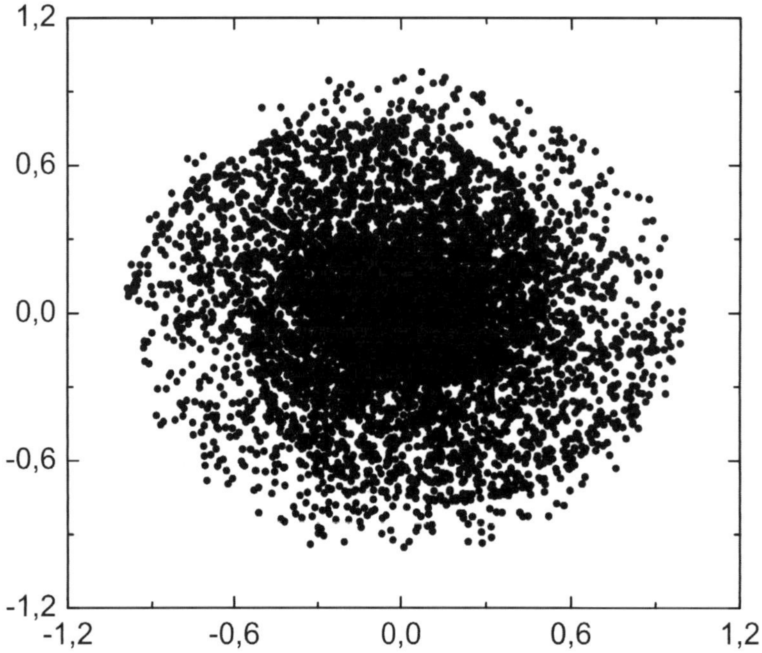

Figure 1. The simulation result.

5. Conclusion

The new cuFFT code "Virtual Planetarium" for astrophysical objects simulation was discussed. The program code is based on the Eulerian-Lagrangian combination of the smoothed particle method and Godunov method. The rationale for transition to such a computational model was also discussed. Using the developed code, the simulation of collapse of a rotating gas cloud was performed.

Acknowledgments

This research was supported by the RFBR grant 18-41-543012 and by the ICMMG SB RAS (project 0315-2019-0009).

References

[1] Kulikov I. GPUPEGAS: A New GPU-accelerated Hydrodynamic Code for Numerical Simulations of Interacting Galaxies // The Astrophysical Journal Supplements Series. 2014. vol. 214, I. 1. Article Number 12.
[2] Springel V. E pur si muove: Galilean-invariant cosmological hydrodynamical simulations on a moving mesh // Monthly Notices of the Royal Astronomical Society. 2010. vol. 401. p. 791-851.
[3] Murphy J., Burrows A. BETHE-Hydro: An Arbitrary Lagrangian-Eulerian Multidimensional Hydrodynamics Code for Astrophysical Simulations // The Astrophysical Journal Supplement Series. 2008. vol. 179. p. 209-241.
[4] Hopkins P. A new class of accurate, mesh-free hydrodynamic simulation methods // Monthly Notices of the Royal Astronomical Society. 2015. vol. 450, I. 1. p. 53-110.
[5] Monaghan J. Smoothed Particle Hydrodynamics // Rep. Prog. Phys. 2005. Vol. 68. p. 1703-1759
[6] Vshivkov V., Lazareva G., Snytnikov A., Kulikov I., Tutukov A. Hydrodynamical code for numerical

simulation of the gas components of colliding galaxies // The Astrophysical Journal Supplement Series. 2011. vol. 194, I. 2. Article Number 47.

[7] Kulikov I., Vorobyov E. Using the PPML approach for constructing a low-dissipation, operator-splitting scheme for numerical simulations of hydrodynamic flows // Journal of Computational Physics. 2016. vol. 317. p. 318-346.

[8] Worrakitpoonpon T. Spherical symmetry breaking in cold gravitational collapse of isolated systems // Monthly Notices of the Royal Astronomical Society. 2015. vol. 446. p. 1335-1346.

[9] Benhaiem D., Sylos Labini F., Joyce M. Long-lived transient structure in collisionless self-gravitating systems // Physics Review E. 2019. vol. 99. Article Number 022125.

IOP Conf. Series: Journal of Physics: Conf. Series **1336** (2019) 012024 doi:10.1088/1742-6596/1336/1/012024

Scalability investigation of parallel algorithms for plasma dynamics problems in open magnetic traps by simulation modeling

D Weins[1], M Boronina[1], V Vshivkov[1], E Mesyats[1], G Dudnikova[1] and I Chernykh[1]

[1] The Institute of Computational Mathematics and Mathematical Geophysics SB RAS (ICM&MG SB RAS) 6, Acad. Lavrentieva Pr., Novosibirsk, 630090, Russian Federation

vins@sscc.ru, boronina@ssd.sscc.ru, vsh@ssd.sscc.ru, mesyats@gmail.com, gdudnikova@gmail.com, chernykh@parbz.sscc.ru

Abstract. This paper describes the use of predictive simulation modeling for investigating the scalability of parallel algorithms. The study is carried out using a multiagent approach. The general principles of designing a model for the execution of a parallel algorithm on a given supercomputer architecture are described along with their application for researching this algorithm. The paper also presents the results of simulating the execution of this algorithm on more than a million cores.

1. Introduction

Modern and promising supercomputers are characterized by high parallelism, hybrid structure, and hierarchical communication medium, so software development for such systems is a complex scientific task. However, even when a program for performing this complex task on a supercomputer is made, a researcher encounters the problem of a boundary of effective scalability of this program, i.e., what is the maximum number of computational nodes required for the best acceleration of this program on a given supercomputer architecture and with given input data? It is possible to solve this problem by multiply running this program on a different number of computational nodes, but the cost of computer time is very high for modern supercomputers, and not all supercomputer centers are ready to provide the required amount of resources.

Within the framework of an integral approach to developing parallel programs [1], created at the Institute of Computational Mathematics and Mathematical Geophysics SB RAS (ICM MG SB RAS), this problem is solved using multiagent modeling methods for simulating multiple runs of the program, which requires a much smaller amount of real resources and time and allows studying the scalability of the algorithm under study on existing and promising supercomputers. This modeling was previously carried out using the AGNES multiagent modeling system [2], but, due to the specific features of this system, it was necessary to apply the Actor model for minimizing the communication overhead of agent communication [3]. Experimental studies particularly show that simplicity of implementation, scalability, and fault tolerance are the features that make the Erlang programming language the most fit for studying the scalability of algorithms by simulation modeling methods [4].

Content from this work may be used under the terms of the Creative Commons Attribution 3.0 licence. Any further distribution of this work must maintain attribution to the author(s) and the title of the work, journal citation and DOI.
Published under licence by IOP Publishing Ltd

This paper illustrates the application of this approach to studying the scalability of numerical modeling of plasma dynamics in axially symmetric open magnetic traps [5].

2. General scheme for the parallel algorithm research

In general form, the computational process is a set of threads executed in parallel on an individual computational node, which interact between each other by exchanging values (messages). The main characteristics of threads are the computation execution time and the time of data transmission to another computational node. The simulation modeling of the execution of a computational process on a supercomputer is carried out by constructing a model of interaction of the computational process threads. In this model, each thread is assigned to a computational unit and communication unit, with a scheme of message interaction between the threads being constructed (Figure 1).

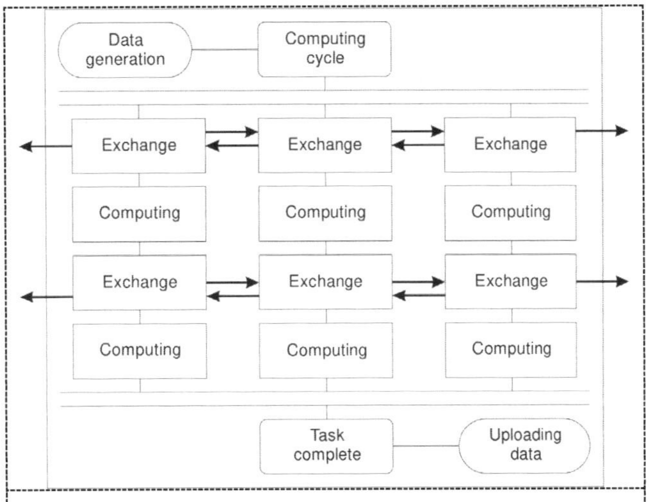

Figure 1. Example of the model of interaction of the computational process thread.

The main efficiency parameter of the computational process under study during its execution on high-performance supercomputers is its scalability. It should be noted that different scalability indicators are important for different computational algorithms. For example, for statistical modeling methods [6], it is weak scalability, i.e., the execution time decreases as the number of computational nodes increases. For network methods of numerical modeling [7], it is strong scalability, i.e., the execution time of the algorithm remains unchanged as the computational domain increases proportionally to the number of computational nodes.

This paper does not touch upon the scalability of the computational processes within the framework of a single central processor or accelerator as the number of cores of computational elements is currently determined and much smaller than the potential number of computational nodes of high-performance supercomputers. Thus, all computations and exchanges between the computational elements of a single computational node are taken as a fixed value that does not affect the general computational scheme.

The modeling of the computational process requires studying the thread interaction model and the threads having similar behavior (performing identical computations and exchanges) are determined. Each group of homogeneous threads is assigned to a class of actors simulating the execution of computational units and message exchange with other actors (threads). The execution of the entire computational process is simulated by forming a necessary number of samples of each class of actors. It should be noted that Erlang allows for message exchange between any actors in the system, thereby making it possible to arrange any kind of interaction between computational threads.

IOP Conf. Series: Journal of Physics: Conf. Series **1336** (2019) 012024 doi:10.1088/1742-6596/1336/1/012024

For each actor, an algorithm for simulating the behavior of a computational thread is programmed and the main execution parameters of a thread are transferred, such as the time it takes to determine each computational unit, the number of transferred values, and the time it takes to transfer each value for all communication units. As soon as all actors are formed and registered in the system, the multiagent model of execution of the computational process is being initialized. Each actor seeks for its "neighbors", i.e., actors with whom it communicates while simulating the thread execution. In accordance with the thread execution algorithm and its own parameters, the actor configures inner gauges for simulating calculations and message exchange-reception.

Upon reaching the criterion for stopping the algorithm execution, the actors finish all the operations from an incomplete cycle and return the results of their work. The main parameters studied after the simulation are the total operating time of the computing process, the execution time of each actor, the number of iterations of the calculation cycle performed by each actor, and the data latency between the actors.

The efficiency of a computational process during its execution on a high-performance supercomputer is based on the concept of its scalability directly dependent on the total execution time, so it is the key parameter. It should be noted that the main parameter affecting the total execution time of the computational process are data latency in communication networks (or in logical lines between actors when dealing with simulation modeling) because the execution time of computational units and the time it takes to transfer an element are fixed values for each computational node. Therefore, this parameter is studied with a special attention in the modeling of execution of computational processes on a large number of computational nodes.

Moreover, only the extensiveness of the computational system is investigated within this scheme. Naturally, with propagation of petascale and following exascale calculations, the number of computers and data transfer rate become higher. At the same time, computational complexity becomes larger too, which complicates algorithms and increases the number of operations per one thread. Therefore, within the framework of this model, the scalability of parallel processes is studied with account for current equipment and, as a consequence, complexity of the computational model.

The proposed method for researching the computational processes has already been successfully used in the study of scalability of several computational algorithms per a large number of computational cores [6,7,8,9].

3. Scalability investigation of parallel algorithms for plasma dynamics problems in open magnetic traps

3.1. Description of the calculation scheme for the algorithm under study

The parallel implementation of the hybrid numerical model of plasma dynamics in open magnetic traps is based on the mixed decomposition of the computational domain, the latter of which is uniformly divided into subdomains, with each subdomain being controlled by a group processor. Particles in subdomains are distributed uniformly among group processors (Figure 2). Boundary nodes for the subdomain of grid nodes are transferred among the main group processors. Particles can be sent to any processor from a neighboring group.

Each step of the computation cycle includes the Lagrangian stage for calculating the particle velocities and coordinates and the Euler stage for computing on a spatial grid.

At the beginning of each step of the computational cycle, the main process sends out an electric and magnetic field in the subdomain through the group using MPI_BCAST.

Moving to the Lagrangian stage requires using bilinear interpolation, calculating the Lorentz force acting on the particle at its location in each process, and determining the particle velocities and coordinates. If a particle flies beyond the subdomain boundaries, then its data is sent to one of the processors of the neighboring group. Using a local rank allows evenly distributing particles in the processes during transfer.

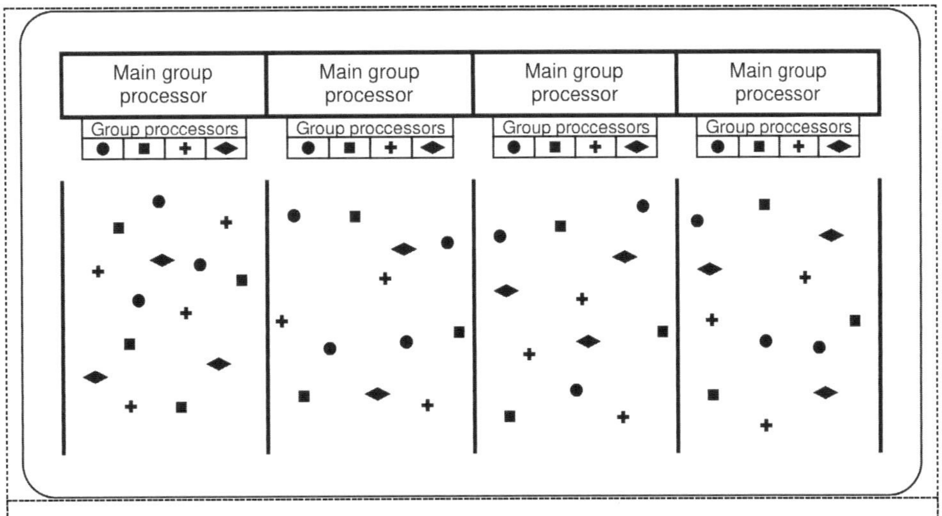

Figure 2. Mixed decomposition of computational domain.

After that, each process calculates the average ion velocities in a cell and densities, after which the Euler stage begins. The main processors of the group collect the group data using MPI_REDUCE and then exchange the boundary data for the grid.

Next, they are solved for electron velocities, for an electric field, for a magnetic field, and for temperature. After solving the equation for each of the listed values, data is exchanged for the boundary grid nodes between neighboring main group processors.

It should be noted that an algorithm with the same decomposition of the computational domain and a similar process interaction scheme has already been studied using the method described [9]. However, the important distinguishing feature is that the numerical modeling algorithm for exactly calculating the characteristics of instabilities excited by an electron beam in plasma at the end of a computational cycle, the processes have an all-to-all data exchange over the entire computational domain using MPI_ALLREDUCE. In the algorithm under study, the main process of the group initially sends data to all processes in the group, then collects data from all the processes following necessary calculations, and it is the generalized data that is exchanged with neighboring main processes of the group afterward.

3.2. Scalability of the parallel algorithm

The execution of the numerical modeling algorithm of plasma dynamics in open magnetic traps is simulated by implementing PlasmaMain and PlasmaSub classes of actors. The actors of the PlasmaMain class simulate the execution of the main processes of the group, which is data transfer and collection from all the processors of the group, as well as data exchange with neighboring actors of the PlasmaMain actors. The actors of the PlasmaSub class simulate the execution of the remaining processes of the group: calculation and exchange cycles. Delays during message exchange between the actors eventually simulate the corresponding delays in message exchange in the real computational system. The algorithm execution model performs a previously specified number of calculation cycles. The time characteristics of execution of all stages of the computational cycles are taken from the results of running a real computational algorithm on the NKS-1P cluster of the of the General-Purpose Computing Center "Siberian SuperComputer Center" of the Institute of Computational Mathematics and Mathematical Geophysics of the Siberian Branch of the Russian Academy of Sciences. The calculations are carried out on Broadwell computational nodes with Intel Xeon E5-2697A v4 processors.

The scalability of the algorithm under study is demonstrated using calculations for a various number of computational nodes (more than 10^6 computational cores). The results of this study are illustrated in figure 3. The differences of the interaction scheme of the processes in the algorithm investigated previously (using the MPI_ALLREDUCE operator [9]) and the algorithm under study (with the MPI_BCAST and MPI_REDUCE operators) are demonstrated in the chart by the scalability of both algorithms.

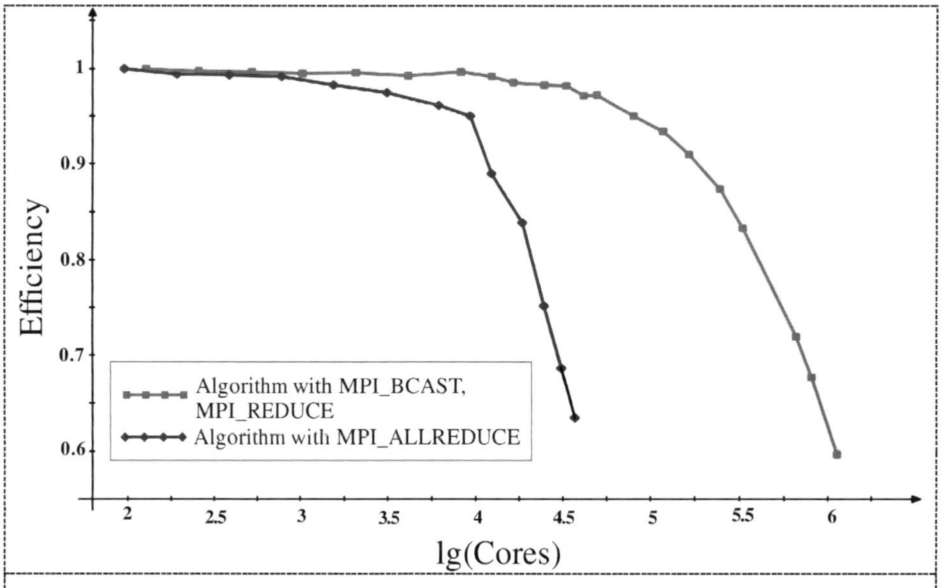

Figure 3. Strong scalability of the studied (MPI_BCAST, MPI_REDUCE) and previously mentioned (MPI_ALLREDUCE) algorithms. The horizontal axis is in logarithmic scale.

It can be concluded from figure 3 that using the calculations of the main process of the group, occupied by data transfer, aggregation, and exchange with neighboring domains can significantly reduce the communication overhead as compared to the algorithm with all-to-all exchange. It should also be noted that this algorithm generally shows high scalability in the case of execution on a large number of computational cores.

4. Conclusion

The method for studying the scalability of parallel algorithms on a large number of computational cores using simulation multiagent modeling is formed within the framework of the integral approach to developing parallel programs. This approach allows predicting the efficiency of using this algorithm on future HTC systems without the help of many computational and material resources.

This method was used to investigate the scalability of the numerical modeling algorithm of the plasma dynamics in axially symmetric open magnetic traps. The study showed that the algorithm could be scaled on more than a million computational cores without any significant efficiency losses. It could also be concluded that the specific features of parallel implementation of this algorithm essentially increased its efficiency due to absence of universal data exchange operations.

Acknowledgements
The work was financially supported by the Russian Science Foundation (Grant No. 19-71- 20026).
The computations were performed by the Siberian Supercomputer Center cluster (ICM&MG SB RAS, Novosibirsk).

References

[1] Glinskiy B M, Kulikov I M, Chernykh I G, Snytnikov A V, Sapetina A F and Weins D V 2017

The Integrated Approach to Solving Large-Size Physical Problems on Supercomputers *CCIS* **793** pp 278-289

[2] Podkorytov D, Rodionov A and Choo H 2012 Agent-based simulation system AGNES for networks modeling: review and researching *Proc. Int. Conf. on Ubiquitous Information Management and Communication* (Kuala Lumpur, Malaysia) P115

Podkorytov D, Rodionov A, Sokolova O and Yurgenson A 2010 Using Agent-Oriented Simulation System AGNES for Evaluation of Sensor Networks *LNCS* **6235** pp 247--250.

[3] Weins D V, Glinskiy B M, Chernykh I G 2019 Analysis of Means of Simulation Modeling of Parallel Algorithms *CCIS* **965** pp 1–11

[4] Cesarini F, Thompson S 2009 *Erlang Programming*.(Boston: O'Reilly Media, Inc) p. 498

[5] Berendeev E, Dudnikova G, Efimova A and Vshivkov V 2017 Computer simulation of plasma dynamics in open plasma trap *LNCS* **10187** pp 227-234.

[6] Glinsky B, Rodionov A, Marchenko M, Podkorytov D and Weins D 2012 Scaling the Distributed Stochastic Simulation to Exaflop Supercomputers *Proc. Int. Conf. on High Performance Computing and Communications* (Liverpool, England, UK) pp 1131–1136.

[7] Glinsky B, Kulikov I, Chernykh I, Weins D, Snytnikov A, Nenashev V, Andreev A, Egunov V and Kharkov E 2016 The Co-design of Astrophysical Code for Massively Parallel Supercomputers *LNCS* **10049** pp 342-353.

Kulikov I, Chernykh I, Glinskiy B, Weins D and Shmelev A 2015 Astrophysics simulation on RSC massively parallel architecture *Proc. Int. Symp. on Cluster, Cloud, and Grid Computing* (Shenzhen, Guangdong, China.) pp 1131-1134.

[8] Glinskiy B, Sapetina A, Martynov V, Weins D and Chernykh I 2017 The Hybrid-Cluster Multilevel Approach to Solving the Elastic Wave Propagation Problem *CCIS* **753** pp 261-274.

[9] Glinskiy B M, Kulikov I M, Snytnikov A, Romanenko A A,Chernykh I G and Vshivkov V A 2017 Co-design of parallelnumerical methods for plasma physics and Astrophysics *Supercomputing Frontiers and Innovations* **1** N3 pp 88–98.

Development of simulation model of HPC system for Super Charm-Tau factory

D Weins[1], V Vorobyev[2], I Chernykh[1] and I Logashenko[2]

[1] The Institute of Computational Mathematics and Mathematical Geophysics SB RAS (ICM&MG SB RAS) 6, Acad. Lavrentieva Pr., Novosibirsk, 630090, Russian Federation

[2] Budker Institute of Nuclear Physics of Siberian Branch Russian Academy of Sciences (BINP SB RAS) 11, Acad. Lavrentieva Pr., Novosibirsk, 630090 Russian Federation

vins@sscc.ru, V.S.Vorobev@inp.nsk.su, chernykh@ssd.sscc.ru, ivan.logashenko@gmail.com

Abstract. The article describes the design of a digital model of an HPC system for processing data from the Super Charm-Tau factory electron-positron collider of the "megascience" class. This model is developed using the AGNES multiagent modeling platform. The model includes intelligent agents that mimic the behavior of the main subsystems of the supercomputer, such as a task scheduler, computing clusters, data storage system, etc. This model allows calculating the parameters of the computing system necessary for processing and storing the results of operation of the Super Charm-Tau factory after its commissioning.

1. Introduction

The «Super Charm-Tau factory» (SCTF) project, which is a symmetric electron-positron collider of ultrahigh luminosity with a beam energy at the mass center system from 2 to 6 GeV [1], is developed at the Budker Institute of Nuclear Physics, Siberian Branch, Russian Academy of Sciences (BINP SB RAS). This project comprises a unique accelerating-storage complex with a luminosity of 10^{35} $cm^{-2}s^{-1}$ and a universal elementary particle detector. The main goal of experiments carried out on the SCTF is to study the properties of tau lepton and charmed particles, subject the existing microworld theory and Standard Model to high-presicion verification, and to search for phenomena not described within the framework of this theory.

In the course of the experiments, about 100 petabytes of "RAW" data is accumulated from the elementary particle detector of the SCTF. An important role in the project is played by the system for data processing and storage, whose tasks include the primary data processing, data transfer to long-term data storage system (decades), data extraction from the storage system for processing and processing using high-tech computing (HPC) systems. Specialized software should allow one to analyze the accumulated data by a collective of about 1000 physicists. The development of the data analysis algorithms and the optimization of the detector structure are carried out using modeling data generated via software for modeling of experiments.

The vast volumes of primary data require large investments into the creation of the localized computational infrastructure of the complex. Within the framework of this paper, it is proposed to

Content from this work may be used under the terms of the Creative Commons Attribution 3.0 licence. Any further distribution of this work must maintain attribution to the author(s) and the title of the work, journal citation and DOI.
Published under licence by IOP Publishing Ltd

solve the problem of designing this computational infrastructure and multilevel systems for storing the experimental data using simulation modeling. This approach allows for preliminary estimation of the structure and amount of data generated by the physical device and determination of the parameters and configuration of computational infrastructure required to solve the experimental problems.

2. Selection of the simuation modeling system

2.1. Multiagent modeling systems

Agent modeling (AM) is an advanced approach to modeling systems containing autonomous and interacting intelligent agents. An intelligent agent is an entity existing in a medium and possessing sensors for perceiving this medium and executive mechanisms for affecting this medium. The agents operate asynchronously according to their laws and interact with other agents in order to reach common goals. During operation, a software agent can change both the outer medium and its own behavior [2]. A distributed, asynchronous behavior is quite important for constructing a simulation model of a supercomputer as it is virtually impossible to ensure the centralized control of tens and hundreds of millions of cores.

A multiagent system (MAS) is a system formed by several interacting intelligent agents in a single medium. Due to such properties as autonomy, limited knowledge, decentralization, and self-organized and intelligent behavior, the agents naturally fit the modeling of inhomogeneous systems with atomic elements, which also include HPC systems [3].

2.2. AGNES

The AGNES system, initially developed for telecommunication and information networks [4], also demonstrated its efficiency in modeling the operation of a distributed control system and executing high-performance parallel programs [5].

The AGNES package is based on Java Agent Development Framework (JADE), serving as a means for creating Java based MAS [3]. JADE is a powerful tool for developing Java based multiagent systems. It is important for modeling large computations that JADE is a FIPA-compatible, distributed agent platform, which can use one or several computers (network nodes), each having only one operationing JAVA machine.

All interaction between JADE agents occurs by exchanging messages according to FIPA specification. The key property of a JADE agent is a set of its «behaviors». An agent's lifecycle ends when this agent has no active behaviors. AGNES uses the advantages provided by JADE and expands the multiagent system to a modeling system. AGNES comprises two types of agents: functional and controlling agents. The primer perform simulation, and the latter control and ensure operation of the primer.

In order to ensure high-level fault tolerance, AGNES implements the basic principles of improved fault tolerance of the modeling medium: storage decentralization and information redundancy.

AGNES is a distributed MAS referred to as a platform. AGNES consists of a system of containers distributed over a network. As a rule, each host has one container. Agents exist within containers. As agents differ in their functions and tasks, AGNES attempts to uniformly redistribute the agents of all types.

Thus, AGNES is a perfect fit for the development and control of the operation model of a HPC system.

3. Description of the model

3.1. General scheme of the HPC system model

The model of the HPC system for data storage and processing with SCTF is developed under the following assumptions:

- The HPC system under consideration is a distributed, heterogeneous, and multicluster computing system (CS) with an MPP architecture. The task flow fed to the HPC is distributed over the clusters according to a certain algorithm. The tasks are implemented without any interaction among the clusters. The HPC system contains a parallel data storage system, with all of the system clusters having access thereto.
- A cluster is regarded as a set of single-type elementary machines (EMs) related to each other by some communication scheme, in which EMs are nodes and the communication lines between them are branches.
- The main resource given to solve the problem is a virtual domain of computational nodes (CNs), satisfying all the stated parameters of the problem. A CN corresponds to an EM in a real system.
- If the main characteristic of solving the problem for a real CS is time, then the number of elementary operations executed by a task is taken a main characteristic in the model.
- Transition from model time and to real time is carried out using a special coefficient

The general scheme of the multiagent model of the HPC system, which accounts for the above-given assumptions, is illustrated in figure 1.

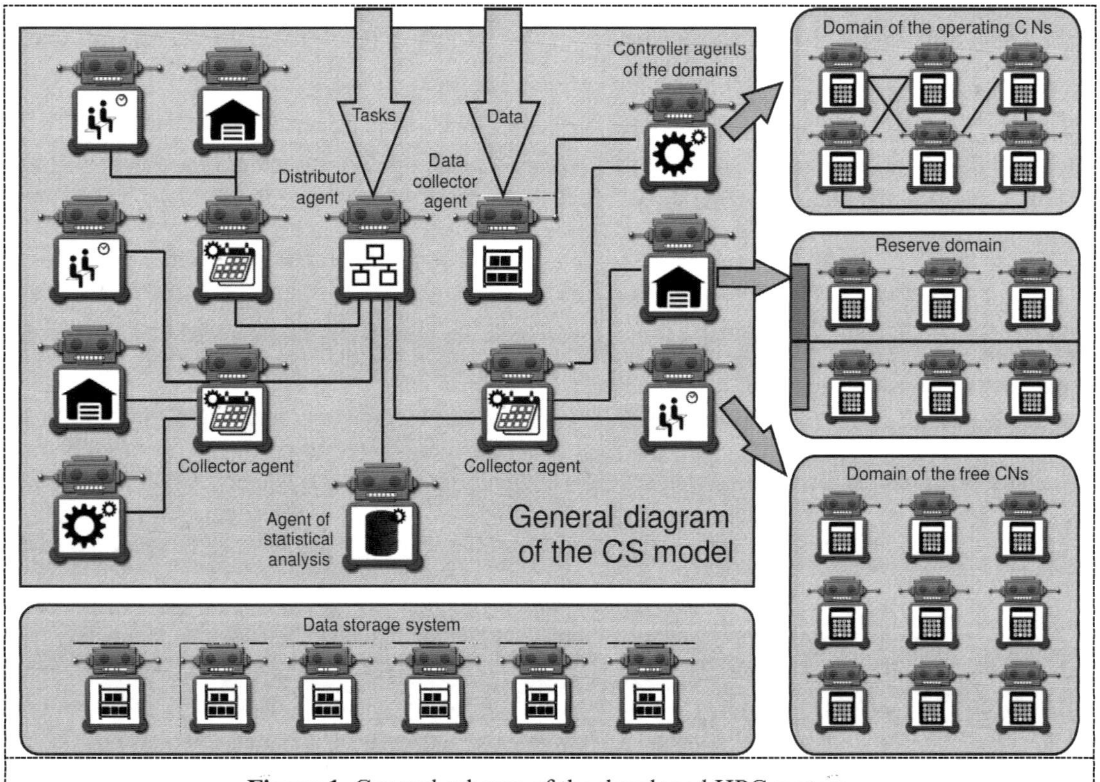

Figure 1. General scheme of the developed HPC system.

All types of objects given in the scheme are intelligent agents possessing their own set of parameters and behavior algorithms.

3.2. Types and functions of agents in the model of the prompt management system of the HPC system

Several types of agents can be seen in the scheme in Fig. 1:

- users at random instances send their parallel tasks to the HPC system;

- the detector with a certain frequency transfers data of certain size to the data storage system of the HPC system;
- the system for data collection and statistical analysis gathers and stores generalized information about tasks solved using the HPC system;
- the distributor distributes the flow of incoming tasks among the clusters included in its composition with account for their state at a current time;
- the collector of the CS schedules the execution of incoming tasks and arranges the implementation of this schedule;
- the data collector ensures that data are saved and read from storage devices of the HPC system;
- the controller of the CS domain arranges correct operation of the CSs under its control;
- the CS simulates the actions of the EMs, executing its branch of the parallel program or being in a stand-by mode.

The multiagent model of the HPC system accounts for combining the agents that perform at a current instance one and the same function into virtual domains (below simply referred to as domains), controlled by the controller agent of the CS regions. Thus, the CS domain is combining the group of agents of computational nodes and the controller agent of the domain, executing at a current instance one and the same function and connected by logical communication lines. During the operation of the HPC system model, the following domains are formed:

- the CN reserve domain contains computational nodes sent into "cold" or "hot" reserve [6], as well as nodes decommissioned for maintenance or repair.
- the domain of the free CNs contains all the CNs of the cluster, which are neither preoccupied by executing a user's task nor in reserve at a given instance;
- the domain of operating CNs contains computational nodes preoccupied by solving one and the same task at a given instance, in which case the controller agent observes both the physical characteristics of the nodes and the course of execution of the parallel program.

3.3. Arranging the execution of parallel tasks in the HPC system model

The general scheme of communication between the model agents during the simulation of the CS operation is given in figure 2.

User's tasks are fed to the distributor input of the high-tech CS. The distributor of the HPC system constantly receives reports from the collector agents about changes in the state of their task queue and available resources. Initially, the switch connects the agent of collection and analysis of statistics. The latter looks through this user's tasks and selects those similar in requested resources. The commuting switch is given a weighted list of preferences of the CS collector agents. Guided by the data received from the agent of collection and analysis of statistics and the available data on the workload of the collector agents, the switch sends the task to the most promising one (the least loaded one) having the largest weigh in the list of preferences.

This, the task is fed to the corresponding collector agent of the CS, and the currently running task schedule is immediately reviewed. If resources are found in this schedule for the task, it is embedded into the schedule, otherwise it is sent to the task queue. As soon as the current task is executed, the collector agent develops a new schedule. Tasks are selected from the queue using one of the planning algorithms.

The task is being executed as soon as its turn arrives. For this purpose, a domain controlling agent is created (or chosen from those already formed). This agent takes control of the CNs from the domain of free CNs. Nodes for the created domain should be maximally similar in the graph of the communication medium of the CS. The controller sends a request to the data controller for a time required to read a certain volume of information for the problem solved and simulates the reading process from the data storage system into the local memory. Then, as the controller agent of the operating domain periodically collects information about the course of the task execution, it waits for the execution to end.

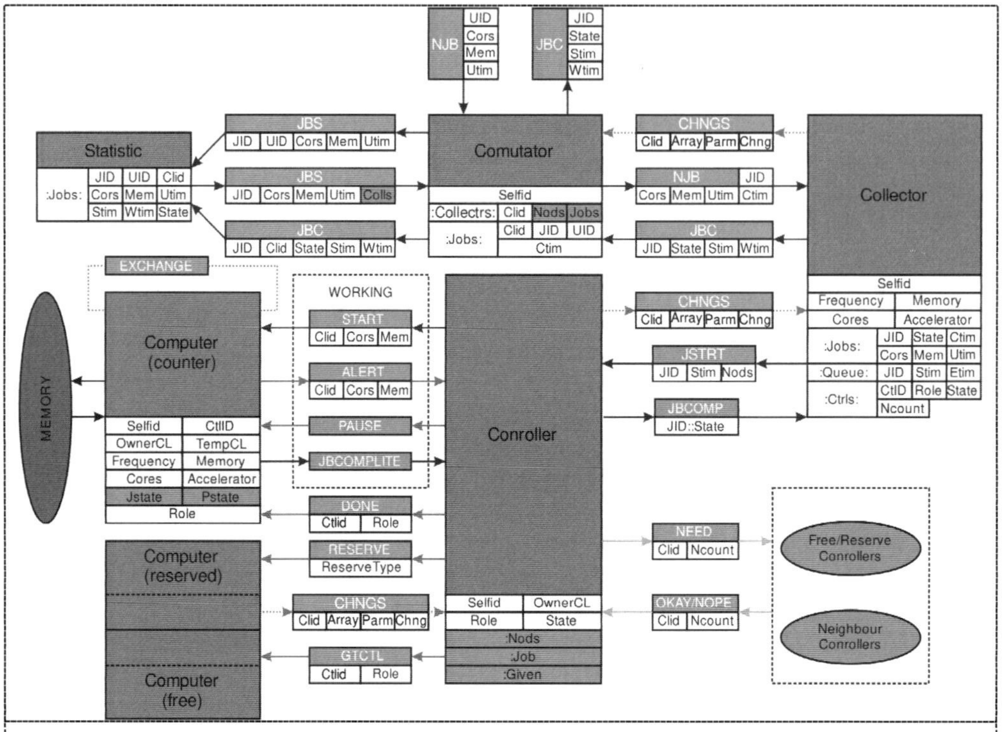

Figure 2. Communication between the model agents in the task arrangement

As the task is executed, its completion parameters are sent back on command until reaching the agent of collection and analysis of statistics, where the final stage of the task execution occurs (its execution statistics is saved). This statistic can be used to estimate the amount and characteristics of computational resources required for the computing system to execute a desired number of user's tasks without serious delays.

3.4. Fault tolerance and energy efficiency

The dynamic fault tolerance system is implemented in this model using the CN agents and domain controlling agents.

The CN agent collects information about the temperature, load, and condition of the communication lines between the nodes and other information that characterizes the CN state. Each CN agent has a set of behavior rules for preventing the CN failures. As soon as one of the parameters is out of range or the trend of changes of this parameter becomes worse (for example, the temperature rapidly rises), the CN agent sends a signal to the controller agent of its domain.

The domain controlling agent receives information about an emergency situation occurring in any node and commands all nodes to stop operating. Next, the received signal is analyzed. If the nodes report looping and deadlock, then the task execution is canceled. If one of the nodes signals violation of its physical parameters (temperature rise or communication line disconnection), the controller agent searches for a node to which the computational load is transferred from the faulty node. Initially, the resources located in a given domain are analyzed. If they are insufficient, the controller connects with the controllers of the domain of free nodes, reserve domain, and "neighboring" operating domains. Next, the computational load is transferred to a faultless node, while the faulty node is transferred to the reserve domain. If the failure cannot be prevented and the node fails during task execution, the domain controlling agent searches for a replacement and then starts the task again from the last execution point.

When it comes to the energy efficiency of cluster computing systems, the following problems may be encountered:

- There are idle cluster nodes that consume significant amounts of electricity;
- Nonuniform temperature distribution over the cluster nodes with a large amplitude, which reduces the reliability and fault tolerance of the cluster and leads to an ineffective use of electricity.

The primer problem is solved in the model using the reserve domain. This domain can be divided into three regions: hot reserve, cold reserve, and nodes requiring repair or maintenance. As the model is launched into the reserve domain (hot reserve), at least 5% of the CNs of the CS are determined. These CNs can be used for the dynamic balancing of the load by the domain controllers of the operating nodes. As soon as the node authorized to perform any task becomes free, it returns to the reserve domain. If the CS collector analyzes the formed schedule and task queue and concludes that there is a certain number of nodes that will remain idle in the foreseeable future, it commands to send them to the reserve domain.

The latter problem is solved by the collector agent of the CS as it forms another schedule. The presence of emergency shutdown thresholds upon reaching a critical temperature is not enough for efficient energy planning [7]. Because of this, the CS collector gathers information about the system nodes and analyzes the temperature trend for the last second. Depending on the fact whether the node processor temperature drops or rises, the trend approaches unity or, conversely, tends to zero. At the instance when the CS collector begins forming a new schedule, it places tasks in the nodes with the lowest value of the trend in a descending order. Thus, the temperature distributes over the cluster nodes more uniformly.

The use of energy efficiency and fault tolerance schemes in the data model allows estimating a required number of additional computing resources that ensure the smooth operation of the HPC system.

3.5. Parallel access to data

The tasks of storage and processing of the data sent from the SCTF in the HPC system model are executed by simulating the operation of the parallel data storage system. Figure 3 shows the scheme of the communication between the data collector agents, the domain controllers executing parallel tasks, and the agents simulating the data coming from the detector and detector event modeling systems. The detector itself and the modeling system are represented by black boxes.

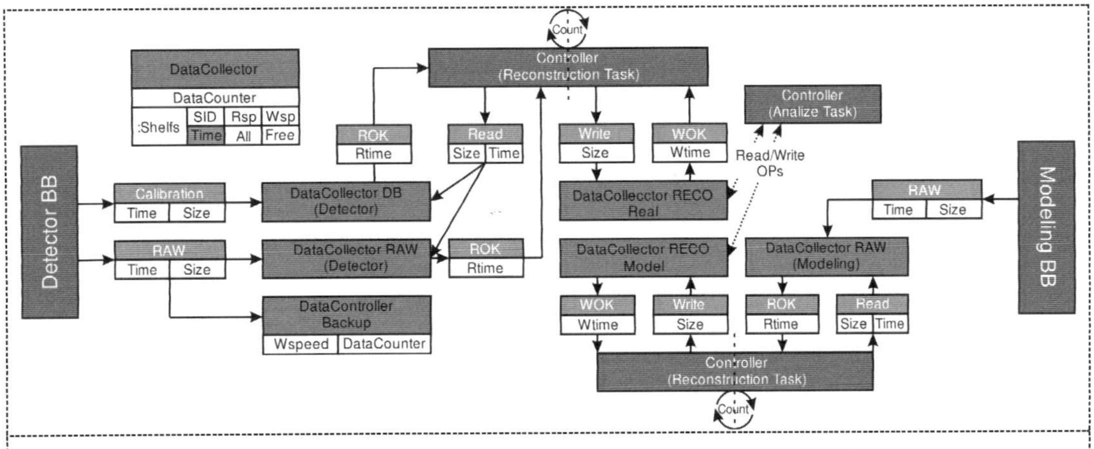

Figure 3. Communication between the model agents when working with a data storage system.

Each data collector agent contains a counter of saved data and information about the state and characteristics of all shelves with data carriers under its control. As the controllers of operating domains send a request to data collector agent, it transfers information about the time required to read and save the desired amount of data.

As seen from the scheme shown in Fig. 3, the HPC system model contains six data controllers, each storing only certain information: the database with detector calibration information, raw data from the detector and from the event modeling system, data from reconstructed real and model events, and raw data backup from the detector. This division is necessary in the model in order determine how many data carriers are needed to ensure the smooth operation of the SCTF.

4. Conclusion

The successful launch of the SCTF requires estimating the computational infrastructure parameters of the complex for storing and processing data of the physical experiment at the stage of design. Using simulation modeling allows for the maximally reliable representation of the exact characteristics and volume of the needed equipment for developing the desired HPC system. The simulation model described in this paper accounts for all the aspects of operation of this system from parallel data storage system to arrangement of the parallel launch of tasks. The developed system for processing software errors and equipment failures, as well as the system for ensuring energy efficiency make it possible to estimate the needed equipment with account for all possible emergency situations.

Thus, the developed simulation model allows calculating the parameters of the computing system required for processing and storing the operation results of the Super Charm-Tau factory after its commissioning.

Acknowledgments

This paper was financially supported by the Russian Science Foundation (Grant No. 19-72-20114) using the resources of the Siberian Supercomputer Center cluster (ICM&MG SB RAS, Novosibirsk).

References

[1] Bondar A 2013 Project of a super charm-tau factory at the Budker Institute of Nuclear Physics in Novosibirsk Phys. At. Nucl.**76** pp 1072–1085.

[2] Chandy K M and Misra J 1978 Distributed Simulation: A Case Study in Design and Verification of Distributed Programs *IEEE Transactions on Software Engineering SE* **5** N5 pp 440-452.

Wooldridge M 2002 *Introduction to MultiAgent Systems* (England: JOHN WILEY & SONS LTD) p 348.

[3] Bellifemine F L, Caire G and Greenwood D 2007 *Developing Multi-Agent Systems with JADE* (Wiley) p 300.

[4] Podkorytov D, Rodionov A and Choo H 2012 Agent-based simulation system AGNES for networks modeling: review and researching *Proc. Int. Conf. on Ubiquitous Information Management and Communication* (Kuala Lumpur, Malaysia) P115

Podkorytov D, Rodionov A, Sokolova O and Yurgenson A 2010 Using Agent-Oriented Simulation System AGNES for Evaluation of Sensor Networks *LNCS* **6235** pp 247--250.

[5] Glinskiy B M, Kulikov I M, Chernykh I G, Snytnikov A V, Sapetina A F and Weins D V 2017 The Integrated Approach to Solving Large-Size Physical Problems on Supercomputers *CCIS* **793** pp 278-289

Glinskiy B, Sapetina A, Martynov V, Weins D and Chernykh I 2017 The Hybrid-Cluster Multilevel Approach to Solving the Elastic Wave Propagation Problem *CCIS* **753** pp 261-274.

Glinsky B, Rodionov A, Marchenko M, Podkorytov D and Weins D 2012 Scaling the Distributed Stochastic Simulation to Exaflop Supercomputers *Proc. Int. Conf. on High*

IOP Conf. Series: Journal of Physics: Conf. Series **1336** (2019) 012025 doi:10.1088/1742-6596/1336/1/012025

Performance Computing and Communications (Liverpool, England, UK) pp 1131–1136.
[6] Kallis J M and Norris M D 1996 Effect of steady-state operating temperature on power cycling durability of electronic assemblies *Tech. report Hughes Aircraft Company* pp 1.
[7] Coskun A K, Rosing T S and Whisnant K 2007 Temperature Aware Task Scheduling in MPSoCs *Proc. Int. Conf. on Design, Automation & Test in Europe Conference & Exhibition* (Nice, France) pp 1659 – 1664.

IOP Conf. Series: Journal of Physics: Conf. Series **1336** (2019) 012026 doi:10.1088/1742-6596/1336/1/012026

Divergence-free finite-difference method for 2D ideal MHD

E.N. Avdeeva[1], **V.V. Lukin**[1,2,*]
[1]Bauman Moscow State Technical University, Moscow, Russia
[2] Keldysh Institute of Applied Mathematics of the RAS, Moscow, Russia
E-mail: *vvlukin@gmail.com

Abstract. The divergence-free finite-difference scheme for 2d ideal MHD using triangular unstructured staggered grids is described. In this approach the density, pressure and velocity fields are attributed to grid cells and magnetic field is defined by its normal components in cells edges. The HLLD approximate Riemann solver is used to compute the numerical flux of gas-dynamics variables at edges. It also provides the electrical field values used to compute the magnetic field. The magnetic field is interpolated into the cells using Raviart — Thomas basis functions. The algorythm is implemented in parallel numerical code. The method is tested on several well-known two-dimensional MHD problems.

1. Introduction

Models based on ideal magnetohydrodynamics equations are high-demanded in geophysical and astrophysical investigations [1]. They are widely applied in modelling of macroscale plasma behavior [2, 3]. The magnetic field divergence-free condition is the crucial for such models [4, 5]. Artificial numerical non-zero divergence of magnetic field leads to non-physical oscillations of solution and the magnetic charge generation. This problem solution demands special numerical techniques to be implemented during computational research [6, 7, 8].

We consider new finite-difference divergence-free numerical scheme for ideal two-dimentional magnetohydrodynamics (MHD). We use the unstructured staggered triangular grids. The grid cells are used as the control volumes for gas variables such as density, velocity and pressure. The plane magnetic field components are associated with cells edges and are defined by normal to edge vector component (like [8]). Under proposed division the gas variables are computed in every time layer using Godunov-type scheme with reduced HLLD numerical flux [9]. Then the magnetic field normal components are computed directly from the Faraday equation using the Stokes theorem by calculating the circulation of electrical field [10] obtained as the components of HLLD flux.

The Stokes theorem exploitation leads to multiple results. On the one hand it guarantees the divergence-free of numerical scheme without any additional steps [6]. On the other hand the direct approximation of the Faraday law using the desribed technique is the general approach applicable for any grid type. The triangular grid is used for example.

Content from this work may be used under the terms of the Creative Commons Attribution 3.0 licence. Any further distribution of this work must maintain attribution to the author(s) and the title of the work, journal citation and DOI.
Published under licence by IOP Publishing Ltd

IOP Conf. Series: Journal of Physics: Conf. Series **1336** (2019) 012026 doi:10.1088/1742-6596/1336/1/012026

2. Governing equations

Ideal MHD equations system includes the conservation laws for gas mass, pulse, energy and Faraday law for magnetic field evolution:

$$\frac{\partial \rho}{\partial t} + \nabla \cdot (\rho \mathbf{v}) = \mathbf{0}, \tag{1}$$

$$\frac{\partial (\rho \mathbf{v})}{\partial t} + \nabla \cdot (\rho \mathbf{v}\mathbf{v} + p\hat{\mathbf{I}}) = (\nabla \times \mathbf{B}) \times \mathbf{B}, \tag{2}$$

$$\frac{\partial e}{\partial t} + \nabla \cdot [(e+p)\mathbf{v}] = ((\nabla \times \mathbf{B}) \times \mathbf{B}) \cdot \mathbf{v}, \tag{3}$$

$$\frac{\partial \mathbf{B}}{\partial t} = \nabla \times (\mathbf{v} \times \mathbf{B}). \tag{4}$$

Here ρ is gas density, $\mathbf{v} = [u, v, w]^{\mathrm{T}}$ is velocity, p is pressure, $\hat{\mathbf{I}}$ is unity tensor, $\mathbf{B} = [B_x, B_y, B_z]^{\mathrm{T}}$ magnetic induction vector, e is the total gas energy excluding the magnetic field energy and including internal and kinetic energy of unity gas volume:

$$e = \frac{p}{\gamma - 1} + \frac{\rho \mathbf{v} \cdot \mathbf{v}}{2},$$

γ is adiabatic index. Let all variables depend only on time t and plane coordinates x and y, all derivatives with respect to z are equal to zero. The system (1) – (4) should be closed by setting the initial and boundary conditions for density, velocity, energy and magnetic field.

The conservation of magnetic field is prescribed by divergence-free condition

$$\nabla \cdot \mathbf{B} = 0. \tag{5}$$

Applying the divergence operator to (4) one can obtain

$$\frac{\partial}{\partial t}(\mathrm{div}\mathbf{B}) = 0.$$

It means that the $\nabla \cdot \mathbf{B}$ will not increase in time and magnetis field comply with condition (5) if initial distribution is divergence-free.

3. Numerical scheme

To construct the numerical scheme we use the explicit first order time integration and the physical processes splitting technique. Under this paradigm on every time step to obtain new variables values one should compute

- the gas fluxes (mass, pulse and energy densities fluxes) on every grid edge using HLLD approximate Riemann solver;
- gas variables inside cells implementing gas fluxes from-edges-to-cell balance condition prescribed in equations (1) – (3) like it usually done in Godunov-type schemes;
- normal components of magnetic field on grid edges using the Stokes theorem;
- magnetic field inside the grid cells by divergence-free interpolation procedure.

Obtained in such approach scheme is monotonic, conservative and divergence-free.

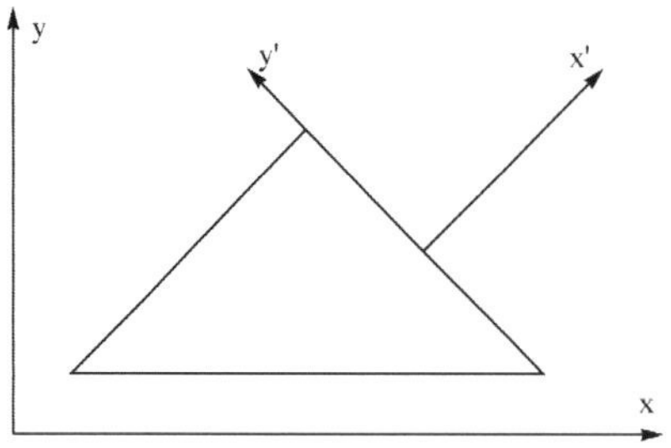

Figure 1. Base grid cell. (x, y) is basic system of axes, (x', y') is local rotated system of axes.

3.1. Numerical fluxes

Consider the triangular cell and its edge with outer normal depicted in Fig. 1. Let x' axis be normal to edge and y' axis be tangential. In standart approach [11] to obtain numerical flux we can locally for edge central point reduce model to 1D approximation and rescribe equations (1-4) in new system of axes (x', y') as:

$$\frac{\partial \mathbf{U}'}{\partial t} + \frac{\partial \mathbf{F}'}{\partial x'} = \mathbf{0}. \tag{6}$$

Here $\mathbf{U}' = [\rho, \rho u', \rho v', \rho w, e', B'_x, B'_y, B_z]^{\mathrm{T}}$ is conservative variables vector including magnetic field, $\mathbf{F}' = [\rho u', \rho(u')^2 - p'_T + (B'_x)^2, \rho u'v' - B'_x B'_y, \rho u'w - B'_x B_z, (e' + p'_T)u' - B'_x(\mathbf{v}' \cdot \mathbf{B}'), 0, B'_y u' - B'_x v', B_z u' - B'_x w]^{\mathrm{T}}$ is flux vector, Notice, that $\mathbf{v}' = [u', v', w]^{\mathrm{T}}$ is velocity vector and $\mathbf{B}' = [B'_x, B'_y, B_z]^{\mathrm{T}}$ is magnetic field vector in rotated system of axis (x', y'). The dependance of total pressure p'_{T} and total energy e' on gas pressure p is given by formulas

$$p'_{\mathrm{T}} = p + \frac{\mathbf{B}' \cdot \mathbf{B}'}{2}, \qquad e' = \frac{p}{\gamma - 1} + \frac{\rho \mathbf{v}' \cdot \mathbf{v}'}{2} + \frac{\mathbf{B}' \cdot \mathbf{B}'}{2}.$$

This hyperbolic system has seven eigenvalues that correspond to entropy, two Alfvenic and four magnetosonic (two fast and two slow) waves [1]:

$$\lambda_4 = u', \quad \lambda_{2,6} = u' \mp c_a, \quad \lambda_{1,7} = u' \mp c_f, \quad \lambda_{3,4} = u' \mp c_s. \tag{7}$$

Here

$$c_a = \frac{|B'_x|}{\sqrt{\rho}}, \qquad c_{f,s} = \left\{ \frac{\gamma p + |\mathbf{B}'|^2 \pm \sqrt{(\gamma p + |\mathbf{B}'|^2)^2 - 4\gamma p (B'_x)^2}}{2\rho} \right\}^{1/2}.$$

All eigenvalues are real and ordered:

$$\lambda_1 \leqslant \lambda_2 \leqslant \lambda_3 \leqslant \lambda_4 \leqslant \lambda_5 \leqslant \lambda_6 \leqslant \lambda_7.$$

Equations system (6) is usually [11] used to construct approximate solver for Riemann problem with two states prescribed by base and neighbouring cells. We use HLLD numerical flux which is described in detail in [9]. It considers six states rising from discontinuity point in cell's edge and divided by five waves depicted in Fig. 2. States U^*_L, U^{**}_L, U^{**}_R and U^*_R are diveded by Alfvenic waves travelling with velocities S^*_L and S^*_R to the left and to the right direction respectively and by contact wave travelling with velocity S_M. Index L designates „left“ side or the base cell and index R designates „right“ side or neighbouring cell for considered edge.

IOP Conf. Series: Journal of Physics: Conf. Series **1336** (2019) 012026 doi:10.1088/1742-6596/1336/1/012026

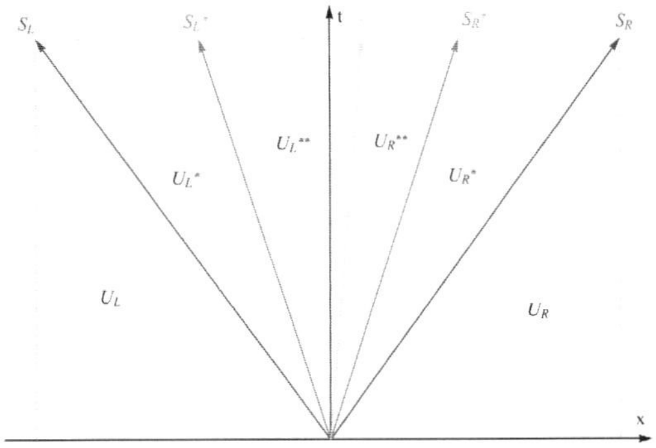

Figure 2. Plasma states after breakdown of a discontinuity for HLLD solver. Dark blue is for fast magnetosonic waves, red is for Alfvenic waves and green is for enthropy wave.

Let normal velocity component and total pressurt are unchanged by Alfvenic and contact waves:

$$u_L^* = u_L^{**} = u_R^{**} = u_R^* = S_M,$$
$$p_{\mathrm{T}_L}^* = p_{\mathrm{T}_L}^{**} = p_{\mathrm{T}_R}^{**} = p_{\mathrm{T}_R}^* = p_{\mathrm{T}}^*.$$

Here

$$S_M = \frac{(S_R - u_R)\rho_R u_R - (S_L - u_L)\rho_L u_L - p_{\mathrm{T}_R} + p_{\mathrm{T}_L}}{(S_R - u_R)\rho_R - (S_L - u_L)\rho_L},$$
$$p_{\mathrm{T}}^* = \frac{(S_R - u_R)\rho_R p_{\mathrm{T}_L} - (S_L - u_L)\rho_L p_{\mathrm{T}_R} + \rho_R\rho_L(S_R - u_R)(S_L - u_L)(u_R - u_L)}{(S_R - u_R)\rho_R - (S_L - u_L)\rho_L}.$$

After S_M and p_{T}^* calculation and using Rankine — Hugoniot condition for discontinuity [1], one can compute states U_α^* with $\alpha = L$ or R in following way:

$$\rho_\alpha^* = \rho_\alpha \frac{(S_\alpha - u_\alpha)}{(S_\alpha - S_M)},$$
$$v_\alpha^* = v_\alpha - B_x' B_{y_\alpha} \frac{(S_M - u_\alpha)}{\rho_\alpha(S_\alpha - u_\alpha)(S_\alpha - S_M) - (B_x')^2},$$
$$B_{y_\alpha}^* = B_{y_\alpha} \frac{\rho_\alpha(S_\alpha - u_\alpha)^2 - (B_x')^2}{\rho_\alpha(S_\alpha - u_\alpha)(S_\alpha - S_M) - (B_x')^2},$$
$$w_\alpha^* = w_\alpha - B_x' B_{z_\alpha} \frac{(S_M - u_\alpha)}{\rho_\alpha(S_\alpha - u_\alpha)(S_\alpha - S_M) - (B_x')^2},$$
$$B_{z_\alpha}^* = B_{z_\alpha} \frac{\rho_\alpha(S_\alpha - u_\alpha)^2 - (B_x')^2}{\rho_\alpha(S_\alpha - u_\alpha)(S_\alpha - S_M) - (B_x')^2},$$
$$e_\alpha^* = \frac{(S_\alpha - u_\alpha)e_\alpha - p_{\mathrm{T}_\alpha} u_\alpha + p_{\mathrm{T}}^* S_M + B_x'(\mathbf{v}_\alpha \cdot \mathbf{B}_\alpha - \mathbf{v}_\alpha^* \cdot \mathbf{B}_\alpha^*)}{S_\alpha - S_M}.$$

Alfvenic waves velocities are computed using formulas

$$S_L^* = S_M - \frac{|B_x|'}{\sqrt{\rho_L^*}}, \qquad S_R^* = S_M + \frac{|B_x|'}{\sqrt{\rho_R^*}}.$$

Then states U_α^{**}, $\alpha = L, R$ are computed using values for U_L^*, U_R^* by formulas:

$$\begin{aligned}
\rho_\alpha^{**} &= \rho_\alpha^*, \\
v_L^{**} &= v_R^{**} \equiv v^{**}, \\
B_{y_L}^{**} &= B_{y_R}^{**} \equiv B_y^{**}, \\
w_L^{**} &= w_R^{**} \equiv w^{**}, \\
B_{z_L}^{**} &= B_{z_R}^{**} \equiv B_z^{**}, \\
e_\alpha^{**} &= e_\alpha^* \mp \sqrt{\rho_\alpha^*}(\mathbf{v}_\alpha^* \cdot \mathbf{B}_\alpha^* - \mathbf{v}^{**} \cdot \mathbf{B}^{**})\text{sign}(B_x').
\end{aligned}$$

Here

$$\begin{aligned}
v^{**} &= \frac{\sqrt{\rho_L^*}v_L^* + \sqrt{\rho_R^*}v_R^* + (B_{y_R}^* - B_{y_L}^*)\text{sign}(B_x')}{\sqrt{\rho_L^*} + \sqrt{\rho_R^*}}, \\
B_y^{**} &= \frac{\sqrt{\rho_L^*}B_{y_R}^* + \sqrt{\rho_R^*}B_{y_L}^* + (B_{y_R}^* + \sqrt{\rho_L^*\rho_R^*}(v_R^* - v_L^*)\text{sign}(B_x')}{\sqrt{\rho_L^*} + \sqrt{\rho_R^*}}, \\
w^{**} &= \frac{\sqrt{\rho_L^*}w_L^* + \sqrt{\rho_R^*}w_R^* + (B_{z_R}^* - B_{z_L}^*)\text{sign}(B_x')}{\sqrt{\rho_L^*} + \sqrt{\rho_R^*}}, \\
B_z^{**} &= \frac{\sqrt{\rho_L^*}B_{z_R}^* + \sqrt{\rho_R^*}B_{z_L}^* + (B_{z_R}^* + \sqrt{\rho_L^*\rho_R^*}(w_R^* - w_L^*)\text{sign}(B_x')}{\sqrt{\rho_L^*} + \sqrt{\rho_R^*}}.
\end{aligned}$$

In formula for energies e_α^* and e_α^{**} plus sign is suitable for $\alpha = L$ and minus one is for $\alpha = R$.

Finally the HLLD numerical flux $\mathbf{F}'$ at the center of edge is computed as

$$\begin{aligned}
&\mathbf{F}_L, && \textbf{if } S_L > 0, \\
\mathbf{F}_L^* &= \mathbf{F}_L + S_L(\mathbf{U}_L^* - \mathbf{U}_L), && \textbf{if } S_L \le 0 \le S_L^*, \\
\mathbf{F}_L^{**} &= \mathbf{F}_L + S_L(\mathbf{U}_L^* - \mathbf{U}_L) + S_L^*(U_L^{**} - U_L^*), && \textbf{if } S_L^* \le 0 \le S_M, \\
\mathbf{F}_R^{**} &= \mathbf{F}_R + S_R(\mathbf{U}_R^* - \mathbf{U}_R) + S_R^*(U_R^{**} - U_R^*), && \textbf{if } S_M \le 0 \le S_R^*, \\
\mathbf{F}_R^* &= \mathbf{F}_R + S_R(\mathbf{U}_R^* - \mathbf{U}_R), && \textbf{if } S_R^* \le 0 \le S_R, \\
&\mathbf{F}_R, && \textbf{if } S_R < 0.
\end{aligned}$$

Values S_L and S_R one can compute as the totally minimal and maximal eigenvalues (7) for MHD equations system

$$S_L = \min(u_L, u_R) - \max(c_{f_L}, c_{f_R}), \quad S_R = \max(u_L, u_R) + \max(c_{f_L}, c_{f_R}).$$

3.2. Faraday equation approximation

During the numerical simulations based on system (6) the divergence-free condition (5) may fail. In this case, non-physical transfer of matter orthogonally to $mathbfB$ and violation of the energy conservation law are possible. To avoid this trouble, we use the direct approximation of Faraday equation (4) with Stokes theorem.

Let's write the approximation for the Faraday equation on base triangular cell depicted in Fig. 1. Consider the unity height triangular prism stretched out of this cell in z direction. For the prism face S based on one of the cell edges we produce the equation (4) by outer normal and integrate it over this face:

$$\int_S \frac{\partial \mathbf{B}}{\partial t} \cdot \mathbf{n} dS = \int_S [\nabla \times (\mathbf{v} \times \mathbf{B})] \cdot \mathbf{n} dS.$$

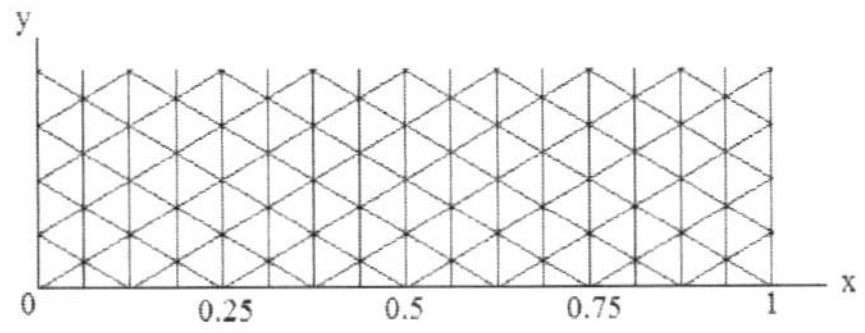

Figure 3. Triangular grid containing equilateral triangles.

Using Stokes formula one can modify right hand side of the equation obtained:

$$\int_S \frac{\partial \mathbf{B}}{\partial t} \cdot \mathbf{n} dS = \oint_{\partial S} (\mathbf{v} \times \mathbf{B}) d\ell.$$

Since we consider pure 2D model, $B_z = 0$ and we write the following formula for magnetic field normal components on cell edges in new time layer computation:

$$\hat{B}_n = B_n + \frac{\tau}{\ell}([\mathbf{v} \times \mathbf{B}]_z^i - [\mathbf{v} \times \mathbf{B}]_z^{i+1}), \quad i = 1, 2, 3, \quad i = 3 \Rightarrow i + 1 \to 1. \tag{8}$$

Here B_n is the normal component of magnetic field in center of cell edge, τ is time step, ℓ is edge length, $[\cdot]_z^i$ is designation for vertical component of the vector defined in ith vertex of triangular cell. Consider this component in the expression (8) in more detail:

$$[\mathbf{v} \times \mathbf{B}]_z^i = [u' B'_y - v' B'_x]^i.$$

The right side of the formula corresponds to the flux component for B'_y in the system of equations (6), let's denote it as F_y^B. Note that previously we have computed fluxes $\mathbf{F}'$ in centers of cells edges. Using these values we compute the average F_y^B by formula

$$[\mathbf{v} \times \mathbf{B}]_z^i = \frac{\sum\limits_{j=1}^{N} (F_y^B)_j}{N},$$

where N being number of edges adjacent to ith vertex.

The magnetic field values $\mathbf{B}$ inside the grid cells now can be calculated aided by the B_n interpolation using Raviart — Thomas vector basis functions [12] with constant divergence:

$$\mathbf{B} = \sum_{j=1}^{3} B_n^j \ell_j \varphi_j(x, y).$$

Since the B_n are calculated using the Faraday law, summary value of div $\mathbf{B}$ is equal to zero in whole the cell.

4. Testing

The presented algorithm for solving the ideal MHD system is implemented as a program in C++ using the OpenMP library for its parallelization. It is tested and verified on several 2D MHD problems described in detail in the literature [6]. In all calculations the discretization of the computational domain is carried out by equilateral triangles (Fig. 3).

4.1. Brio — Wu test

First test is Brio — Wu shock tube [6]. It is the variant of Riemann problem and illustrates process of complex discontinuities evolution. Initial conditions are

$$\begin{gathered}\rho = 1, \quad u = 0, \quad v = 0, \quad w = 0, \\ p = 1, \quad B_x = 0.75, \quad B_y = 1, \quad B_z = 0, \qquad x < 0.5,\end{gathered}$$

$$\begin{gathered}\rho = 0.125, \quad u = 0, \quad v = 0, \quad w = 0, \\ p = 0.1, \quad B_x = 0.75, \quad B_y = -1, \quad B_z = 0, \qquad x > 0.5.\end{gathered}$$

Adiabatic index is $\gamma = 2$. Computational domain is $\Omega = [0, 1] \times [0, 0.01]$.

Boundary conditions (BC) for magnetic field are defined using the field **B** normal components values in the computational domain boundary edges. Both magnetic and gas BC implement free flow condition.

Fig. 4 presents the solution distribution along the x axis. The grid contains $n_x = 500$ triangles along the x axis (along the waves travelling direction). The results show good resolution of

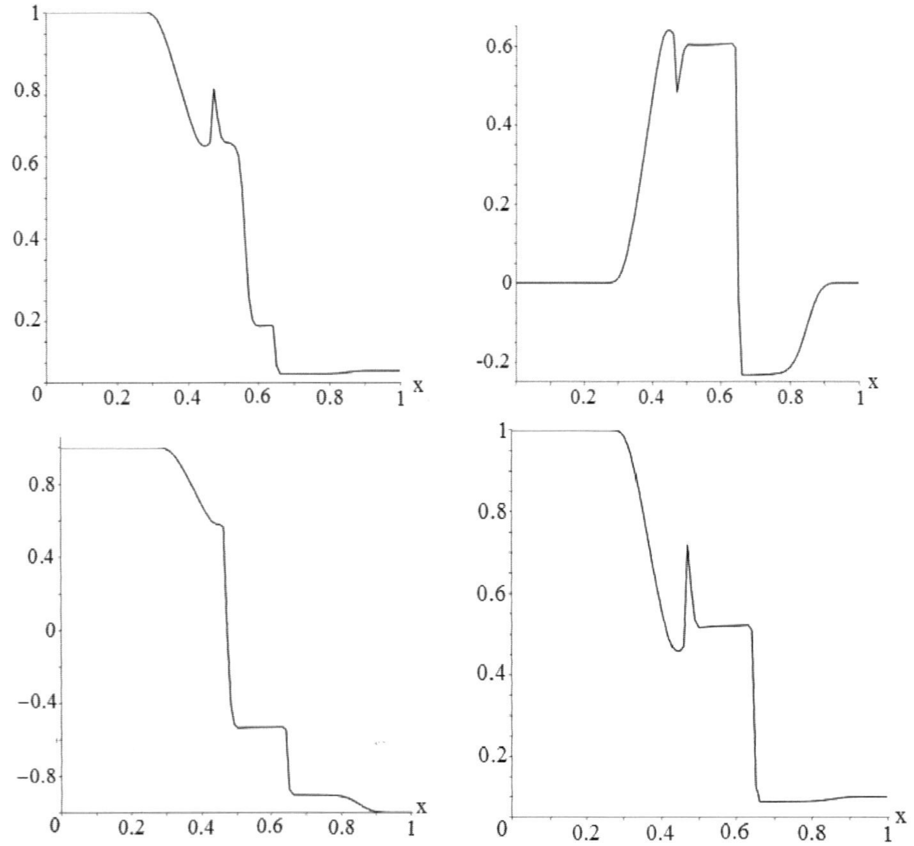

Figure 4. Brio — Wu shock tube test. From top to bottom and from left to right panels present density, velocity component u, perpendicular magnetic field B_y and pressure at time moment $t = 0.1$.

presented method. Note, that shock wave front contains only two layers of cells and contact discontinuity is smeared between 3–4 layers.

IOP Conf. Series: Journal of Physics: Conf. Series **1336** (2019) 012026 doi:10.1088/1742-6596/1336/1/012026

4.2. Cylinder rotation in magnetized medium

Let the computational domain be $\Omega = [0,1] \times [0,0.01]$. The initial gas pressure and magnetic field are uniform. Initial magnetic field is $\mathbf{B} = [B_x, 0, 0]^{\mathrm{T}}$. There is the dense rotating disk of plasma in the center of computational domain with radius $r_0 = 1$. The point radius-vector is computed as $r = \sqrt{(x-0.5)^2 + (y-0.5)^2}$.

If $r < r_0$ values of gas variables are

$$\rho = 10, \quad u = -v_0(y-0.5)/r_0, \quad v = v_0(x-0.5)/r_0, \quad w = 0.$$

If $r < r_1 = 0.115$ values are

$$\rho = 1, \quad u = 0, \quad v = 0, \quad w = 0.$$

The plasma between rotating and ambient zones $r_0 < r < r_1$ has linear distribution of variables:

$$\rho = 1 + 9f, \quad u = -fv_0(y-0.5)/r_0, \quad v = fv_0(x-0.5)/r_0, \quad w = 0,$$

where $f = (r_1 - r)/(r_1 - r_0)$.

At the initial moment centrifugal forces are not balanced, so the system set in this way is not an equilibrium configuration. The rotating substance will be gradually redistributed over the computational domain, invading its ambient part, while the magnetic field will keep the rotating substance in a flattened form.

We used following values:

$$r_1 = 0.115, \quad v_0 = 2, \quad p = 1, \quad B_x = 5/\sqrt{4\pi}, \quad \gamma = 1.4, t = 0.15.$$

The solution obtained on a grid with 400 computational cells along one of the axes is shown in Fig. 5. The distribution of density and pressure in the section along the diagonal $x = y$ is shown in Fig. 6.

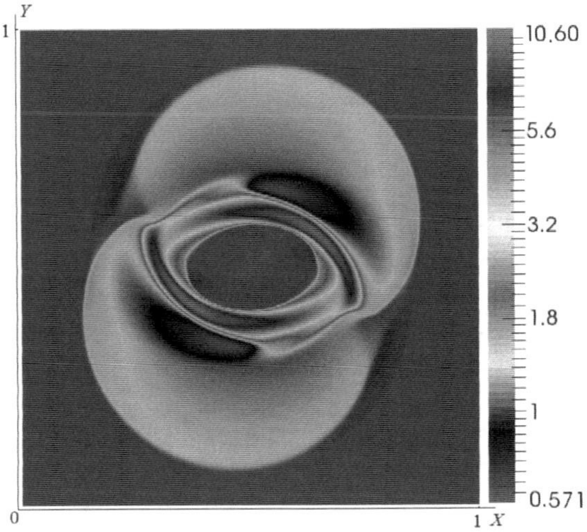

Figure 5. Cylinder rotation in magnetized medium. The density distribution at time $t = 0.15$, the number of cells along the axis is 400.

IOP Conf. Series: Journal of Physics: Conf. Series **1336** (2019) 012026 doi:10.1088/1742-6596/1336/1/012026

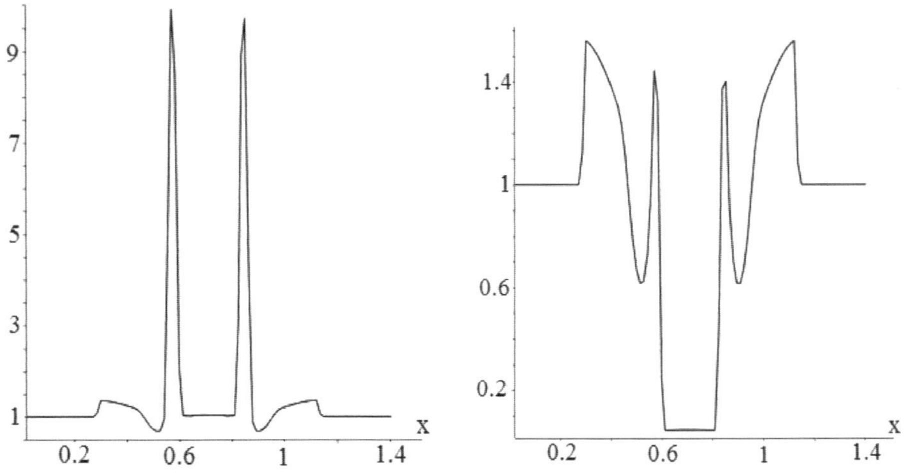

Figure 6. Cylinder rotation in magnetized medium. From left to right, the cross section along the diagonal $x = y$, the density and pressure distributions at the time instant $t = 0.15$.

4.3. Orszag — Tang vortex

The Orszag — Tang vortex problem [6] is one of the standart 2D tests for ideal MHD. It's solution describes evolution of smooth variables distribution into the complex combination of shock waves, contact discontinuities and other. Orszag — Tang vortex problems gives information about numerical scheme resolution of such waves and efficiency in waves interaction exposition.

The computational domain is square $[0, 1] \times [0, 1]$. Initial conditions are

$$\begin{aligned} u &= -\sin(2\pi y), \quad & v &= \sin(2\pi x), \quad & w &= 0. \\ B_x &= -\sin(2\pi y), \quad & B_y &= \sin(4\pi x), \quad & B_z &= 0. \end{aligned}$$

Adiabatic index is $\gamma = 5/3$. Density and pressure are equal to $25/(36\pi)$ and $5/(12\pi)$ respectively. Boundary conditions are periodical.

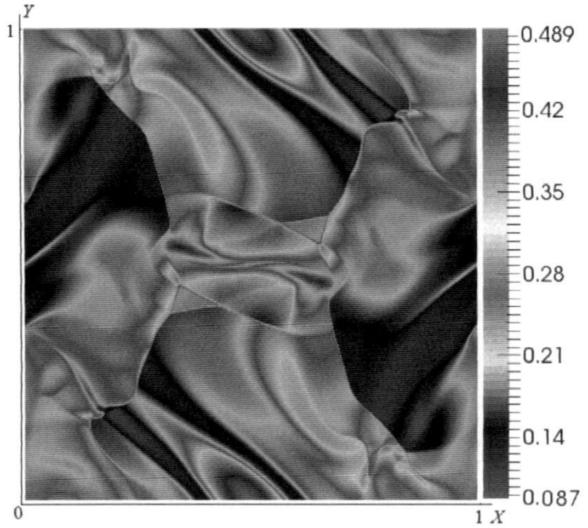

Figure 7. Orszag — Tang vortex. Density distribution at the moment $t = 0.5$, number of cells along horizontal axis is 800.

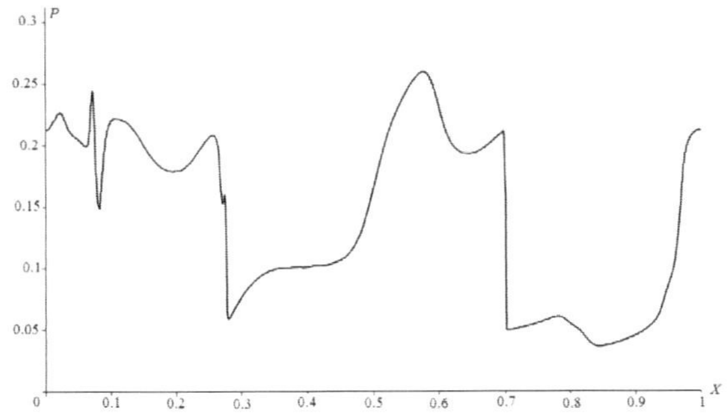

Figure 8. Orszag — Tang vortex. Pressure distribution along line $y = 0.3125$ at the time moment $t = 0.5$.

Fig. 7 shows the problem solution at time moment $t = 0.5$. The number of grid cell along axis is 800. Fig. 8 shows the pressure distribution along line $y = 0.3125$. Notice, that finite-difference used doesn't lead to spurious oscillations and has good resolution of discontinuities (only 2 cells on shock wave front).

5. Conclusions

The numerical scheme for 2D ideal MHD based on physical processes splitting is considered. The unstructured triangular staggered grids technique is used. The matter fluxes (mass, pulse, energy) are calculated using Godunov-type finite-difference scheme and HLLD numerical flux. Evolution of magnetic field is modelled in divergence-free way using direct Faraday law approxiation and Stokes theorem application. The algorythm is implemented in parallel program package. Program is verified using standart test problems. The tests performed showed the effectiveness of the developed technique, which provides suppression of the growth of magnetic field divergence. They also showed that proposed method gives a good resolution for both smooth and discontinuous solutions.

Acknowledgments

The reported study was funded by RFBR, projects numbers 18-31-20020, 18-01-00252, 18-502-12025.

References

[1] Kulikovskij A, Pogorelov N and Lubimov G 2001 *Mathematical questions of hyperbolic equations systems numerical solution* (Moscow: Nauka. Fizmatlit)
[2] Kulikovskij A and Lubimov G 1962 *Magnetohydrodynamics* (Moscow: Fizmatgiz)
[3] Galanin M and Popov Y 1995 *Quasistationary magnetic fields in inhomogeneous media* (Moscow: Nauka. Fizmatlit)
[4] Galanin M and Lukin V 2015 *Computational mathematics and mathematical physics* **55** 1298 – 1309
[5] Galanin M and Lukin V 2007 *Keldysh IAM Preprints* **50** 1 – 29
[6] Toth G 2000 *J. Comp. Phys.* **161** 605–652
[7] Samarskii A and Popov Y 2004 *Finite-diffrence methods for gas dynamics problems solution* (Moscow: Editorial URSS)
[8] Balsara D and Spicer D 1999 *J. Comp. Phys.* **149**
[9] Miyoshi T and Kusano K 2005 *Journal of Computational Physics* **208** 315–344
[10] Landau L, Lifshitz E and Pitaevskii L 1984 *Electrodynamics of Continuous Media. Vol. 8* (Oxford: Butterworth-Heinemann)
[11] Toro E F 2009 *Riemann Solvers and Numerical Methods for Fluid Dynamics* (Berlin: Springer)
[12] Dubois F, Greff I and Pierre C 2017 *arXiv e-prints* arXiv:1710.04395 (*Preprint* 1710.04395)

AUTHOR INDEX